Functional Foods of the Future

Edited by

Vijai Kumar Gupta
SRUC, University of Edinburgh, UK
Email: vijaikumar.gupta@dcu.ie

Minaxi Sharma
CARAH ASBL, Belgium
Email: minaxi86sharma@gmail.com

Smriti Gaur
Jaypee Institute of Information Technology, India
Email: taru10@gmail.com

and

Ramesh Chander Kuhad
Shree Guru Gobind Singh Tricentenary University, India
Email: kuhad85@gmail.com

ROYAL SOCIETY
OF **CHEMISTRY**

Food Chemistry, Function and Analysis No. 44

Hardback ISBN: 978-1-83767-000-0
PDF ISBN: 978-1-83767-329-2
EPUB ISBN: 978-1-83767-330-8
Print ISSN: 2398-0656
Electronic ISSN: 2398-0664

A catalogue record for this book is available from the British Library

The Royal Society of Chemistry is a charity, registered in England and Wales, Number 207890, and a company incorporated in England by Royal Charter (Registered No. RC000524), registered office: Burlington House, Piccadilly, London W1J 0BA, UK, Telephone: +44 (0)20 7437 8656.

For further information see our website at www.rsc.org

For general enquiries, please contact books@rsc.org

For EU product safety enquiries, please email books@rsc.org or contact Royal Society of Chemistry Worldwide (Germany) GmbH, Römischer Hof, Unter den Linden 10, 10117 Berlin.

Preface

Biobased compounds and their derivatives have been used worldwide for many years. Many current-day medical drugs are derived either from biobased compounds or their derivatives. Such compounds may affect a diverse range of molecular targets and be used to treat diseases such as cancer, cardiovascular diseases, gastrointestinal diseases, and diabetes. Moreover, today's consumer has become more concerned about their health; as a result, we are increasingly consuming food products that serve the dual purpose of being nutritious and (potentially) providing health benefits such as prevention from modern lifestyle diseases. Consequently, the development of novel functional foods has gained increased attention from researchers over the past decade.

This book discusses how the fields of bioproduction, biochemistry, and processing of biobased therapeutics can be used for current and future functional foods. This book comprises 16 chapters that focus on the relevant information regarding the possible health impacts of using biobased therapeutics as functional foods. Chapter 1 describes the current scope and recent trends of functional foods focusing on sustainable development goals (SDGs). This chapter emphasizes that the consumption of functional foods in a regular diet, within limits, is a viable option to meet SDGs, especially in terms of reducing hunger and malnutrition. The second chapter explains the significance of nanotechnology in the development of functional foods and current research trends along with toxicity and regulatory concerns, especially on the applications of nanotechnology in the development of functional foods. Chapter 3 shows recent advancements in fermentation-based functional foods. In this book, to discuss the therapeutic effects of different biobased compounds, Chapter 4 focuses on the significance of omega-3s, Chapter 5 discusses the importance of conjugated linoleic acids, Chapter 6 provides a concise overview of curcumin's medicinal potential,

Food Chemistry, Function and Analysis No. 44
Functional Foods of the Future
Edited by Vijai Kumar Gupta, Minaxi Sharma, Smriti Gaur and Ramesh Chander Kuhad

Published by the Royal Society of Chemistry, www.rsc.org

and Chapter 7 elaborates resveratrol's medicinal and therapeutic significance and its use in food formulations. Chapter 8 focuses on the fortification of polyphenolic compounds in dairy products, whereas Chapter 9 highlights different types of prebiotics, their food formulations and therapeutic significance towards human health. Chapter 10 discusses microbes as bio-factories in production of different metabolites like organic acids, enzymes, proteins, vitamins, antibiotics and hydrocolloids, and their importance in functional food development. Chapter 11 comprehensively discusses recent insights into the development of edible packaging utilizing essential biomolecules, namely carbohydrates, proteins and lipids. This chapter further describes the potential applications and recent insights into the development of smart/intelligent films in food industries. Chapter 12 discusses formulation, consumption, safe dosage, label claims, promotions and regulations related to functional energy drinks. Chapter 13 highlights the applications of lactobacilli in functional food formulations and their therapeutic role in gastrointestinal health. Chapter 14 discusses the critical significance that religious and cultural influences have on food choices to clarify the complex effects of religious doctrines and cultural norms on people's dietary preferences and behaviours across varied countries. This chapter thoroughly analyses the food customs of various religions, including Hinduism, Sikhism, Islam, Judaism, and Jainism. The chapter ends with a summary of the main conclusions, highlighting the importance of understanding religious and cultural viewpoints while creating functional foods. Chapter 15 deals with the major natural pigments, their chemistry, bioactive/therapeutic potential, and their importance in sustainable functional food production. Chapter 16 covers the therapeutic significance of vitamins and carotenoids manufactured using microbial cell factories and their impact on human health and food sustainability.

Overall, this book fills a gap by linking the existence of biobased therapeutic compounds and functional foods.

Minaxi Sharma
Smriti Gaur
Ramesh Chander Kuhad
Vijai Kumar Gupta

Contents

Food Chemistry, Function and Analysis No. 44
Functional Foods of the Future
Edited by Vijai Kumar Gupta, Minaxi Sharma, Smriti Gaur and Ramesh Chander Kuhad

Published by the Royal Society of Chemistry, www.rsc.org

CHAPTER 1

Scope and Current Status of Functional Foods: The SDGs Perspective

SHIKHA MAHAJAN,*[a] KIRANDEEP,[b] MINAXI SHARMA[c] AND KANDI SRIDHAR*[d]

[a] Department of Food and Nutrition Punjab Agricultural University, Ludhiana 141004, India; [b] Krishi Vigyan Kendra, Langroya, Distt SBS Nagar, 144516, India; [c] Research Centre for Life Science and Healthcare, Nottingham Ningbo China Beacons of Excellence Research and Innovation Institute (CBI), University of Nottingham Ningbo China, Ningbo-315000, China; [d] Department of Food Technology, Karpagam Academy of Higher Education (Deemed to be University), Coimbatore 641021, India
*Emails: shikha_bathla@yahoo.com; sridhar4647@gmail.com

1.1 Brief Overview

The concept of functional food began in Japan in 1980 with different food items being enriched or fortified with special constituents or bio-active components that exert some sort of positive health effect. Functional foods are also known as nutraceuticals, pharma, designer or therapeutic food items and aim to treat and prevent diseases. Functional foods have many positive health effects due to their multivariate properties like antibacterial, antimicrobial and antiviral characteristics used in treating, for example, cardiovascular issues, diabetes mellitus, immune system-related disorders and osteoporosis. The term 'functional food' is already in use by various scientific institutions but is still not yet legislative terminology or with a

Food Chemistry, Function and Analysis No. 44
Functional Foods of the Future
Edited by Vijai Kumar Gupta, Minaxi Sharma, Smriti Gaur and Ramesh Chander Kuhad

Published by the Royal Society of Chemistry, www.rsc.org

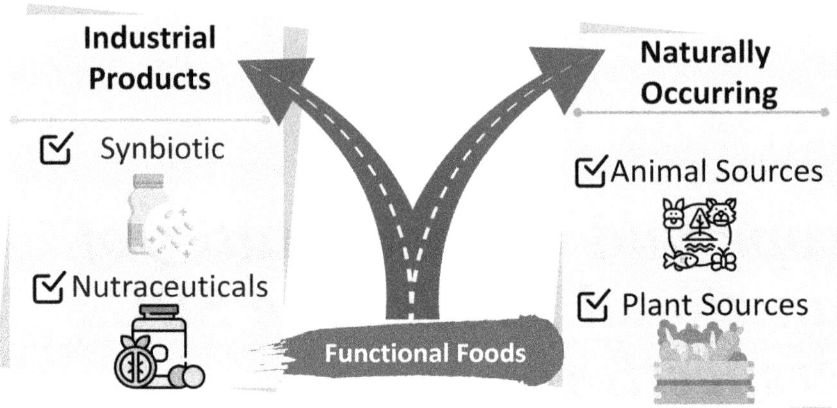

Figure 1.1 Classification of functional foods. Reproduced from ref. 1 with permission from Springer Nature, Copyright 2018.

universally accepted definition. The definitions of the term 'functional food' defined by different scientific and regulatory bodies are listed below:

IFT (Institute of Food Technologies) has described functional foods as any food items or part of a food item that may be fortified or enriched with a particular bioactive component that is necessary to meet physiological requirements, and beyond basic nutrition also has a certain positive health effect that is necessary for proper growth and development of the human body. IFIC (International Food Information Council) has also defined functional food is any food component that gives a health benefit apart from the basic nutrition. Another definition provided by the Food and Nutrition Board of the National Academy of Science (USA) has discussed the concept of functional food as any edible ingredient of food or customized food that gives some sort of health benefits beyond meeting the basic food and nutritional requirements. In India, FSSAI (Food Safety Standard Authority of India) has defined the term functional food as any food item that is derived naturally and when consumed as part of a daily meal has a defined health effect. Moreover, functional foods are naturally occurring food items and not in the form of a capsule, tablet or powder.

So, the different types of functional food are derived naturally from different sources and thus can be used in food products for their health benefits (Figure 1.1).

1.2 Overview of SDGs

The sustainable development goals (SDGs) were designed and implemented by United Nation General Assembly (UNGA), covering 193 nations under one roof, with title *Transforming Our World: the 2030 Agenda*. They are built on the simple and novel principle: "The future we want". The SDGs were framed in

2015 to succeed the Millennium Development Goals (MDGs) which were going to terminate in that year. UNGA formally enunciated and embraced an Agenda in July 2017 and termed it the 2030 Agenda, with 17 specific goals to be achieved by the year 2030.[2] By keeping sustainability at its center, the first five SDGs highlight different aspects of the economy to focus on the overall well-being of humanity with the aims of eradicating poverty and recurrent episodes of hunger and increasing quality education and gender equality. The next five SDGs focus on basic facilities of personal hygiene and quality of life, including availability of clean drinking water, healthy hygiene and sanitation practices, and accessible clean energy sources.[3] Moreover, the right to equal wages for both sexes is also one of the SDGs, being necessary for the growth of the economy and to decrease inequality. The SDGs are not limited to these basic things but also place emphasis on there being peace and equal justice to all.[4–6] They also focus on new innovation and the latest technologies in the industrial sector for growth of the current economy. The UNGA set the goal of achieving all 17 SDGs by 2030, but it does not guarantee that all the goals will be fully achieved by that deadline.

1.2.1 Brief Description and Relevance of the 17 SDGs

1. **No poverty:** There are seven targets that are aim to eradicate poverty and provide equal wages to all individuals. Gender biases in terms of employment opportunities should be avoided completely in all sectors. Moreover, they also ensure implementation of social protection systems for security or facilitation of human rights (Ritche and Ortiz 2018).[7]
2. **Zero hunger:** Each and every person in this world shall have access to daily nutritious meals. This goal can be achieved when agricultural production at the global level is sufficient to meet the food and nutritional requirements of every individual. Moreover, to eliminate the incidence of malnutrition especially, hidden hunger can be eliminated by production of fortified crops such as genetically modified foods and adoption of integrated food system models. Use of the latest research methodologies and technological application of innovative ideas are needed to accomplish to achieve this target. Trade restrictions can also play a vital role to decrease the incidence of both hunger and malnutrition.
3. **General health and wellbeing:** This SDG focuses on reducing mortality, whether child mortality or any disease-related mortality. Road accidents and death from certain hazards or incidental attacks should be prevented. Moreover, consumption of any dangerous or addictive substance and use of tobacco or nicotine should be banned completely. Basically, this goal focuses on the general health and well-being of individuals to improve quality of life and also to enhance the life expectancy.

4. **Quality of education:** This SDG focusses on the quality of education to all people to improve the economic growth of nations on a global level. It also focusses on imparting skill-based vocational education to give financial independence. This SDG also aims to improve current existing education patterns and to increase the number of academic scholarships to generate more job opportunities in under-developed and developing nations.
5. **Gender equality:** Both genders should have equal opportunities and rights in all sectors and across all the world. This SDG also demands that there should be a legal scaffold so that both sexes have equal opportunities in all working sectors. Moreover, this SDG also aims to reduce episodes of child marriages and forced marriages with the aim to "leave no one behind". Restrictions on mutilation and female genital cutting should be monitored exactingly. So, gender equality is also one of the SDGs to help to create a peaceful working environment for both sexes along with respect of females in society.
6. **Clean water and sanitation:** Availability of fresh clean drinkable water along with healthy sanitation facilities is an important target of an SDG that is to be achieved by 2030.[7] It also targets having a closed toilet system for every household so that open defecation is prevented and diseases occurring from open defecation are reduced. Moreover, waste water treatment should be done so that the ecosystem and water channels can be restored.[8]
7. **Affordable and clean energy:** Production of energy should be affordable and clean. Moreover, access to electricity and use of renewable energy channels should be expanded in a number of countries, especially India, Bangladesh and Kenya.[9] This SDG also aims to improve energy services in developing countries, along with sufficient funds for affordable and clean energy production and the infrastructure associated with it.
8. **Decent work and economic growth:** This SDG targets economic growth by reducing youth unemployment, occupational hazards or injuries, and more options of employment for women. Equal employment opportunities and equal rights to all individuals are the main basis of this SDG. Moreover, this target also aims to ensure a greater number of job opportunities and to encourage enterprise so that diversity in trade and economic growth can be achieved by 2030. It also aims to have equal wages to both sexes and promotes skill-based training to young people. This target also focuses on improving work culture and promotes healthy work environments for all individuals at all work places. Child labor should be avoided in all spheres and safe working environments for women created. Economic growth can also be affected by tourism so emphasis on tourism is also one of the objectives of this SDG.[2]
9. **Industry, innovation and infrastructure:** This SDG is aimed at increasing access to IT services by applying the latest research and

development technologies. Development of resilient infrastructure and fiscal services to promote industrialization is also one of the missions of this SDG to be achieved by 2030.

10. **Reduced inequalities:** This SDG demands reduced income inequalities and discrimination in all spheres related to the economic growth of a country. Reducing all sorts of inequalities or discrimination in terms of social, economic, political or any other inequality that hinder or obstruct economic growth. This SDG also aims at reduction of the cost transportation services and promotion of global finance markets. It also aims to develop social policies that can initiate equality in all spheres.
11. **Sustainable cities and communities:** This SDG targets meeting the basic requirements of reasonable and sustainable transport systems. This SDG also focuses on the reduction of adverse environmental conditions of urban areas and natural disasters. Moreover, it also ensures protection of our cultural heritage parks, museums and other natural sites to attract tourism. It also targets sustainable policies and regulations to preserve natural resources to decrease disasters occurring.
12. **Responsible consumption and production:** This target aims to have efficient use of natural resources along with good practices of post-harvest management of horticultural produce, so that food with good quality and quantity is accessible. Wastage of any food from production to final consumption should be avoided. The triple R approach of reduce, recycle and reuse can be promoted through activities at grass root level. This SDG also targets strengthening public distribution systems to give equal access to basic food to meet nutritional security at the household level. Moreover, it also focus on the development and strengthening of the research and development sector to monitor the achievement of these goals.
13. **Climate action:** This SDG aims integrate climate change-based policy and procedures to reduce the risk of climate change. For this purpose, 2023 was declared International Year of Millets (IYM) by UNGA in March 2021 to initiate cultivation of different millets under the change of climatic conditions around the globe.
14. **Life below water:** This SDG aims to reduce water pollution at all levels, for example associated with marine farming, acidity of the ocean and plastic pollution. Moreover, it also focuses on decreasing excessive subsidies of fish rearing or marine businesses so that over fishing can be prevented.[10] Rules and regulations associated with international maritime law and to stop marine pollution should be followed strictly.
15. **Life on land:** This SDG aims to conserve and maintains the ecosystem by prohibition of practices related to cutting of forests and destruction of natural resources. This SDG also aims to have a healthy life on Earth by protecting the habitat of natural species and conserving the resources of our ecosystem in a planned and scientific way.

16. **Peace, justice and strong institutions:** This SDG aims to target frequent episodes of attack, violence and other abuse at domestic and work places. Moreover, children should be protected from any sort of exploitation, violence and trafficking. This SDG also demands equal justice and liberty for all individuals and reduction of corruption.
17. **Partnerships for the goals**: This SDG targets implementation and revival of global partnerships to achieve the above-stated sixteen goals. Moreover, this goal also focuses on reinforcing co-ordination between different developed and developing nations. Further, this goal also aims to improve public–private partnerships between different zones likes north–south and south–south in five major spheres of finance, technology, infrastructure and trade.[11]

1.3 Current Status of Functional Foods

1.3.1 Global Market Size and Growth Trends of Functional Foods

The functional foods available in the global market are of various types such as for sports nutrition, management of weight, health, immunity boosting, *etc.* There is an increased demand of functional foods for treating various health issues like heart diseases, cardiovascular diseases, congenital heart defects, *etc.* Sports enthusiasts require food that is rich in vitamins, protein, and increases muscular endurance and strength, which in turn impacts the demand for functional foods. Some kinds of functional foods improve digestion and also lower the risk of cardiovascular disease as well as diabetes. The demand for functional foods in the sports nutrition sector is expected to increase at CAGR 10% between 2022 and 2030.[12]

According to statistics for the year 2021, the market for functional foods in the immunity application sector was valued at 57.6 billion USD. A person's immune system defends them against various infections. Functional foods are a rich source of antioxidants, a variety of vitamins, nucleotides, fatty acids, along with L-arginine that promotes immune response. Preventative medicine and immunity boosters became popular worldwide during the COVID-19 pandemic. The aforementioned products also suppress cellular immunity in addition to lowering plasma triglyceride levels. Owing to these factors the market is anticipated to grow rapidly.

The North American functional food market accounted for 24.7% of global revenues in 2021. Immunity, weight management, digestive health, and cardio health make up the major segments of the North American market. As fitness interests grow among consumers in the U.S., Mexico, and Canada, functional foods and drinks are expected to be in high demand. Millennials' preference for protein bars and cookies as well as shakes as healthy snack choices is also expected to increase the demand for functional foods in

North America over the coming years. As a result of a burgeoning population and rising disposable incomes, the Asia–Pacific market for functional foods during the year 2021 was 35.8%. Developing economies are experiencing rapid growth in the consumption of processed food and also increased awareness regarding health and fitness. Due to the COVID-19 worldwide pandemic, consumer awareness and concerns regarding food safety are increasing, which is driving market growth in the region.

A highly competitive market exists for functional foods, with both public and private companies are focusing their attention on research and development so as to have innovative products. Public companies follow a proactive approach in order to gain advantage in the market. There is a likelihood that privately owned market participants will follow suit in the near future. Recent years have seen significant advancements in the functional foods market in terms of product innovation as well as portfolio expansion. Arla Foods Ingredients, for instance, launched a juice drink which was enriched with protein during March 2021 by using Lacprodan ISO for fortifying the beverages. The global functional food market has many important companies, which include the Coca-Cola Company, Amway Corp, Arla Foods amba, Inc., Standard Functional Foods Group, Herbalife International of America, Inc. and many others.[12]

1.3.2 Popular Functional Food Ingredients and Products

The definition of functional foods varies from organization to organization. All definitions of functional food agree that they are food products that have been altered to impart additional health benefits and reduce disease risk. Various sources and ingredients can be used to make functional foods, including plants, fruits and vegetables, animals, probiotics, and prebiotics. Therefore, functional foods may be classified based upon the product and also the ingredients present in them.

1.3.2.1 Animal-based Products

In addition to amino acids (like glycine, cysteine and methionine, *etc.*), meat also contains vitamins such as linolic acid, taurine, and creatine.[13] A functional food can be made from it by adding nutritional components like fish oil, natural extracts, and fiber from nuts. For example, a functional beverage that promotes joint health can be made by extracting chondroitin sulfate and glucosamine from bovine skin. There is also abundant use of eggs in the market, and their enrichment with omega-3 can reduce the risk of ischemic heart disease. It has been recommended to add linseed or flaxseed to the hens' diet in order to increase the concentration of linolic acid in the egg. In addition, fish oil can also be added to the hens' diet to increase its linolic acid content. As a result of this technique, the egg yolk may have a fishy taste, which is regarded as an undesirable characteristic.[14]

1.3.2.2 Dairy Products

There are many nutrients in dairy products, including proteins, vitamins, and minerals. Proteins and peptides in dairy-based products can help decrease the risk of cardiovascular diseases such as hypertension, myocardial infarction, and stroke. In some products, functional peptides having the sequences of valine–proline–proline and isoleucine–proline–proline have been added to alter their properties, including β-lactoglobulin and glycomacropeptides (GMP), which provide enhanced protection against viral and bacterial infections.[15]

1.3.2.3 Herbal Products

Many diseases have been treated with herbs for thousands of years.[16] In addition to polyphenols, they can contain caffeic acids and tannic acids, which can chelate iron, which may reduce the effects of neurodegenerative diseases such as Alzheimer's.[17] In addition to scavenging free radicals, propolis has been shown to protect against myocardial damage. The neuroprotective properties of curcumin have been demonstrated through its ability to modify inflammatory cytokines, reduce oxidized proteins, and increase interleukin-1β. As reported by Agarwal and Harikumar (2009), curcumin inhibited oxidative damage and inflammation in neurons as well as cognitive deficits due to oxidative damage.[18] Hence, incorporating herbal extracts into food products can prevent diseases and promote health.

1.3.2.4 Fruits and Vegetables

Various fruits and vegetables contain antioxidants and anti-inflammatory substances that have cardioprotective properties that promote health. There are many polyphenols found in berry fruits, such as flavonols, catechins, anthocyanins, phenolic acids such as hydroxycinnamic acids, and also tannins such as proanthocyanins and ellagitannins, which are among the most significant because of their anti-inflammatory properties.[19] A study also reported that the phytochemical resveratrol in red grapes was effective at modifying lipids and enhancing antioxidant activity. The carotenoid lycopene can be found primarily in tomatoes and serves as an antioxidant, reducing oxidative stress, hypertension, and atherosclerosis.[20]

1.3.2.5 Marine Products

There are an incredible number of species in the marine environment, including marine plants, microorganisms, sponges, fish, and many others. Aquatic environments provide a wealth of functional compounds such as omega-3 polyunsaturated fats, glucosamine, chitosan/chitosan oligosaccharides, xanthophylls, carotenoids as well as marine enzymes and protein hydrolysates.[21]

1.3.2.6 Probiotics and Prebiotics

A probiotic is a live microorganism, usually a bacterium, that when consumed in enough numbers can be beneficial to the consumer. Intestinal bacteria consume prebiotics or oligosaccharides which cannot be digested by the human enzymes. They promote bacterial growth and have health benefits for the host.[22] Researchers have investigated *Lactobacillus* and *Bifidobacteria* as probiotic strains for use in functional foods.[19,23]

1.4 Food Technology and Its Impact on Functional Food Development

In order to enhance the nutritional value of foodstuffs, different strategies and technologies have been used. There has recently been a lot of interest in vacuum impregnation as an emerging technology in food science and technology. In this technique, a liquid medium having bioactive compounds is introduced inside solid porous food through capillary pressure replacing internal gas.[24] This method has been used to increase the nutritional quality of a variety of foods, such as adding polyphenols to ready-to-eat sweet potatoes,[25] adding ascorbic acid to potato tubers,[26] adding calcium, vitamin C, and vitamin E to potato snacks[27] and adding *Lactobacillus casei* to apple cylinders.[28]

Conditions such as low pH and gastrointestinal issues can have detrimental effects on bioactive ingredients. Consequently, encapsulation is a practical approach that encloses bioactive compounds in a variety of nontoxic materials in order to protect them against various adverse conditions.[29] Proteins,[30] lipids,[31] polysaccharides,[32,33] hydrogels[34] and metal–organic frameworks, or a combination of them, are commonly used for encapsulation and controlled release in the field of food science.[35,36] Recently, many new and innovative technologies have been studied by various researchers for the encapsulation of bioactive compounds. For example, spray chilling and electrospinning are widely used for the process of encapsulation.[37] There are two types of carriers: nanocarriers and microcarriers.[38,39] Nanocarriers are mostly used for bioactive compounds whereas microcarriers can be used for probiotics as well as bioactive compounds.[40]

1.5 Regulatory Frameworks for Functional Foods in Different Regions/Countries

The U.S. Food and Drug Administration (FDA) does not recognize functional foods as a unique product class and does not have a legal definition. Although functional foods are mainly a marketing tool, they have received considerable attention from researchers. As defined by various sources, functional foods refer to foods that are enhanced to produce some health benefit. Generally, the U.S. Federal Food, Drug, and Cosmetic Act 1938 (the Act) applies to all foods including functional foods; therefore these foods are regulated by the FDA as conventional foods. As a matter of regulatory policy, conventional foods (including functional foods) and dietary supplements must prove their safety. As a result, the Act prohibits the marketing of any food product unless it is determined to be safe. Food safety has traditionally not been weighed in favor of risk or benefit, as it is for drugs. Food products must meet statutory safety standards regardless of their effectiveness.[41–43]

1.5.1 Specific Provisions

There were no safety standards governing dietary supplements or functional foods before 1994. As a food additive, ingredients have to be generally recognized as safe (GRAS) or undergo a pre-marketing approval process if they are not GRAS. As a result, either the FDA or the manufacturer determined that the ingredients were GRAS or, if the food additive was newly approved, the manufacturer submitted relevant safety data. Regulations and guidance documents outline the review process and data requirements for food additive petitions. Manufacturers are generally required to convene expert review boards to determine the GRAS status, as recommended by the agency.

The Food and Drug Administration has kept a close eye on the incorporation of dietary supplement ingredients into conventional foods. Considering that dietary supplements have a different regulatory framework than traditional foods in terms of safety and other provisions, the agency is concerned that dietary supplement ingredients meet the appropriate safety standards for conventional foods when they are added to conventional foods. This means that they are GRAS for their intended use or approved as food additives. The standards for additions to conventional foods must be met for functional foods, which means that functional foods cannot simply be made by adding a dietary supplement ingredient to a conventional food without ensuring that the ingredient is safe in consumption of the food. Functional foods can only be made by adding an ingredient that is safe for food consumption. As with conventional foods, functional foods must adhere to all the specific requirements. In order to maintain a sufficient knowledge-base to ensure safety, the FDA and manufacturers should keep pace with the growing interest in this area.[44]

1.5.2 Label Claims and Product Boundaries

Regulations determine which regulatory category a product falls into based on its intended use, not its chemical composition or other characteristics. The FDA considers all products intended for diagnosing, treating, preventing, curing, or mitigating diseases to be drugs regardless of the ingredient's origin. Manufacturers' claims about their products determine how the FDA reviews and regulates their products, based on the nature of their claimed effect. As far as advertising is concerned, primarily the FTC (U.S. Federal Trade Commission) regulates it, and does not make any distinctions. Keeping aside the claims about the nutrient content, mainly two types of claims are examined for functional foods and dietary supplements. One claim is related to health which focuses on reducing the risk of disease in the context of a daily diet plan. The second claim is related to effects on the structure or function of the body. A product which claims its role in curing or preventing a disease is being considered as a drug.

Traditionally, food-like effects have been required for structure or function claims for conventional as well functional foods. The 1994 U.S. Dietary Supplement Health and Education Act (DSHEA) expanded the labeling of dietary supplements to include non-food like claims as well as food-like claims. According to these 1994 provisions, manufacturers of dietary supplements are required to use the standard label disclaimer ("This statement has not been assessed by FDA"). Whenever a structure/function claim is used on a product label, the FDA must receive a notification detailing how the claim is worded within 30 days of the product's marketing and that it is not intended to diagnose, treat, cure, or prevent any disease. When manufacturers submit claims that imply a treatment for a disease, the FDA often informs them that their claims should not be used. A specific restriction of the DSHEA is that dietary supplements cannot masquerade as conventional foods. Unlike conventional foods (such as cereals and candy bars), dietary supplements cannot represent themselves as conventional foods, despite the fact that they can take the shape of such foods. Therefore, a dietary supplement bar product with a dietary supplement label (*i.e.* with the dietary supplement facts panel rather than the nutrition facts panel) would be acceptable from a regulatory standpoint. The product would be subject to regulation as a conventional food if, for example, the labeling indicates that it is a snack food or a substitute for candy bars.

The FDA has the authority to take action for false or misleading claims under current regulations, but its substantiation approach is case-by-case. Furthermore, there is the misconception that structure/function claims are simply "emerging" health claims. It is not relevant to structure/function claims to provide data about the effects of disease, since structure/function claims are by definition not about disease, either implied or inferred. Generally, structure/function claims focus on maintenance, support, and the promotion of health. As dietary supplements and functional foods become more widely available, appropriate regulatory strategies and supporting scientific underpinnings are emerging. To provide consumers with safe and effective products that have meaningful claims, many components of society must contribute to the system over the next few years. A critical aspect of public health policy in maintaining consumer trust in the food supply should also to be taken care of when formulating claims.

1.6 Conclusions

This chapter highlights that major SDGs, that is SDG 1 (no poverty) and SDG 2 (zero hunger), should be met by 2030. The consumption of different types of functional foods from naturally derived sources exerts a positive health effect to meet the SDGs pertaining to especially hunger and malnutrition, in terms of under-nutrition, over-nutrition, specific nutrient deficiency and imbalance of nutrition. Moreover, label claims for functional foods products should also be taken care off.

References

1. S. Mohanty and K. Singhal, Functional Foods As Personalised Nutrition: Definitions and Genomic Insights, in *Functional Food and Human Health*, ed. V. Rani and U. C. S. Yadav, Springer Singapore, Singapore, 2018, pp. 513–535.
2. J. Schleicher, M. Schaafsma and B. Vira, Will the Sustainable Development Goals address the links between poverty and the natural environment?, *Curr. Opin. Environ. Sustainability*, 2018, **34**, 43–47.
3. F. Biermann, N. Kanie and R. E. Kim, Global governance by goal-setting: the novel approach of the UN Sustainable Development Goals, *Curr. Opin. Environ. Sustainability*, 2017, **26–27**(Open issue, part II), 26–31.
4. United Nations, Resolution adopted by the General Assembly on 25 September 2015, Transforming our world: the 2030 Agenda for Sustainable Development (A/RES/70/1 Archived 28 November 2020 at the Wayback Machine), 2015.
5. United Nations, Resolution adopted by the General Assembly on 6 July 2017, Work of the Statistical Commission pertaining to the 2030 Agenda for Sustainable Development (A/RES/71/313 Archived 28 November 2020 at the Wayback Machine, 2017.
6. United Nations, Sustainable development goals report Archived 30 December 2020 at the Wayback Machine, New York, 2020.
7. H. Ritchie, M. Roser, J. Mispy and E. Ortiz-Ospina, Measuring Progress towards the Sustainable Development Goals, SGD-Tracker.org, United Nations Sustainable Development Goals, 2018, https://www.un.org/sustainabledevelopment/.
8. WHO and UNICEF, Progress on Drinking Water, Sanitation and Hygiene: 2017 Update and SDG Baselines Archived 25 July 2019 at the Wayback Machine. Geneva: World Health Organization (WHO) and the United Nations Children's Fund (UNICEF), 2017.
9. J. Bartram and P. Hunter, Bradley classification of disease transmission routes for water-related hazards, in *Routledge handbook of water and health*, ed. J. Bartram, R. Baum, P. A. Coclanis, D. M. Gute, D. Kay, S. McFadyen, *et al.*, Routledge, London and New York, 2015, pp. 20–37.

10. F. Biermann, N. Kanie and R. E. Kim, Global governance by goal-setting: the novel approach of the UN Sustainable Development Goals, *Curr. Opin. Environ. Sustainability*, 2017, **26–27**(Open issue, part II), 26–31.
11. Ocean acidification (2007) (Issues Brief) (PDF), IUCN (International Union for Conservation of Nature), November 2017, Archived (PDF) from the original on 30 December 2020, Retrieved 3 November 2020.
12. Functional Foods Market Size, Share and Trends Analysis Report By Ingredient (Carotenoids, Prebiotics and Probiotics, Fatty Acids, Dietary Fibers), By Product, By Application, By Region, And Segment Forecasts, 2022–2030, https://www.grandviewresearch.com/industry-analysis/functional-food-market.
13. W. H. Górska, W. Laskowski, O. Kulykovets, A. Kudlińska-Chylak, M. Czeczotko and K. Rejman, Food products as sources of protein and amino acids—The case of Poland, *Nutrients*, 2018, **10**(12), 1977.
14. P. F. Surai and N. H. C. Sparks, Designer eggs: From improvement of egg composition to functional food, *Trends Food Sci. Technol.*, 2001, **12**(1), 7–16.
15. Z. M. S. Hosseini, M. Sarhadi, M. Zarei, R. Thilagavathi and C. Selvam, Synergistic effects of curcumin and its analogs with other bioactive compounds: A comprehensive review, *Eur. J. Med. Chem.*, 2021, **210**, 113072.
16. C. K. B. Ferrari, Functional foods, herbs and nutraceuticals: Towards biochemical mechanisms of healthy aging, *Biogerontology*, 2004, **5**(5), 275–289.
17. S. Chopra, K. K. Pillai, S. Z. Husain and D. K. Girl, Propolis protects against doxorubicin induced myocardiopathy in rats, *Exp. Mol. Pathol.*, 1995, **62**(3), 190–198.
18. B. B. Aggarwal and K. B. Harikumar, Potential therapeutic effects of curcumin, the anti-inflammatory agent, against neurodegenerative, cardiovascular, pulmonary, metabolic, autoimmune and neoplastic diseases, *Int. J. Biochem. Cell Biol.*, 2009, **41**(1), 40–59.
19. S. Kaur and M. Das, Functional foods: An overview, *Food Sci. Biotechnol.*, 2011, **20**(4), 861–875.
20. K. W. Kong, H. E. Khoo, K. N. Prasad, A. Ismail, C. P. Tan and N. F. Rajab, Revealing the power of the natural red pigment lycopene, *Molecules*, 2010, **15**(2), 959–987.
21. F. Shahidi and P. Ambigaipalan, Novel functional food ingredients from marine sources, *Curr. Opin. Food Sci.*, 2015, **2**, 123–129.
22. C. Stanton, R. P. Ross, G. F. Fitzgerald and S. D. Van, Fermented functional foods based on probiotics and their biogenic metabolites, *Curr. Opin. Biotechnol*, 2005, **16**(2), 198–203.
23. L. F. Rincón, *Encyclopedia Food Science Nutrition*, 2003, pp. 2827–2832.
24. E. Betoret, N. Betoret, P. Rocculi and R. Mdalla, Strategies to improve food functionality: Structure–property relationships on high pressures homogenization, vacuum impregnation and drying technologies, *Trends Food Sci. Technol.*, 2015, **46**(1), 1–12.

25. R. A. Abalos, E. F. Naef, M. V. Aviles and M. B. Gómez, Vacuum impregnation: A methodology for the preparation of a ready-to-eat sweet potato enriched in polyphenols, *LWT*, 2020, **131**, 109773.
26. K. Hironaka, M. Kikuchi, H. Koaze, T. Sato, M. Kojima and K. Yamamoto, *et al.*, Ascorbic acid enrichment of whole potato tuber by vacuum-impregnation, *Food Chem.*, 2011, **127**(3), 1114–1118.
27. C. Y. Duarte, R. D. Granda, M. Cortés and C. O. Vega, Potato snacks added with active components: Effects of the vacuum impregnation and drying processes, *J. Food Sci. Technol.*, 2020, **57**(4), 1523–1534.
28. N. Betoret, L. Puente, M. Diaz, M. Pagan, M. Garcia and M. Gras, *et al.*, Development of probiotic-enriched dried fruits by vacuum impregnation, *J. Food Eng.*, 2003, **56**(2–3), 273–277.
29. C. P. Champagne and P. Fustier, *Microencapsulation for Delivery of Probiotics and Other Ingredients in Functional Dairy Products*, Woodhead Publishing Limited, Sawston, Cambridge, 2nd edn, 2007, pp. 404–426.
30. J. Burgain, M. Corgneau, J. Scher and C. Gaiani, Encapsulation of probiotics in milk protein microcapsules, in *Microencapsulation and Microspheres for Food Applications*, 2015, pp. 391–406.
31. I. Katouzian, E. A. Faridi, S. M. Jafari and S. Akhavan, Formulation and application of a new generation of lipid nano-carriers for the food bioactive ingredients, *Trends Food Sci. Technol.*, 2017, **68**, 14–25.
32. A. S. Akbari, R. Shaddel and S. M. Jafari, Encapsulation of food bioactives and nutraceuticals by various chitosan-based nanocarriers, *Food Hydrology*, 2020, **105**, 105774.
33. A. Taheri and S. M. Jafari, Gum-based nanocarriers for the protection and delivery of food bioactive compounds, *Adv. Colloid Interface Sci.*, 2019, **269**, 277–295.
34. W. Hennink and C. Van Nostrum, Novel crosslinking methods to design hydrogels, *Adv. Drug Delivery Rev.*, 2002, **54**(1), 13–36.
35. Z. Moussa, M. Hmadeh, M. G. Abiad, O. H. Dib and D. Patra, Encapsulation of curcumin in cyclodextrin-metal organic frameworks: Dissociation of loaded CD-MOFs enhances stability of curcumin, *Food Chem.*, 2016, **212**, 485–494.
36. Y. Chen, K. Tai, P. Ma, J. Su, W. Dong and Y. Gao, *et al.*, Novel γ-cyclodextrin metal–organic frameworks for encapsulation of curcumin with improved loading capacity, physicochemical stability and controlled release properties, *Food Chem.*, 2021, **347**(17), 128978.
37. S. Klettenhammer, G. Ferrentino, K. Morozova and M. Scampicchio, Novel technologies based on supercritical fluids for the encapsulation of food grade bioactive compounds, *Foods*, 2020, **9**, 1395.
38. D. J. McClements, Delivery by Design (DbD): A standardized approach to the development of efficacious nanoparticle- and microparticle-based delivery systems, *Compr. Rev. Food Sci. Food Saf.*, 2018, **17**(1), 200–219.
39. D. J. McClements, *Future Foods: How Modern Science is Transforming the Way We Eat*, 2019, pp. 1–392.

40. M. Martínez-Ballesta, Á. Gil-Izquierdo, C. García-Viguera and R. Domínguez-Perles, Nanoparticles and controlled delivery for bioactive compounds: Outlining challenges for new "smart-foods" for health, *Foods*, 2018, **7**, 72.
41. R. J. Head, I. R. Record and A. K. Roger, Functional foods: approaches to definition and substantiation, *Nutr. Rev.*, 1996, **54**, S17–S20.
42. M. F. Jacobson and B. Silverglade, Functional foods: health boon or quackery?, *BMJ*, 1999, **319**(7204), 205–206.
43. L. E. Batson, Food Additive Petition Agency: Food and Drug Administration, *Fed. Regist.*, 1994, **59**(2), 406–409.
44. C. L. Taylor, Regulatory Frameworks for Functional Foods and Dietary Supplements, *Nutr. Rev.*, 2004, **62**(2), 55–59.

CHAPTER 2

Nanotechnology: A Boon for Sustainable Development of Functional Foods

NIMISHA TEHRI,[a] SONIA SANGWAN,[b] TEJPAL DHEWA[c] AND ANIL KUMAR PUNIYA*[d]

[a] Department of Food Safety and Quality Assurance, CoPHT & FP, Sardar Vallabhbhai Patel University of Agriculture and Technology, Meerut- 250110, U.P., India; [b] Department of Food Technology, Bhai Jaita Subharti Engineering College, Swami Vivekanand Subharti University, Meerut – 250005, UP, India; [c] Department of Nutrition Biology, Central University of Haryana, Mahendergarh – 123031, Haryana, India; [d] Dairy Microbiology Division, ICAR-National Dairy Research Institute, Karnal – 132001, Haryana, India
*Email: akpuniya@gmail.com

2.1 Introduction

The food industry is consistently growing globally, 'with a compound annual growth rate of over 24% projected from 2019 to 2023' according to market analysis. Functional foods constitute an important component of the food industry and have a crucial function in supplying essential nutrients and preventing and treating various ailments, including cancer, diabetes, heart disease, respiratory problems, arthritis, and safeguarding against COVID-19. However, traditional approaches used for functional food development have drawbacks in sustaining the nutritional value, safeguarding against environmental conditions, and ensuring long-term stability. In the past few decades, functional food nanotechnology has emerged as an exciting field of

Food Chemistry, Function and Analysis No. 44
Functional Foods of the Future
Edited by Vijai Kumar Gupta, Minaxi Sharma, Smriti Gaur and Ramesh Chander Kuhad

Published by the Royal Society of Chemistry, www.rsc.org

research. Nanotechnology operates at the nanoscale, dealing with atoms, molecules, and macromolecules sized ranging roughly from 1–100 nm, to develop and use materials with distinct properties. These nanomaterials exhibit one or more dimensions or internal structures ranging from 1–100 nm, enabling the use and manipulation of matter at the nanoscale. Due to their high surface-to-volume ratio and novel physiochemical properties such as colour, diffusivity, solubility, strength, toxicity, and magnetic, optical and thermodynamic characteristics, these materials possess unique characteristics that differ from their macroscale counterparts.[1,2]

The applications of nanotechnology in the food industry can be categorized primarily into two groups, *i.e.* food nanostructured ingredients and food nanosensing. Food nanostructured ingredients are utilized in a number of aspects of food production, ranging from the processing to packaging. In food processing, nanostructures have the potential to serve as food additives, carriers for nutrient delivery, anti-caking agents and antimicrobial agents. Nano-enabled delivery systems enhance the nutraceutical properties of food components. Food packaging empowered by nanotechnology represents a significant advancement compared to traditional packaging methods reliant on plastic barriers. Nanomaterials are also used as fillers to enhance the mechanical strength and durability of packaging materials. The functional elements of nanotechnology-based packaging, including antimicrobial properties, contribute to extended shelf life for food products and thus make a significant contribution through intelligent and innovative packaging solutions.[3] In contrast, the implementation of food nanosensors in intelligent food packaging also plays a significant role in the valuation of food safety by monitoring the quality of stored food.[4,5] Nanomaterials have revolutionized the food industry, enhancing not only food quality and safety but also the health benefits of food. Cutting-edge nanomaterials and nanoparticles enable precise and sustainable action, manipulating particle size to facilitate cluster formation and alter surface charge, resulting in improved solubility, bioavailability, taste, texture, and consistency of food products.

A number of organizations, researchers, and industries are actively developing innovative techniques, methods, and products that directly apply nanotechnology in the field of food science and technology.[6] This chapter covers the current trends and prospects of functional food nanotechnologies with a meticulous description of state-of-the-art nanomaterial applications in the food industry, encompassing processing, nano-delivery, nano-sensing and packaging.

2.2 Functional Foods

Functional foods contain functional ingredients that can be defined as dietary ingredients that positively influence the consumer's health. Functional foods have health-improving qualities in addition to their nutritional value. Preferably, the functional food should fulfil the following conditions:

- Functional foods should not be in the form of powders, granules, tablets, capsules, liquids, or jellies if the purpose is to maintain their natural state or minimize processing

- Does not include a drug as defined in clause (b) or ayurvedic, Siddha and Unani drugs as defined in clauses (a) and (h) of Section 3 of the Drugs and Cosmetics Act, 1940 and rules constituted thereunder
- Does not assert to cure or diminish any specific disease, disorder or condition (except for certain health benefits or promotion claims) as may be allowed by the regulations made under this Act
- Does not include a narcotic or psychotropic substance as defined in the Schedule of the Narcotic Drugs and Psychotropic Substances Act, 1985, and rules created thereunder or substances listed in Schedules E and EI of the Drugs and Cosmetics Rules, 1945.

Functional foods contain bioactive ingredients that protect consumers from various diseases. Diet, food habits, and lifestyle are major factors influencing the health of individuals worldwide. Over 2500 years ago, Hippocrates recognized the medicinal importance of food when he stated "Let food be thy medicine and medicine be thy food." Adequate nutrition is essential for national development and the health of families. Food nutrients are crucial for improving the overall health of children and adults, including their intelligence, educational performance, and economic status. Phytochemicals and zoochemicals, physiologically active ingredients in foods, can help reduce the risk of many chronic diseases. Moreover, all the foods that are functional provide taste, aroma, or nutritional value.[7]

Functional foods are usually studied under two different categories: conventional and modified. Conventional foods are natural, whole-food ingredients rich in various nutrients like vitamins, minerals, antioxidants, and heart-healthy fats. On the other hand, modified foods have been enriched with ingredients like vitamins, minerals, probiotics, or fibre to increase the benefits of functional foods. Different examples of functional foods include fruits (*e.g.* kiwifruit, pears, berries, oranges, apples and bananas), vegetables (*e.g.* spinach, kale, cauliflower and zucchini), nuts (*e.g.* cashews, Brazil nuts, macadamia nuts and almonds), seeds (*e.g.* pumpkin seeds, flax seeds and chia seeds), whole grains (*e.g.* brown rice, buckwheat and oats), seafood (*e.g.* cod, sardines and salmon), herbs and spices (*e.g.* pepper, ginger, cinnamon and turmeric), beverages (*e.g.* green tea, coffee, black tea), fortified dairy products (*e.g.* milk and yoghurt), fortified grains (*e.g.* pasta and bread, fortified cereal and granola) and fortified eggs, *etc.*[8]

2.3 Nanotechnology for Functional Foods Development

Nanotechnology is one of the most promising, multidisciplinary technological advancements to revolutionize the conventional food industry. Functional food nanotechnology uses nanotechnology and nanomaterials to enhance food processing, nutrient delivery, food safety, and packaging.

2.3.1 Approaches for the Formulation of Nanoparticles

The methodologies used for the formation of nanoparticles are known as the top-down and bottom-up ones.[9] In the top-down methodology, based on emulsification, solvent evaporation, or high-pressure homogenization, nanoparticle synthesis is accomplished by the infusion of energy to disrupt bulk materials, thereby facilitating the reduction in size and the shaping of their structure for specific purposes. On the other hand, the bottom-up methodology, using coalescence, nanoparticle precipitation, inclusion bonding, or supercritical gas, entails the construction of nanoparticles from discrete atoms that possess the capability to naturally self-assemble in a self-regulated manner.[10] The top-down approach exhibits a slower production rate, whereas the bottom-up approach yields a more uniform chemical composition when contrasted with the top-down approach.[11]

2.3.2 Bioavailability and Controlled Release of Nano-encapsulated Bioactive Ingredients

Nutrients must go through the gastrointestinal tract to reach their final absorption location. The gastrointestinal tract process can be divided into three stages: oral processing, gastric phase and intestinal phase. During oral processing, the food is mixed with saliva, broken down into smaller particles, and swallowed. In the stomach, a highly acidic environment and enzymes like pepsin transform the food further. The partially digested food then enters the small intestine, where it is absorbed through active and passive transport systems. Nanoencapsulation protects bioactive ingredients from degradation in both acidic and alkaline conditions. Their minute size contributes to enhanced uptake and absorption by facilitating the controlled release of these bioactive components at their designated target sites.[12,13] There are two major mechanisms for controlled release of bioactive compounds: delayed release and sustained release. In delayed release, the bioactive ingredient is shielded during the gastric phase and only released when it reaches the intestinal phase. Conversely, in the sustained release, the concentration of the bioactive ingredient is maintained consistently over an extended period.[14] Various elements, including the type, configuration and size of the carrier, play a role in influencing the release of nano-encapsulated bioactive ingredients.[15] Consequently, the design of nano-encapsulation serves as a critical factor that can impact the accessibility of the encapsulated ingredient. Multiple studies have indicated increased bioavailability of nano-encapsulated bioactive compounds when assessed through *in vivo* models.

2.3.3 Nano-delivery Systems for Functional Foods

Nanotechnology represents a burgeoning domain with the potential to provide an efficient means of delivering functional foods. The nano-delivery systems enhance the functional characteristics of nutraceuticals, encompassing bioactive peptides, vitamins, antioxidants, shell/wall materials, and probiotics, all at submicron dimensions (1–200 nm). The increased surface area of these nanomaterials amplifies their function as carriers, shielding the core components from degradation, while also extending their stability, solubility, and capacity for effective delivery. This, in turn, enhances their bioavailability at specific target sites.[7] Incorporating beneficial elements for health, such as minerals, vitamins, essential fatty acids, antioxidants, carotenoids, dietary fibers and phytosterols, has garnered substantial attention in the domain of functional foods.[16,17] Nonetheless, introducing these elements into food formulations poses a formidable challenge because they are prone to instability throughout food processing and storage. This instability arises from their rapid degradation, often induced by various environmental factors such as pH, temperature, moisture, light, oxygen, *etc.* Moreover, these can engage in unwanted interactions with other components within the food matrix.

Certain bioactive substances, including polyphenols, peptides, and phytochemicals, may carry unappealing aromas or flavours that can lead to an unpleasant taste and astringency. Such bioactive compounds tend to be vulnerable as they journey through the gastrointestinal tract, where they encounter a wide pH range and various digestive enzymes. Consequently, they sometimes degrade in the stomach before reaching the small intestine, leading to reduced bioavailability. Additionally, some bioactives, such as essential oils, curcumin, and bioactive lipids, exhibit poor solubility in aqueous solutions, which limits their utility in functional foods.[15,18,19] The utilization of nano-encapsulation for safeguarding bioactive compounds sensitive to heat and oxygen presents a promising strategy for integrating them into functional foods. This method effectively safeguards these compounds throughout the processing and storage phases, masks undesirable tastes and flavours, enhances their water dispersibility, enables controlled release, and consequently enhances their bioavailability.[16,20]

Various biomolecules, such as casein micelles and fat globule networks, naturally exist in the form of nanoparticles within milk systems. In contrast, food nanotechnology pertains to the deliberate introduction of engineered nanoparticles into food. These nanoscale materials exhibit unique characteristics, presenting significant potential for the food industry, mainly in the advancement of functional foods. Most bioactive compounds, encompassing carbohydrates, lipids, vitamins, and proteins, tend to be fragile when exposed to the highly acidic conditions and enzymatic activity present in the duodenum and stomach. The encapsulation of these diverse bioactives not only enables them to withstand these challenging conditions but also facilitates their seamless integration into food products. Achieving this effect proves to be a challenging task when using non-encapsulated bioactive compounds with limited solubility. To improve the delivery of delicate micronutrients, vitamins, and medications, nanoparticle-encapsulated small edible capsules are presently employed in everyday food products to

Table 2.1 Food-grade nano-delivery systems.

Delivery system	Description	Applications	Ref.
Casein micelles	Colloidal nanostructures (~20–300 nm)	Delivery of vitamins, proteins, and minerals	23–29
Films/edible coatings	Thin layers or coatings similar to membranes	Preservation of fresh-food quality during extended storage	
Liposomes	Vesicles formed by amphipathic lipid molecules (*i.e.* phospholipids)	Delivery of hydrophilic and hydrophobic biomolecules	
Microemulsions	Thermodynamically stable systems (particles with radii of about 2 to 50 nm)	Hydrophobic and hydrophilic molecule delivery and solubility	
Nanoemulsions	Conventional emulsion (average droplet size between 10 and 100 nm)	Delivery and better physical stability and bioavailability of lipophilic compounds	
Nanoparticles	Nanoparticles (dimensions between 1 and 100 nm)	Encapsulation of bioactive ingredients and allow controlled functionalization	
Solid lipid nanoparticles	Submicron emulsion	Deliver and stabilize hydrophobic materials by preventing degradation, improving solubility, and ensuring controlled release	

impart advantageous health benefits.[21,22] A variety of functional nanostructured elements can serve as foundational units for developing innovative structures and introducing new features into food items. Numerous nanoscale materials, including nanocapsules, nano-emulsions, liposomes, nanofibers, and nanotubes, can find application in functional foods. Both organic and inorganic nanomaterials hold the potential for use in food-related applications. Various forms of nano-delivery systems commonly employed across the food industry to transport bioactive substances within food products are described with applications in Table 2.1.

2.3.4 Functional Food Quality and Safety

Presently, nano-delivery systems are also employed within the nutraceutical industry to control and ward off illnesses. The stability of green tea, caffeine, eugenol and curcumin, and their ability to be effectively absorbed by the body, enhance their healing capabilities. Zerumbone, originally obtained from wild ginger, is famous for its wide array of therapeutic benefits, encompassing antibacterial, anti-platelet, anti-inflammatory, antifungal, cytotoxic, and cancer-preventive attributes.[30] Nonetheless, its limited ability to dissolve in water and its low bioavailability when taken orally pose constraints on its utility.[31] Due to advancements in nano-delivery systems, the problematic oral bioavailability of zerumbone has been effectively enhanced. Multiple research studies have documented a rise in both the oral bioavailability and therapeutic benefits of zerumbone when administered through nano-delivery systems.[31–33] Furthermore, as well as its function as a food additive and colouring agent, curcumin has also been noted for its anti-inflammatory properties.[34] Nonetheless, its use in pharmaceuticals was limited due to its poor solubility. In another study, curcumin was encapsulated within redox nanoparticles containing silica, and the outcomes indicated enhanced water solubility and anti-inflammatory impact.[35]

Vitamins and essential dietary components like antioxidants play a crucial role in supporting the proper operation of diverse metabolic processes within our body. Insufficient levels of these substances can result in severe health complications such as anaemia, cardiovascular disorders, and the development of cancer. Unfortunately, these vital dietary components are susceptible to degradation when present in various processed foods. To monitor their deterioration, numerous reports have highlighted the use of various nanosensors combined with different nanomaterials. Several reports highlight the potential use of 'ionic liquid nanocomposites' comprised of 'carbon nanotubes and nickel oxide nanoparticles' to detect ascorbic acid and folic acid within a range of samples, including milk, fruit juices, and wheat.[36–38]

The assessment of food quality often involves examining the levels of citric acid, succinic acid, fructose, L-malic acid, sucrose, D-sorbitol, glucose, L-glutamic acid and hydrogen peroxide in stored food items.[39,40] To accomplish this, various nanosensors, combined with nanoparticles like zirconium dioxide, iron silver, nickel–platinum, chitosan, tin dioxide, gold, cuprous oxide and Prussian blue–gold have been employed to monitor the quality of these food components within different food items.[41–44]

2.3.5 Functional Food Packaging

The utilization of nanotechnology in food packaging offers a broad spectrum of probable commercial applications, particularly in active, intelligent, and enhanced packaging.[7,45–47] Active packaging, as defined by European regulation (EC) No. 450/2009, encompasses a packaging system that actively interacts with the food it contains by "intentionally incorporating elements capable of releasing or absorbing substances into or from the packaged food or the surrounding environment".[48] This approach often incorporates metals like gold, silver, titanium dioxide, and copper, and their unique semiconductor properties make them well-suited for tasks such as adsorption, stain prevention, and catalytic reactions.[47] Nanotechnology has improved the packaging of various food products. In the realm of smart packaging, nanoparticles are employed to monitor the development of chemical, biochemical or microbiological changes within the food itself or in its surrounding environment.[1,49,50] These nanoparticles in packaging serve the purpose of restricting the growth of microbial groups that might exist on the surfaces of food items.[51] For example, Ag nanoparticles have been widely applied as antimicrobial agents in food and beverage preservation.[52,53] A study was conducted where the authors developed a biocomposite film using gelatin extracted from the skin of fish and incorporated Ag–Cu nanoparticles to investigate the impact of Ag–Cu NPs on the antibacterial and thermal properties of the gelatin-based film.[54] The findings revealed that these films exhibited enhanced thermal stability and demonstrated increased antibacterial effectiveness.

Moreover, a study observed the inhibition of microbial growth when metal and metal oxide nanoparticles were introduced.[54] This inhibition was attributed to the production of reactive oxygen species, which leads to oxidative stress and cell death. Additionally, the nanoparticles were found to bind to DNA and enzymes, thereby disrupting cellular functions. Nanotechnology contributes to the reduction of environmental pollution by enabling the creation of biodegradable packaging materials. In the food industry, a nanostructure known as a silicate layer is utilized for packaging food products. Advances in sensor technology incorporated into smart food packaging offer valuable insights into material quality, safety, and shelf life.[55] Nanocomposites also find applications in food product packaging, offering improved attributes to withstand thermal stress during food processing, transportation, and storage. Presently, nanocomposites are being utilized in beer bottles, effectively extending their shelf life by as much as six months. Carbon-based graphene nanoplates, known for their heat resistance, hold promise for use in food packaging within the food industry.

2.4 Toxicological and Regulatory Concerns

Nanotechnology holds great potential in enhancing the efficiency and efficacy of delivering nutraceuticals and bioactive substances within functional foods to enhance human well-being, as previously discussed. Nevertheless, despite the numerous benefits and prospects presented by nanotechnological advancements, several hurdles persist for their integration into the development of food products. Although nanoscale delivery systems enhance the bioavailability, permeability, and absorptive capabilities, at the same time they also evoke concerns of potential toxicity,[24,56] due to poor understanding of the toxicity issues related to the use of food-grade nanomaterials. The small size of nanoparticles is linked to potential risks since they can cross cellular barriers. Extensive research has been conducted to thoroughly examine the mechanisms of nanomaterial toxicity. The primary source of nanotoxicity stems from an overabundance of reactive oxygen species, imposing oxidative stress and disruption of regular cellular functions. Consequently, nanotoxicity can lead to genetic material damage, unregulated cellular signaling, cell movement, cellular toxicity, programmed cell death, and the initiation of cancer. Therefore, the safety of nanomaterials in foods is a matter of significant importance and relies on various factors, including the type of nanotechnology utilized, its application (*i.e.* particular food product, food-contact materials, or nutritional supplement), and additional considerations like temperature.[57]

Currently, there are no specific guidelines or regulatory documents tailored to nanotechnology, and there are no testing requirements outlined in existing regulations that specifically address nanoparticles based on their size or unique physicochemical characteristics for functional food development. This lack of specific guidance could present a significant obstacle to the commercialization of food products that incorporate nano-encapsulation techniques.[58] Moreover, consumer perception, acceptance, and familiarity with nanotechnology could pose challenges to its implementation in the food industry.[59] A significant majority of respondents (67%) expressed a limited grasp of nanotechnology, nanoproducts, and nanomaterials with around 42% indicating a lack of understanding regarding their impact on food. Low awareness leaves consumers ill-equipped to make informed assessments of probable risk associated with health and environment.[60,61]

2.5 Summary and Future Outlook

Inspired by nanotechnology, the popularity of using nanostructures in the functional food industry is increasing, and consequently interest and activities in this research area have greatly increased in recent times. Use of nanotechnology in the areas of food packaging and maintaining food safety are well known. Furthermore, potential application has been attained in food preservation by means of nanomaterials which might protect the food from lipids, moisture, gases, off-flavours, and odours. Developments in of new delivery systems for functional foods offer significant scope for improving the bioavailability or absorption of bioactive compounds in specific parts of the gastrointestinal tract to further improve human health. The delivery systems integrate nutraceuticals and bioactive compounds in a food matrix and protect them from adverse food-processing and storage conditions and thus enhance their effectiveness and efficiency. No doubt this technology offers numerous advantages but the negative effects of nanometre-scale particles on human health must be addressed to alleviate consumer concerns. Furthermore, issues concerned with environmental safety should be thoroughly examined before the release of any food in the market.

Conflicts of Interest

The authors do not have any conflict of interest.

Acknowledgements

The authors are grateful to the authorities of the respective organizations for their support in writing this manuscript.

References

1. M. Rai, A. Yadav and A. Gade, Silver nanoparticles as a new generation of antimicrobials, *Biotechnol. Adv.*, 2009, **27**(1), 76–83.
2. A. Gupta, H. B. Eral, T. A. Hatton and P. S. Doyle, Nanoemulsions: formation, properties and applications, *Soft Matter*, 2016, **12**(11), 2826–2841.
3. S. H. Nile, V. Baskar, D. Selvaraj, A. Nile, J. Xiao and G. Kai, Nanotechnologies in food science: applications, recent trends, and future perspectives, *Nano-Micro Lett.*, 2020, **12**, 1–34.
4. P. N. Ezhilarasi, P. Karthik, N. Chhanwal and C. Anandharamakrishnan, Nanoencapsulation techniques for food bioactive components: a review, *Food Bioprocess Technol.*, 2013, **6**, 628–647.
5. T. Singh, S. Shukla, P. Kumar, V. Wahla, V. K. Bajpai and I. A. Rather, Application of nanotechnology in food science: perception and overview, *Front. Microbiol.*, 2017, **8**, 1501.
6. N. Dasgupta, S. Ranjan, D. Mundekkad, C. Ramalingam, R. Shanker and A. Kumar, Nanotechnology in agro-food: from field to plate, *Food Res. Int.*, 2015, **69**, 381–400.
7. C. Han, A. Zhao, E. Varughese and E. Sahle-Demessie, Evaluating weathering of food packaging polyethylene-nano-clay composites: Release of nanoparticles and their impacts, *NanoImpact*, 2018, **9**, 61–71.
8. Sonia, Role of functional food in therapeutic nutrition: brief review, *Asian J. Sci. Technol.*, 2019, **10**(01), 9221–9227.
9. N. Tehri, A. Vashishth, A. Gahlaut and V. Hooda, Biosynthesis, antimicrobial spectra and applications of silver nanoparticles: current progress and future prospects, *Inorg. Nano-Met. Chem.*, 2022, **52**(1), 1–19.
10. A. J. Paredes, C. M. Asensio, J. M. Llabot, D. A. Allemandi and S. D. Palma, *Nanoencapsulation in the food industry: manufacture, applications and characterization*, 2016.
11. P. Y. H. Hsieh and J. A. Ofori, Innovations in food technology for health, *Asia Pac. J. Clin. Nutr.*, 2007, **16**(S1), 65–73.
12. E. Acosta, Bioavailability of nanoparticles in nutrient and nutraceutical delivery, *Curr. Opin. Colloid Interface Sci.*, 2009, **14**(1), 3–15.
13. M. R. Mozafari, C. J. Reed and C. Rostron, Prospects of anionic nanolipoplexes in nanotherapy: transmission electron microscopy and light scattering studies, *Micron*, 2007, **38**(8), 787–795.
14. M. Fathi, M. R. Mozafari and M. Mohebbi, Nanoencapsulation of food ingredients using lipid based delivery systems, *Trends Food Sci. Technol.*, 2012, **23**(1), 13–27.
15. D. J. McClements and Y. Li, Structured emulsion-based delivery systems: Controlling the digestion and release of lipophilic food components, *Adv. Colloid Interface Sci.*, 2010, **159**(2), 213–228.
16. K. O. Aboalnaja, S. Yaghmoor, T. A. Kumosani and D. J. McClements, Utilization of nanoemulsions to enhance bioactivity of pharmaceuticals, supplements, and nutraceuticals: Nanoemulsion delivery systems and

nanoemulsion excipient systems, *Expert Opin. Drug Delivery*, 2016, **13**(9), 1327–1336.
17. D. J. McClements, Utilizing food effects to overcome challenges in delivery of lipophilic bioactives: structural design of medical and functional foods, *Expert Opin. Drug Delivery*, 2013, **10**(12), 1621–1632.
18. F. L. Sousa, M. Santos, S. M. Rocha and T. Trindade, Encapsulation of essential oils in SiO_2 microcapsules and release behaviour of volatile compounds, *J. Microencapsulation*, 2014, **31**(7), 627–635.
19. L. Zou, W. Liu, C. Liu, H. Xiao and D. J. McClements, Designing excipient emulsions to increase nutraceutical bioavailability: emulsifier type influences curcumin stability and bioaccessibility by altering gastrointestinal fate, *Food Funct.*, 2015, **6**(8), 2475–2486.
20. S. M. Jafari and D. J. McClements, Nanotechnology approaches for increasing nutrient bioavailability, *Adv. Food Nutr. Res.*, 2017, **81**, 1–30.
21. A. Haider and I. K. Kang, Preparation of silver nanoparticles and their industrial and biomedical applications: a comprehensive review, *Adv. Mater. Sci. Eng.*, 2015, **2015**, 1–16.
22. Q. Chaudhry, M. Scotter, J. Blackburn, B. Ross, A. Boxall and L. Castle, *et al.*, Applications and implications of nanotechnologies for the food sector, *Food Addit. Contam.*, 2008, **25**(3), 241–258.
23. J. Zhu and Q. Huang, Chapter Four – Nanoencapsulation of functional food ingredients, in *Advances in Food and Nutrition Research*, ed. L.-T. Lim and M. Rogers, Academic Press, 2019, vol. 88, pp. 129–165.
24. Y. D. Livney, Nanostructured delivery systems in food: Latest developments and potential future directions, *Curr. Opin. Food Sci.*, 2015, **3**, 125–135.
25. H. Singh, Nanotechnology applications in functional foods; opportunities and challenges, *Prev. Nutr. Food Sci.*, 2016, **21**(1), 1.
26. D. J. McClements, Edible nanoemulsions: fabrication, properties, and functional performance, *Soft Matter*, 2011, **7**(6), 2297–2316.
27. Q. Lin, S. Ge, D. J. McClements, X. Li, Z. Jin and A. Jiao, *et al.*, Advances in preparation, interaction and stimulus responsiveness of protein-based nanodelivery systems, *Crit. Rev. Food Sci. Nutr.*, 2023, **63**(19), 4092–4105.
28. D. J. McClements and B. Öztürk, Utilization of nanotechnology to improve the handling, storage and biocompatibility of bioactive lipids in food applications, *Foods*, 2021, **10**(2), 1–17.
29. E. Assadpour and S. M. Jafari, Nanoencapsulation: Techniques and developments for food applications, in *Nanomaterials for food applications*, Elsevier, 2019, pp. 35–61.
30. M. A. Haque, I. Jantan and H. Harikrishnan, Zerumbone suppresses the activation of inflammatory mediators in LPS-stimulated U937 macrophages through MyD88-dependent NF-κB/MAPK/PI3K-Akt signaling pathways, *Int. Immunopharmacol.*, 2018, **55**, 312–322.
31. S. Md, B. C. Kit, S. Jagdish, D. J. David, M. Pandey and L. A. Chatterjee, Development and *in vitro* evaluation of a zerumbone loaded nanosuspension drug delivery system, *Crystals*, 2018, **8**(7), 286.

32. M. G. Dorrington and I. D. Fraser, NF-κB signaling in macrophages: dynamics, crosstalk, and signal integration, *Front. Immunol.*, 2019, **10**, 705.
33. J. N. Foong, G. T. Selvarajah, A. Rasedee, H. S. Rahman, C. W. How and C. Y. Beh, *et al.*, Zerumbone-loaded nanostructured lipid carrier induces apoptosis of canine mammary adenocarcinoma cells, *BioMed Res. Int.*, 2018, **2018**, 8691569.
34. D. D. Clarence, K. R. Paudel, B. Manandhar, S. K. Singh, H. P. Devkota and J. Panneerselvam, *et al.*, Unravelling the therapeutic potential of nano-delivered functional foods in chronic respiratory diseases, *Nutrients*, 2022, **14**(18), 3828.
35. D. J. McClements, Recent advances in the production and application of nano-enabled bioactive food ingredients, *Curr. Opin. Food Sci.*, 2020, **33**, 85–90.
36. H. Karimi-Maleh, M. Moazampour, M. Yoosefian, A. L. Sanati, F. Tahernejad-Javazmi and M. Mahani, An electrochemical nanosensor for simultaneous voltammetric determination of ascorbic acid and Sudan I in food samples, *Food Anal. Methods*, 2014, **7**, 2169–2176.
37. S. Wei, F. Zhao, Z. Xu and B. Zeng, Voltammetric determination of folic acid with a multi-walled carbon nanotube-modified gold electrode, *Microchim. Acta*, 2006, **152**, 285–290.
38. F. Xiao, C. Ruan, L. Liu, R. Yan, F. Zhao and B. Zeng, Single-walled carbon nanotube-ionic liquid paste electrode for the sensitive voltammetric determination of folic acid, *Sens. Actuators, B*, 2008, **134**(2), 895–901.
39. S. Vermeir, B. M. Nicolai, K. Jans, G. Maes and J. Lammertyn, High-throughput microplate enzymatic assays for fast sugar and acid quantification in apple and tomato, *J. Agric. Food Chem.*, 2007, **55**(9), 3240–3248.
40. K. J. Verstrepen, D. Iserentant, P. Malcorps, G. Derdelinckx, P. Van Dijck and J. Winderickx, *et al.*, Glucose and sucrose: hazardous fast-food for industrial yeast?, *Trends Biotechnol.*, 2004, **22**(10), 531–537.
41. E. Filippo, A. Serra and D. Manno, Poly (vinyl alcohol) capped silver nanoparticles as localized surface plasmon resonance-based hydrogen peroxide sensor, *Sens. Actuators, B*, 2009, **138**(2), 625–630.
42. K. Z. Liang and W. J. Mu, ZrO_2/DNA-derivated polyion hybrid complex membrane for the determination of hydrogen peroxide in milk, *Ionics*, 2008, **14**, 533–539.
43. Y. J. Liu, D. J. Yao, H. Y. Chang, C. M. Liu and C. Chen, Magnetic bead-based DNA detection with multi-layers quantum dots labeling for rapid detection of *Escherichia coli* O157: H7, *Biosens. Bioelectron.*, 2008, **24**(4), 558–565.
44. J. S. Ye, Y. Wen, W. De Zhang, L. M. Gan, G. Q. Xu and F. S. Sheu, Nonenzymatic glucose detection using multi-walled carbon nanotube electrodes, *Electrochem. Commun.*, 2004, **6**(1), 66–70.
45. S. Kumar, A. Shukla, P. P. Baul, A. Mitra and D. Halder, Biodegradable hybrid nanocomposites of chitosan/gelatin and silver nanoparticles for

active food packaging applications, *Food Packag. Shelf Life*, 2018, **16**, 178–184.
46. S. D. F. Mihindukulasuriya and L. T. Lim, Nanotechnology development in food packaging: A review, *Trends Food Sci. Technol.*, 2014, **40**(2), 149–167.
47. K. Neme, A. Nafady, S. Uddin and Y. B. Tola, Application of nanotechnology in agriculture, postharvest loss reduction and food processing: Food security implication and challenges, *Heliyon*, 2021, 7(12), 149–167.
48. S. Yildirim, B. Röcker, M. K. Pettersen, J. Nilsen-Nygaard, Z. Ayhan and R. Rutkaite, *et al.*, Active packaging applications for food, *Compr. Rev. Food Sci. Food Saf.*, 2018, **17**(1), 165–199.
49. B. Kuswandi and M. Moradi, Improvement of food packaging based on functional nanomaterial, *Nanotechnology: applications in energy, drug and food*, 2019, 309–344.
50. A. Ashfaq, N. Khursheed, S. Fatima, Z. Anjum and K. Younis, Application of nanotechnology in food packaging: Pros and Cons, *J. Agric. Food Res.*, 2022, 7, 100270.
51. M. Rossi, D. Passeri, A. Sinibaldi, M. Angjellari, E. Tamburri and A. Sorbo, *et al.*, Nanotechnology for food packaging and food quality assessment, *Adv. Food Nutr. Res.*, 2017, **82**, 149–204.
52. A. S. da Costa Ribeiro, J. de Lima Gomes, M. M. Kushida and R. R. Petrus, Sugarcane juice stability in plastic bottle treated with silver and zinc oxide, *Packag. Technol. Sci.*, 2019, **32**(3), 155–161.
53. A. E. Fadiji, D. M. N. Mthiyane, D. C. Onwudiwe and O. O. Babalola, Harnessing the known and unknown impact of nanotechnology on enhancing food security and reducing postharvest losses: Constraints and future prospects, *Agronomy*, 2022, **12**(7), 1657.
54. Y. A. Arfat, J. Ahmed, N. Hiremath, R. Auras and A. Joseph, Thermo-mechanical, rheological, structural and antimicrobial properties of bionanocomposite films based on fish skin gelatin and silver-copper nanoparticles, *Food Hydrocolloids*, 2017, **62**, 191–202.
55. D. Schaefer and W. M. Cheung, Smart packaging: opportunities and challenges, *Proc. CIRP*, 2018, **72**, 1022–1027.
56. P. M. Reque and A. Brandelli, Encapsulation of probiotics and nutraceuticals: applications in functional food industry, *Trends Food Sci. Technol.*, 2021, **114**, 1–10.
57. J. C. Buzby, Nanotechnology for food applications: more questions than answers, *J. Consum. Aff.*, 2010, **44**(3), 528–545.
58. R. Singh, B. Mann, R. Sharma and S. Singh, Application of Nanotechnology in Functional Foods, in *Nanoscience for Sustainable Agriculture*, ed. R. Pudake, N. Chauhan and C. Kole, Springer, Cham, 2019.
59. C. L. Cummings, J. Kuzma, A. Kokotovich, D. Glas and K. Grieger, Barriers to responsible innovation of nanotechnology applications in food

and agriculture: A study of US experts and developers, *NanoImpact*, 2021, **23**, 100326.
60. A. S. Ummi and S. Siddiquee, Nanotechnology applications in food: opportunities and challenges in food industry, *Nanotechnology: Applications in Energy, Drug and Food*, 2019, pp. 295–308.
61. Z. H. Arabeyyat, M. M. Jamaliah and M. A. Khalaf, Public awareness of nanotechnology and its implications for health in Jordan, *Sustainability*, 2022, **14**(10), 5786.

CHAPTER 3

Recent Advances in Fermented Functional Foods

M. L. MOHEDANO,[a] K. ZAROUR,[b] I. DIEZ-OZAETA,[a,c,†]
M. T. DUEÑAS,[c] P. LÓPEZ*[a] AND P. RUSSO[d]

[a] Departamento de Biotecnología Microbiana y de Plantas, Centro de Investigaciones Biológicas Margarita Salas (CSIC), Madrid, Spain; [b] Laboratoire de Microbiologie Appliquée (LMA), Faculté des Sciences de la Nature et de la Vie, Université Oran 1 Ahmed Ben Bella, Oran, Algeria; [c] Departamento de Química Aplicada, Facultad de Química, Universidad del País Vasco (UPV/EHU), San Sebastián, Spain; [d] Department of Food Environmental and Nutritional Sciences (DeFENS), University of Milan, Italy
*Email: plg@cib.csic.es

3.1 Introduction

Fermentation, an ancient food preservation and enhancement technique, has served as a milestone for the development of functional foods (see Figure 3.1). The transformation of substrates through microbial metabolism not only imparts unique flavors and textures but also produces biochemical changes that can modulate the nutritional and health attributes of the final product.[1] Functional foods can be defined as dietary products that, besides providing nutrients and energy, beneficially modulate one or more targeted functions in the body, by enhancing certain physiological responses and/or by reducing the risk of disease.[2] Fermented functional foods are more than just gastronomic delights; they are evidence of the potential of food as a

† Current address: BCC Innovation, Technology Center in Gastronomy, Basque Culinary Center, 20009 Donostia-San Sebastián, Spain.

Food Chemistry, Function and Analysis No. 44
Functional Foods of the Future
Edited by Vijai Kumar Gupta, Minaxi Sharma, Smriti Gaur and Ramesh Chander Kuhad

Published by the Royal Society of Chemistry, www.rsc.org

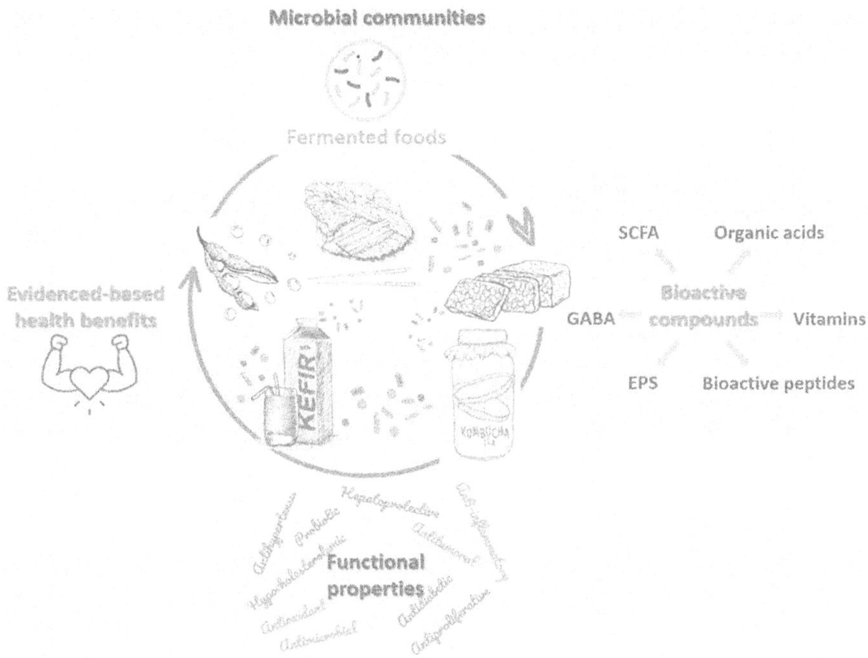

Figure 3.1 Schematic representation of the beneficial effects that microbial communities can exert through food fermentation.

source of holistic wellbeing. One of the outstanding aspects of the fermented functional foods is their capability to introduce or enhance the occurrence of bioactive compounds that are often limited or absent in their non-fermented counterparts. This phenomenon is intricately linked to the metabolic activities of the diverse microbial consortia involved in fermentation processes. Bioactive compounds, characterized by their capacity to exert specific physiological effects when consumed, have become a focal point of research in the quest to understand the mechanisms that underpin the health benefits attributed to fermented foods. The spectrum of these compounds includes, but is not limited to, probiotics, prebiotics, vitamins, peptides, polyphenols, and other metabolites synthesized during microbial fermentation.[3] These compounds offer a multifaceted array of functionalities, from the modulation of gut microbiota to the direct interaction with host cells and tissues.[4,5] In particular, their potential to exert beneficial effects on digestive health, immune system modulation, antioxidant capacity, and the prevention of chronic diseases is of great interest to the food industry.[6]

In recent years, the understanding of these intricate processes has evolved significantly, with an array of advanced analytical tools enabling the in-depth characterization of the microbial consortia and metabolites responsible for these transformations. Concurrently, the genomic and metagenomic analysis of these microbial communities has expanded our comprehension of the intricate web of interactions occurring during fermentation.[7]

3.2 Bioactive Ingredients for Fermented Functional Food

3.2.1 Fermentation: A Tool to Increase the Content of B-group Vitamins

Vitamins are essential micronutrients involved as coenzymes of critical metabolic reactions that can be synthesized only by certain plants and microorganisms. Water-soluble B-complex vitamins include thiamin (vitamin B_1), riboflavin (vitamin B_2), niacin (vitamin B_3), pantothenic acid (vitamin B_5), pyridoxine (vitamin B_6), biotin (vitamin B_7), folate (vitamin B_9) and cobalamin (vitamin B_{12}). The ability to synthesize vitamins of the B-group is a strain-dependent trait shared among different microorganisms involved in food fermentation, including both yeasts and bacteria.[8]

Fermentation is considered a good way to increase the level of some vitamins of the B-group through bacterial metabolism. However, spontaneous fermentation could lead to a loss of these compounds due to the detrimental effect of non-producing microorganisms. Therefore, microbial-based approaches can be employed to improve and optimize the microbial vitamin production during fermentation.[9] Overall, a screening of the vitamin-overproducing strains is the main strategy to select starters for the biofortification of different matrices, since significant differences in terms of quantity and distribution, either intracellularly or extracellularly, have been reported.[10]

Moreover, several conditions related to the fermentation process, including pH and the occurrence of reactive oxidative species, could cause the degradation and/or interconversion among different vitamers, *i.e.* closely related forms of the same vitamins but with different physiological activities and bioavailability.[11] Therefore, the selection of starter cultures mainly producing more stable vitamers under the technological conditions used for food processing would be useful to enhance the vitamin content of some fermented foods. Finally, it is important to consider that different processes can be responsible for the loss of high amounts of B-group vitamins including thermolabile, water-soluble and photosensitive compounds.[12] For example, submerged fermentation has been reported to drastically reduce the concentration of niacin, thiamin, and riboflavin in pumpkin and kale leaves, respectively.[13,14] Furthermore, from a technological point of view, it should be noted that optimizing some fermentation conditions, such as increasing the fermentation time, could be a useful approach to prevent vitamin losses as well as to encourage their biosynthesis.[15]

3.2.1.1 Riboflavin

Riboflavin is the precursor of the coenzymes flavin mononucleotide (FMN) and flavin adenine dinucleotide (FAD), playing a central role in essential reactions including the Krebs cycle, respiratory chain, fatty acid oxidation, purine catabolism, and biosynthesis of pyridoxine. Deficiency of vitamin B_2 is due not only to malnutrition, but also to some dietary lifestyle and physiological or pathological conditions. The current solutions to overcome riboflavin deficiency are supplementation or fortification by using exogenous riboflavin produced by chemical or biotechnological synthesis. The biotechnological production is mainly obtained using strains of *Ashbya gossypii* or genetically modified *Bacillus subtilis*.[16] In recent years, there has been a growing interest in the concept of *in situ* biofortification as an economic and eco-friendly alternative to the industrial production of riboflavin. However, the question is whether a bacterium can produce this micronutrient at levels for adequate fortification. It has been reported that food-grade riboflavin-overproducing spontaneous mutant strains can be selected under the selective pressure of roseoflavin, a toxic analogue of riboflavin.[17] This phenotype is due to an impaired regulation of the *rib* operon.[18] By using this biotechnological strategy different species of interest to the food industry have been selected including *L. plantarum*, *Limosilactobacillus fermentum*, *Lactococcus lactis*, *Leuconostoc mesenteroides*, *Propionibacterium freudenreichii* and *Weisella cibaria*.[18–21] These strains have been employed for the production at laboratory scale of biofortified bread,[15,22] yogurt,[19] cereal-based beverages,[21,23] soy-milk and okara.[24,25] The higher levels of riboflavin production in culture media reported for different species of food-grade bacteria selected by treatment with roseoflavin are shown in Figure 3.2. The highest yield was synthesized by the *W. cibaria* BAL3C-5 C120T riboflavin-overproducing strain (1.5×10^{11} colony forming units per L produced about 6.8 $mg L^{-1}$) showing the yellow color of riboflavin in cultures in solid and liquid media (Figure 3.2).

Recently, riboflavin-overproducing and dextran-producing strains of *Weissella cibaria*, which is still without qualified presumption of safety (QPS), have also been proposed to produce bread with enhanced nutritional and technological qualities.[26] Interestingly, *Limosilactobacillus reuteri* AMBV339 isolated from human vagina was found to be the lactic acid bacterium synthesizing the highest level of riboflavin (6×10^{12} colony forming units per L produced about 18 $mg L^{-1}$) in culture medium reported so far. This strain has been employed to fortify plant-based and dairy foods to different extents, indicating the important role of the food matrix to induce riboflavin production.[27]

In order to evaluate the potential production of riboflavin-overproducing strains directly in the gut of the host, their probiotic aptitudes have been evaluated using both *in vitro* and *in vivo* models.[27–30] In particular, *in vitro* evidence showed a passive and an active transport of riboflavin by using a gastrointestinal dialysis model and Caco-2 cell lines, respectively.[27,28]

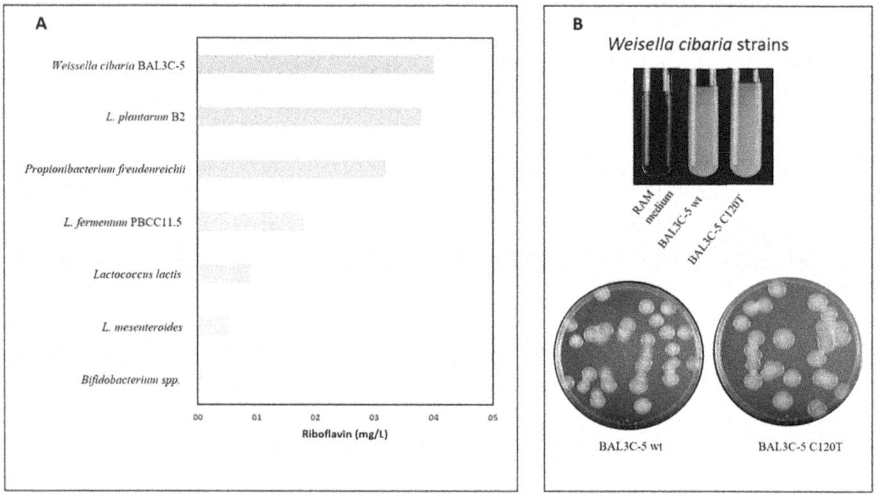

Figure 3.2 (A) Levels of riboflavin produced by spontaneous-overproducer roseoflavin-resistant strains belonging to species of interest in the food industry. (B) Detection of riboflavin production by *W. cibaria* BAL 3C-5 wild type (wt) and its isogenic riboflavin-overproducing BAL3C-5 C120T mutant strains in liquid and solid media.

Accordingly, *in vivo* assays using a murine model have demonstrated that administration of biofortified foods or riboflavin-overproducing probiotics is able to revert ariboflavinosis (a deficiency disease due to lack of vitamin B_2), as well to ameliorate some pathological conditions including inflammatory, cancer and neurodegenerative diseases.[31–35]

3.2.1.2 Cobalamin

Cobalamin is the only vitamin of microbial origin, since it can be synthesized only by some bacteria and archaea. Its deficiency in human beings might lead to disturbance of cell division, neuropathy, nervous system disease and pernicious anemia. This vitamin is industrially produced by microbial fermentation, as chemical synthesis is very costly. However, among food-grade bacteria of interest in the field of food fermentation, only some strains belonging to a few species of lactic acid bacteria (LAB) have been reported to carry the complete gene clusters required for vitamin B_{12} biosynthesis.[36] In particular, *L. reuteri* was the first LAB reported for this ability.[37] Over the years, other strains belonging to the LAB species *L. plantarum*,[38] *L. fermentum*,[39] *Furfurilactobacillus rossiae*,[40] and *Loigolactobacillus coryniformis*[41,42] have also been described as cobalamin producers. It should be remarked that *L. plantarum* and *L. fermentum* are usually employed as starters to ferment different matrices including sourdoughs and vegetables. Moreover, several strains belonging to these species as well as to *L. reuteri* have been reported for their probiotic properties. In contrast,

F. rossiae and *Loigolactobacillus coryniformis* are underexploited LAB species with sporadic applications in the food industry. Nonetheless, to the best of our knowledge, until now only soy milk fermented by *L. reuteri* or *L. plantarum* strains has been proposed as a prospective approach to provide the daily intake with nutritionally-relevant and biologically-active B_{12}.[43–45] However, the ability of LAB strains to produce vitamin B_{12} in nutritionally relevant amounts rather than other corrinoids not available for humans remains controversial.[46]

Unlike LAB, *Propionibacterium* spp. have been widely reported for their capability to synthesize biologically active cobalamin.[47] Propionibacteria are mainly associated with a dairy environment and are routinely utilized as starters in Swiss-type cheese making. Recently, a pro-bioactive ricotta cheese whey containing cobalamin derivatives in the range 0.49–1.31 $mg\,L^{-1}$ was obtained by fermentation with *Propionibacterium freudenreichii*.[48] Also, in recent years *P. freudenreichii* has been employed to ferment sourdough-related matrices including cereals (*i.e.* wheat, oat, rice, and rye bran, as well as millet and sorghum flours), pseudocereals (*i.e.* buckwheat bran, amaranth and quinoa flours), and legumes (*i.e.* soybean, faba bean, and lupine flours) suggesting new applications for the *in situ* biofortification of this vitamin.[49–51]

3.2.1.3 Folate

Folate is mainly involved in nucleotide biosynthesis and DNA replication and repair. Folate deficiency is relevant to the etiologies of chronic diseases, and an adequate intake during pregnancy reduces the occurrence of megaloblastic anemia and birth defects.

Several species and strains among food-grade bacteria and yeasts exhibit folate biosynthesis capability. For example, *Lactiplantibacillus* spp. LP1 and *Saccharomyces cerevisiae* Y-J-1 synthesize folate levels up to 1940 μg per 100 g and 2115 μg per 100 g, respectively, thus increasing 1.6- and 1.8-fold the levels of the vitamin in *Moringa oleifera* leaf powder,[52] while a co-fermentation performed by *L. lactis* N8 and *Saccharomyces boulardii* SAA6[53] increased riboflavin and folate levels of Indian idli batter by 40–90%.[54] In the last decade, several studies reported on the bio-enrichment of yogurt and fermented milk. Thus, *P. freudenreichii* subsp. *shermanii* enhanced by 49% (39.2 $\mu g\,mL^{-1}$) the total folate content when inoculated together with traditional yogurt starter cultures.[55] Similarly, a mix of *Lactobacillus delbrueckii* subsp. *bulgaricus*, *Streptococcus thermophilus*, and *Bifidobacterium bifidum* showed a satisfactory potential (59% increase, about 105.4 $\mu g\,kg^{-1}$) to synthesize folate in fermented milk.[53] *L. plantarum* 15HN produced high total folate levels (1487 $\mu g\,L^{-1}$) in yogurt,[56] while *L. plantarum* GSLP-7 V was able to produce 3720 $\mu g\,L^{-1}$ and to modulate the gut microbiota, suggesting its potential utility as a treatment for folate-deficiency related diseases.[57]

The occurrence of folate production has been reported as a promising functional trait of African cereal-based fermented foods.[58] Thus, selected

LAB strains have been used to produce sorghum *motoho* (a sour soft porridge) or *injera* (a sour fermented flatbread made with teff flour) with the final amount of folate reaching up to 22% of the recommended nutrient intakes of folate.[59–61] *Pichia kudriavzevii* in mixed fermentation with *L. fermentum* was a strategy to ferment pearl millet-based gruel to obtain folate-enriched foods,[62] while folate fortified *ragi* gruel (Indian finger millet porridge) obtained by using *Lacticaseibacillus rhamnosus* has been recommended for the diet of pregnant and breast-feeding women.[63] However, folate can be lost during food processing, and strict management of the fermentation should be implemented in homemade products, as suggested for *ben-saalga* biofortification, a typical porridge consumed in Burkina Faso.[12]

Recently, growing interest in the food industry has been devoted to the formulation of functional products for specific dietary lifestyles. A *Latilactobacillus sakei* strain employed to ferment potato substrates supplemented with amaranth and chia flour was able to increase folate levels up to 89–95%. Furthermore, its ability to survive under simulated gastrointestinal tract (GIT) conditions suggested its probiotic potential.[64] The same strain was also employed to ferment Andean potato and amaranth achieving a folate concentration of 1900 $ng\,g^{-1}$, and demonstrating the feasibility of bio-enriched novel vegan spread products through lactic fermentation with no changes in technological parameters.[65]

3.2.1.4 Other B-group Vitamins

Less investigated is the role of food-grade microorganisms in the biosynthesis of the other B-group vitamins in fermented foods. However, a positive effect on the content of thiamine and riboflavin during the fermentation of quinoa seeds with *Lacticaseibacillus casei* has been reported.[66] In amaranth leaves submitted to spontaneous fermentation, the amount of thiamin, riboflavin, and vitamin B_6 was higher compared to the fresh unfermented air-dried leaves.[67] Intriguingly, a yogurt-like product from African yam bean fermented with *S. thermophilus* and *L. delbrueckii* subsp. *bulgaricus* resulted in a 75% reduction of riboflavin levels, while thiamine content increased about 20%.[68] However, studies on thiamine-producing LAB are still pioneering, and preliminary evidence of a neuroprotective effect in a murine Parkinsonian model has been reported.[69,70]

3.2.2 Exopolysaccharides

Exopolysaccharides (EPS) are high-molecular-weight extracellular polysaccharides produced by a wide range of microorganisms. Among them, EPS produced by LAB (LAB-EPS) have been extensively studied[71] and they have a fundamental role in maintaining bacterial homeostasis. They provide protection against adverse environmental conditions, such as changes in temperature, extreme pH and osmotic stress.[72] These metabolites also improve the interaction between bacteria and their surroundings and safeguard them from toxic substances, including antibiotics, enzymes and heavy metals.[73–75] Additionally, they play a crucial role in biofilm formation and serve as a source of energy.[76]

The LAB-EPS can be classified as homopolysaccharides (HoPS) or heteropolysaccharides (HePS) based on the monosaccharide composition.[77] The HoPS[78] are composed of a single type of monosaccharide, mainly D-glucose or D-fructose. The HoPS can be further categorized into a α-glucans, β-glucans and fructans, depending on the type of glycosidic bond, either a single linkage or a combination of a limited number of linkages. The HePS[79] possess a range of at least two different monosaccharides with varying linkage patterns and may also include non-sugar molecules.

These LAB-EPS have been thoroughly studied due to their high production yield (in the case of α-glucans), considerable variation, and generally recognition as safe (GRAS) status by the Food and Drug Administration (FDA) of the USA, and some species are included in the QPS list of the European Food Administration (EFSA) in Europe.[72,73] These polymers have garnered widespread interest in the food industry, including their use as gelling agents, viscosifiers, stabilizers and emulsifiers for improvement of the organoleptic and rheological properties of food.[74–76] Some of the commercially available LAB starter cultures used for yoghurt production are EPS producers.[80] Moreover, LAB-EPS have been found to offer various health benefits, such as acting as immunomodulators[81,82] and anti-inflammatories,[83,84] reducing cholesterol level,[85] serving as anticancer agents,[86] exhibiting antioxidant activity,[87] or modulating the intestinal microbiota.[88] These properties of LAB-EPS enhance their potential for producing fermented functional foods, which should contribute to improved human health.[89]

3.2.2.1 Immunomodulatory Activity

Numerous studies have shown that LAB-EPS can have immunomodulatory effects, which depend on their physicochemical properties, such as monosaccharide composition, molecular weight and electrical charges.[90] Recent studies suggest that there are two clearly differentiated factors that regulate the immunomodulatory capacity of the EPS. The first factor is the EPS electrical charge, *i.e.* the content of phosphate groups (negative charge) in their structure, which activates different immune cells (*i.e.* macrophages, lymphocytes) and stimulates the immune response.[91] The second factor is the molecular mass: small EPS activate immune system cells, whereas large EPS trigger a suppressive immune response.[92] The complexity of their ramifications can also influence their immunomodulatory properties.[93]

The HePS from *L. delbrueckii* subsp. *bulgaricus* strains, which are a mixture of acidic phosphopolysaccharides (negatively charged) and neutral polymers, were analyzed and only the acidic polymers were able to stimulate the immune system response.[94] In a related study,[95] treatment with the acidic HePS produced by *L. delbrueckii* subsp. *bulgaricus* OLL1073R-1 strain generated, in mouse splenocytes, a slight increase in interferon gamma (IFN-γ) levels and natural killer cell activity, as well as a minor decrease in interleukin 4 (IL-4) production. Furthermore, after separation of the high and low molecular weight fractions of the acidic HePS, it was observed that only high molecular weight EPS exhibited these effects. The functional food potential of the OLL1073R-1-EPS[95] was confirmed by the observation of a similar effect when a yogurt prepared by fermentation with *L. delbrueckii* subsp. *bulgaricus* in combination with *S. thermophilus* OLS3295 was administered to mice and its immunomodulation activity was tested.[94]

Currently, intolerance to lactose in milk and allergy to milk proteins are increasingly diagnosed disorders, which prevent or hinder the consumption of dairy products by people who suffer from them. In addition, more and more people are following a vegan diet, and not consuming dairy products because they are derived from animals. In all these cases, people opt for plant-based beverages, such as soy, which are usually made without dairy ingredients to avoid these problems. Studies on EPS production in various fermented matrices (soymilk compared to cow's milk) have been carried out using strains of *S. thermophilus* (*S. thermophilus* SBC8781[96] and *S. thermophilus* 05-34[97,98]). The results revealed that each strain produced different EPS, the soybean matrix being where the LAB produced a higher yield of EPS, with a greater molecular weight and containing a more acidic structure. In addition, when the EPS from the plant matrix were analyzed for their immunomodulatory effect, it was found that they were able to regulate the expression of inflammatory factors in intestinal epithelial cells, reducing the expression of interleukins 6 and 8 (IL-6 and IL-8) as well as IFN-γ.[96]

Kefir is a functional fermented beverage which contains kefiran, a HePS produced by *Lactobacillus kefiranofaciens*, and has been shown to stimulate cytokine production, resulting in increased levels of IL-6, interleukin 10 (IL-10) and interleukin 12 (IL-12) in the small intestine and an increase in serum levels of IL-4, IL-6, IL-10 and IFN-γ.[99,100]

Furthermore, a β-glucan produced by *Pediococcus parvulus* 2.6,[101,102] which was isolated from Basque Country cider[103] and used to produce functional fermented foods,[104] was shown to activate human macrophages with an anti-inflammatory response.[105] Additionally, the exposure of gnotobiotic zebrafish larvae to this HoPS resulted in the inhibition of gene expression of the pro-inflammatory cytokines, TNF-α and IL-8.[82] In addition, β-glucan suppressed the activation of pro-inflammatory cytokines through the nuclear factor kappa light chain enhancer of activated B cells (NF-κB) pathway by inhibiting the protein adaptor MyD88, as well as the recruitment and proliferation of the neutrophils.[82] Moreover, this polymer was found to have anti-inflammatory properties, leading to a decrease in both the gene expression and secretion of IL-8, in an *ex vivo* model of human biopsies from patients with Crohn's disease.[106] On the other hand, the HoPS dextrans produced by three strains of *L. mesenteroides* (CM9, SMS34 and CM30) isolated from mammalian milk, and characterized as high molecular weight glucans (ranging from 1.74×10^8 Da to 4.41×10^8 Da), were able to

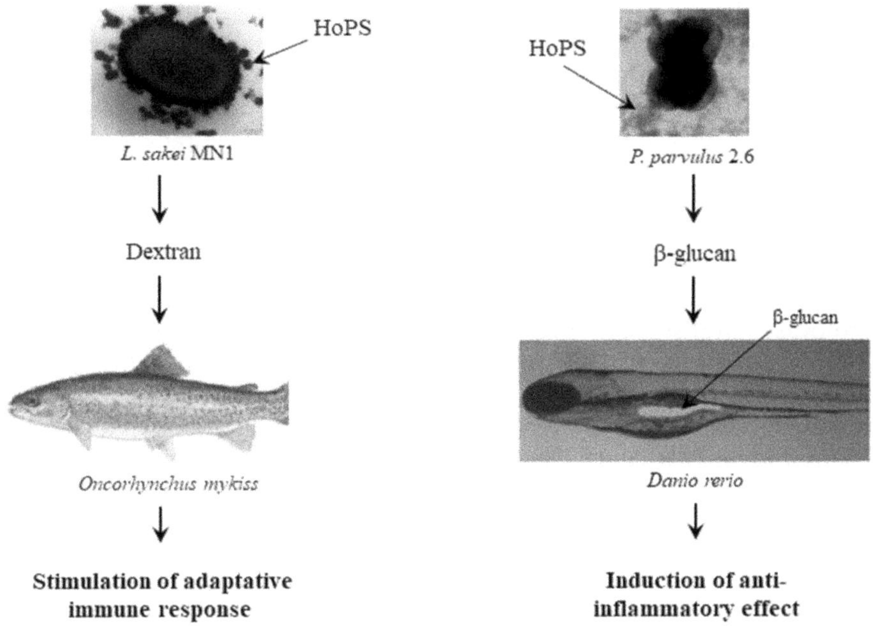

Figure 3.3 Schematic representation of *in vivo* evaluation of HoPS immunomodulation in rainbow trout (dextran of *L. sakei* MN1)[83] and zebrafish (β-glucan of *P. parvulus* 2.6).[82]

immunomodulate *in vitro* human macrophages, decreasing the levels of TNF-α and increasing the levels of IL-10, partially counteracting the inflammatory effect of *Escherichia coli* O111:B4 lipopolysaccharide.[81] Moreover, the HoPS dextran immunomodulates trout *in vivo*, since fish feeding with the polymer of *L. sakei* MN1 generated an increase of interferon gamma after 15 days of treatment.[83]

Consequently, the above-detailed *in vivo* experiments performed with fish models support that the β-glucan produced by *P. parvulus* 2.6 has an anti-inflammatory effect and the dextran from *L. sakei* MN1 stimulates the adaptive immune response (Figure 3.3).

3.2.2.2 Antioxidant Activity

Reactive oxygen species (ROS) play a crucial role in maintaining immune system function and redox balance, including apoptosis, cell signaling, and ion transport. However, the accumulation of ROS can also induce oxidative stress in living organisms, leading to a range of adverse health effects, such as cancer, liver cirrhosis, fatty liver and degenerative systemic diseases (including inflammatory bowel disease, Alzheimer's disease and diabetes).[107] Various studies reported that some LAB-EPS possess potent natural antioxidant properties, effectively combating oxidative stress associated with free radicals.[108] Among them, the HePS produced by *L. plantarum* RO30 isolated from romi cheese can scavenge hydroxyl and superoxide anion radicals.[109] Furthermore, certain EPS can increase the levels of antioxidant enzymes such as catalase and superoxide dismutase.[110,111] In addition, some EPS, such as the dextran produced by *L. mesenteroides* WiKim32, have the potential to chelate metal ions, thus preventing the formation of free radicals.[112] Thus, some LAB-EPS offer an alternative to synthetic antioxidants,[113] which may be carcinogenic or have adverse effects on cell viability.

3.2.2.3 Cholesterol-lowering Abilities

Cardiovascular diseases (CVD) are the leading cause for worldwide mortality, claiming an estimated 17.9 million lives each year. The incidence of heart diseases may decrease with a reduction in serum cholesterol and blood pressure levels. Administration of fermented functional foods that contains EPS-producing LAB could have beneficial effects for managing CVD, since several studies have demonstrated that some of these polymers can decrease serum cholesterol levels (*i.e.* the HePS produced by *Lactiplantibacillus paraplantarum* NCCP 962 and *Lacticaseibacillus paracasei* M7[85,114]). Administration of fermented milk containing acidic HePS produced by *L. lactis* subsp. *cremoris* SBT0495 to hypercholesterolaemic rats demonstrated a favorable outcome in reducing serum cholesterol and increasing the ratio of HDL cholesterol to total cholesterol.[115] Furthermore, the administration of kefiran to spontaneously hypertensive, overweight-prone rats, which had consumed a high-fat diet, caused a lower cholesterol level and reduced blood pressure.[116,117] This last effect may be related to the inhibition of the angiotensin-converting enzyme, and a similar reduction in systolic blood pressure was observed in hypertensive mice, to which the HePS produced by *L. casei* LC2W was administered.[118]

3.2.2.4 Anticancer Activity

Several studies have demonstrated the presence of various anti-tumor mechanisms for LAB-EPS. The primary mechanism is outlined as follows: the EPS can interrupt the cell cycle, thereby inhibiting the proliferation of tumor cells. Thus, the HePS from *L. plantarum* 70810 inhibited the proliferation of human cancer cell lines from liver (HepG-2), stomach (BGC-823) and colon (HT-29).[119] Also, some LAB-EPS possess the potential to stimulate apoptosis of cancer cells, as is the case of the HePS116 produced by *L. plantarum* NCU116, which activates the c-Jun dependent Fas/Fasl-mediated pathway in intestinal cancer cells.[120] In addition, it has been demonstrated that the HoPS dextran synthesized by *L. mesenteroides* SN-8 promotes apoptosis of HepG-2 cells.[121] The cell-bound EPS of *Lactobacillus acidophilus* 606 also exhibits similar properties, inducing apoptosis of HT-29 cells and inhibiting their growth, although only by 20%.[122] Additionally, the EPS can inhibit the expression of genes involved in angiogenesis (*e.g.* the polymer of *L. acidophilus* 10307[123]). Finally, they can enhance the immune response against tumor cells. This was demonstrated in a study where subcutaneous inoculation of tumor cells was performed on mice fed with food containing HePS from *L. delbrueckii* subsp. *bulgaricus* OLL1073R-1.[124]

3.2.2.5 Regulation of Gut Microbiota

LAB-EPS can act at different levels to improve the gut microbiota, thereby improving human health. On one hand, they can act as prebiotics, stimulating and increasing the population of beneficial bacteria in the microbiome. For example, the dextran produced by *L. mesenteroides* XG-5 can resist the hostile conditions of the GIT system and thus reach the gut where it acts as a prebiotic, increasing the concentration of bifidobacteria in the gut microbiome.[125] The same applies to levan-type EPS from *Fructilactobacillus sanfranciscensis* strains,[126] the HePS from *L. rhamnosus* E/N[127] and the dextran from *L. plantarum* DM5,[128] which exert a bifidogenic effect on the microbiome. Furthermore, some EPS, such as the HePS produced by *L. rhamnosus* ZFM231, not only have a bifidogenic effect, but also regulate the biodiversity of the gut microbiota increasing the abundance of other genera such as *Ruminococcus*, *Dorea*, *Butyricoccus* and *Blautia*.[129] On the other hand, the EPS can support the interaction of the probiotic producing bacteria with the intestinal mucus. Thus, HePS generated by the LAB *L. rhamnosus* GG, as well as by *Bifidobacterium longum* NB667 and *Bifidobacterium animalis* IPLA-R1, have this positive effect.[130]

Finally, it has been shown that LAB-EPS can enhance the intestinal barrier. They differentially impact the adherence of pathogenic and beneficial bacteria to the intestinal epithelium cells. This has been demonstrated in studies analyzing the attachment of probiotic and enteropathogenic bacteria to human intestinal mucosal surface of the GIT composed of epithelial cells that act as the initial physical barrier against external factors. Tight junctions, that seal these epithelial cells together, may become compromised causing intestinal barrier dysfunction.[131] In this context, Chen *et al.*[132] demonstrated that the HePS generated by *S. thermophilus* MN-BN-A01 could restore the intestinal barrier in Caco-2 cells treated with *E. coli* lipopolysaccharide, raising the intestinal permeability by inducing changes in the expression of multiple tight junction genes. This HePS was assessed in a murine colitis model, exhibiting the potential to hinder colitis progression in mice and enhance the integrity of the intestinal barrier. Furthermore, the HePS synthesized by *L. plantarum* NCU116 demonstrated efficacy in restoring the intestinal barrier in a mouse model of colitis.[120]

3.2.3 Polyols

Increasing public awareness of dietary health is leading consumers to seek healthier alternatives to traditional sugars and sweeteners. Amidst this growing interest, polyols have emerged as an appealing class of sugar alcohols or polyhydric alcohols, offering a promising solution to address the global health challenges associated with excessive sugar consumption. Polyols, also known as sugar substitutes, present a wide range of applications in various food products and have garnered attention for their potential benefits.[133,134]

They occur naturally in certain fruits and vegetables, but they can also be produced through various processes, including both chemical methods and natural fermentation by microorganisms, including yeast, fungi and bacteria such as LAB, as part of their metabolic processes. Some well-known polyols include mannitol, sorbitol, erythritol, xylitol, and maltitol.[135,136] These compounds are derived from saccharides through various processes, involving the reduction of the aldehyde or ketone group to an alcohol group[137] (Figure 3.4). Enzymatic production of polyols can be achieved through various approaches, including designing bioconversion pathways using the Izumoring strategy and primarily focusing on expressing and co-expressing specific enzymes in engineered microorganisms.[138] The use of functional lactic starter cultures capable of producing polyols opens up new possibilities for creating naturally sweetened fermented foods with several benefits.[139]

Heterofermentative LAB (HeLAB) produce polyols (Figure 3.4) and several studies have demonstrated their spontaneous production of mannitol and erythritol in various fermented foods, including Romanian mixed vegetables, *kimchi*, soybean fermented food, cocoa pulp and cocoa beans, gluten-free sourdough, and sorghum and wheat sourdoughs. This phenomenon highlights the natural ability of LAB to synthesize polyols *in situ* during the fermentation process, leading to the presence of low-calorie sugars in the final products.[140]

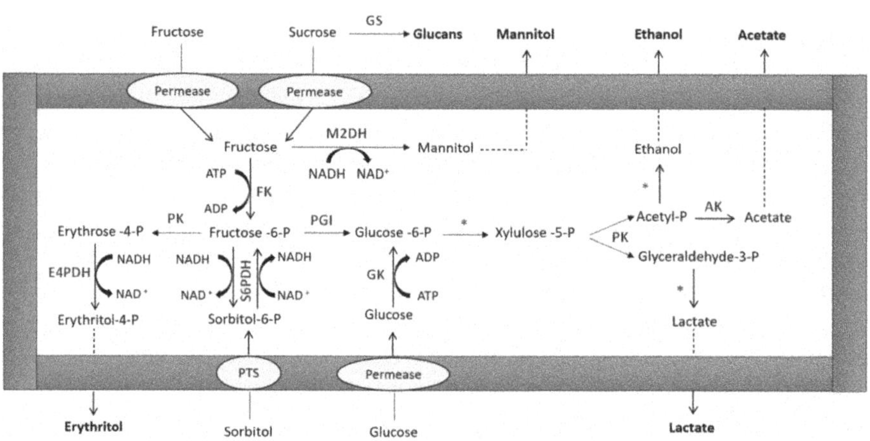

Figure 3.4 Proposed pathway for polyol metabolism of HeLAB in the presence of other carbohydrates. AK, acetate kinase; E4PDH, erythrose-4-phosphate dehydrogenase; FK, fructokinase; GK, glucokinase; GS, glucansucrase; M2DH, mannitol-2-dehydrogenase; PGI, phosphoglucose isomerase; PK, phosphoketolase; S6PDH, sorbitol-6-phosphate dehydrogenase; *, several enzymatic reactions.

One of the key attributes that distinguishes polyols from traditional sugars is their reduced caloric content. With caloric contribution ranging from 0.2 to 2.8 kcal g^{-1}, polyols offer a significant calorie reduction compared to regular sugars, making them appealing to individuals seeking to manage their calorie intake or follow low-calorie diets.[141] Additionally, their impact on blood sugar levels is considerably lower, making polyols a valuable choice for individuals with diabetes or those looking to maintain stable blood glucose levels.[142]

In recent years, the application of polyols has expanded far beyond their role as sugar substitutes. Notably, polyols have found a significant presence in the development of functional fermented foods, a category of products known for their beneficial effects on the gut health and its overall wellbeing. Incorporating polyols into these products not only improves taste and texture but also enhances their nutritional value, creating a win-win scenario for health-conscious consumers.[143,144]

Polyols can be partially metabolized in the colon by gut bacteria. When ingested, polyols are not fully absorbed in the small intestine and reach the colon intact. There, certain bacteria can ferment and metabolize polyols, leading to the production of short-chain fatty acids (SCFA), which are considered beneficial because they play important roles in promoting gut health as well as gases (such as hydrogen, methane, and carbon dioxide). However, it should be pointed that the gas production during this process could generate in some people GIT discomfort.[145] In addition, moderate doses of polyols have been found to induce a shift in the microbiome of healthy individuals, leading to an increase in probiotics such as bifidobacteria, suggesting potential benefits as prebiotics.[146]

D-Mannitol, (2*R*,3*R*,4*R*,5*R*)-hexane-1,2,3,4,5,6-hexol, $C_6H_{14}O_6$, occurs naturally in substantial quantities in olives, carrots, figs, pineapples, sweet potatoes, and larches. This sugar alcohol can be produced by yeasts (such as *Candida* spp. and *Yarrowia* spp.),[147] filamentous fungi (such as *Aspergillus* spp. and *Penicillium* spp.)[148] and bacteria (such as LAB).[140] On the contrary to homofermentative LAB (HoLAB), which are not commonly used for mannitol production, heterofermentative LAB (HeLAB), such as *Leuconostoc* and *Oenococcus* genera and several lactobacilli, are considered the most efficient microorganisms in producing mannitol.[149]

LAB known for their high mannitol production showed high potential to ameliorate the losses of specific volume, denser crumb structure, prolonged shelf life, brown crust color, and sweetness, which are typical attributes of burger buns and challenging to maintain during sugar reduction. Particularly, sourdough formulated with added fructose and fermented by *Leuconostoc citreum* TR116 contains higher amounts of mannitol and lower amounts of lactate, positively influencing dough properties and significantly enhancing the overall quality of the burger buns.[150] The fermentation of apple juice using the mannitol-producing HeLAB, *Levilactobacillus brevis* TR052, *Leuconostoc fallax* TR111, *L. mesenteroides* TR154, *Weissella paramesenteroides* TR212 and *Leuconostoc pseudomesenteroides* TR080, resulted in sugar reduction ranging from 33% to 72%, and the best producer was *L. citreum* TR116 with a high rate of mannitol yield (46.24 g L^{-1}).[151] In addition, the production of mannitol in a quinoa-based milk substitute was investigated using two LAB as starter cultures, *L. citreum* TR116 and *L. brevis* TR055. Quinoa proved to be a suitable substrate, supporting the strains' growth with high viability. The process involved starch hydrolysis, further conversion of glucose to fructose through the enzyme glucose isomerase, and subsequent reduction of fructose to mannitol *via* fermentation, reducing the glucose content by 40% and glycaemic load by 35%. These findings suggested that the two strains can serve as promising functional starter cultures applicable to various fermented food products, including sourdough and beverages.[152]

D-Sorbitol, (2*R*,3*R*,4*R*,5*S*)-hexane-1,2,3,4,5,6-hexol, $C_6H_{14}O_6$, is also known as D-glucitol. Sorbitol production is a rare trait found in only a limited number of microorganisms. Yeasts such as *Candida boidinii*, *Candida famata*, and *S. cerevisiae*, along with the Gram-negative bacteria *Zymomonas mobilis*, possess this ability.[153] While no LAB strain is known to naturally produce sorbitol, certain LAB types, including HeLAB and HoLAB strains, possess catabolic pathways for sorbitol metabolism. This means that they have the capability to metabolize sorbitol as part of their cellular processes.[154]

Erythritol (*meso*-erythritol, $C_4H_{10}O_4$) is a naturally occurring four-carbon sugar derived from glucose through fermentation. It can be found in various fruits, including pears, melons, and grapes, as well as in fermented foods like wine, soy sauce, and cheese. Apart from being utilized as a sugar substitute in baking, erythritol is also known for its antioxidant properties.[133] Erythritol does not significantly affect blood sugar levels, making it suitable for individuals with diabetes or those following a low-carbohydrate diet.[155] It is primarily produced industrially through the fermentation of glucose or cornstarch by certain yeast and bacteria strains including LAB.[156] HeLAB, such as *Oenococcus oeni*, *L. mesenteroides*, and *F. sanfranciscensis*, have the capacity to produce erythritol.

Polyols stand as a compelling player in the quest for healthier alternatives to traditional sugars. From their role as low-calorie sweeteners to their application in functional fermented foods, polyols showcase their potential to revolutionize the way people approach sweetness, taste, and overall nutrition.

3.2.4 Short-chain Fatty Acids

In recent years, the gut microbiota has been the focus of numerous in-depth studies. Its pivotal roles protecting against pathogens, as a metabolic regulator, its endocrine and immune functions, as well as influencing drug metabolism and absorption, have begun to be elucidated. Moreover, it has recently come to light that the influence of the microbiota extends beyond the GIT, since it plays a significant role in the bidirectional communication between the GIT and the central nervous and other systems.[157] Gut microbiota in the colon can produce some beneficial metabolites, which play a crucial role in maintaining gut health and overall wellbeing such as short-chain fatty acids (SCFA).[158]

SCFA are a group of organic acids with a relatively small number of carbon atoms in their chemical structure, typically ranging from 2 to 6 carbon atoms. The discovery that SCFA act as natural ligands for free fatty acid receptors 2 and 3, present in various cell types including enteroendocrine and immune cells, has sparked a renewed interest for the potential impact of SCFA on human health.[159] They are the end products of the fermentation of dietary fiber, *i.e.* non-digestible carbohydrates (NDC). In addition, other components contribute to SCFA production such as resistant starch, plant-derived NDC including non-starch polysaccharides, oligosaccharides, oligofructose, lactose, stachyose, raffinose and polyols (sorbitol, mannitol).[160] Acetate, propionate and butyrate are major SCFA found in the colon and serum.[161] Their concentration and proportion produced in the colon depend not only on the composition and abundance of the gut microbiome but also on the types of dietary fibers provided as substrates for the fermentation process. Consequently, the SCFA profile is influenced by individual microbial populations and the specific dietary choices, making the diet a critical factor in determining their production.[162] Probiotic LAB and bifidobacteria such as *L. rhamnosus*, *Lactobacillus gasseri*, *B. longum*, *B. bifidum*,[163] *Ligilactobacillus salivarius* subsp. *salcinius*, *Ligilactobacillus agilis*,[164] *L. acidophilus*,[165] are involved directly or indirectly in producing SCFA through their metabolic activities in the gut. While LAB can produce some amount of SCFA (acetic, propionic, and butyric acids), the majority of these compounds in the gut is synthesized by other groups of bacteria, which can be modulated by LAB. In this context, SCFA are produced as a result of cross-feeding interactions and modulation of the gut environment. Primary fermenters such as LAB, are efficient at breaking down and metabolizing complex carbohydrates (dietary fiber and NDC) to produce various compounds as byproducts such as lactic acid and EPS.[166] Other bacterial groups, as secondary fermenters, can utilize the metabolic end products produced by primary fermenters as a nutrient source to produce SCFA such as propionate and butyrate.[167] Several key functions of SCFA in the human body have been reported. They represent an essential source of energy for the cells lining the colon: (i) contributing to a balanced gut microbiota, which is crucial for digestive health and overall wellbeing;[168] (ii) playing a role in enhancing the integrity of the intestinal barrier, which helps to prevent the translocation of harmful substances from the gut into the bloodstream;[169] (iii) modulating the immune response in the gut, reducing inflammation and supporting a healthy immune system;[170] (iv) influencing appetite regulation and energy metabolism. Furthermore, SCFA have been associated with improved insulin sensitivity[171] and cardiovascular benefits, including the potential to lower blood pressure and to improve lipid profiles.[172]

Thus, the consumption of fiber-rich foods and prebiotics is essential in promoting SCFA production, contributing to a healthy gut environment and positively impacting various aspects of human health.

3.2.5 Bioactive Peptides

Bioactive peptides, produced through protein hydrolysis by proteases synthesized by fermenting microorganisms, have attracted significant attention due to their numerous health benefits. These peptides have demonstrated various health-promoting properties, including anti-hypertensive, anti-cancer, anti-inflammatory, anti-diabetic, angiotensin-1-converting enzyme (ACE)-inhibitory, anti-microbial, anti-adipogenic, anti-mutagenic, antithrombotic, and anti-atherogenic effects.[173,174] Among these, the most extensively studied bioactive peptides are ACE inhibitor peptides,[175] which are particularly effective in the control of hypertension, a severe medical condition associated with increased risks of heart, brain, and kidney diseases, and affecting around 1.13 billion people worldwide.

Hypertension regulation is largely attributed to the renin-angiotensin-aldosterone system (RAA), with ACE catalyzing the conversion of angiotensin I to angiotensin II, a key player in raising blood pressure.[176] ACE inhibition reduces the formation of angiotensin II, thereby lowering blood pressure. Moreover, the significance of the RAA system has grown due to its connection with ACE2 and its role in SARS-COV-2 virus cell entry,[177] making inhibitors of this system a potential treatment for viral infections. ACE-inhibitory peptides are commonly found in milk fermentation, such as valyl-prolyl-proline and isoleucyl-prolyl-proline, offering a strategy for hypertension control.[178] They have also been identified in clinical studies, with milks fermented with *L. lactis* DIBCA2 and *L. lactis* NRRL B-50571 showing their effectiveness in lowering blood pressure.[178,179]

While most studies have been focused on the antihypertensive properties of bioactive peptides, other multiple health benefits have been also confirmed. For instance, antithrombotic and cholesterol-lowering activities of peptides derived from milk fermented with specific *L. lactis* strains[180] have been demonstrated. Additionally, fermented foods like soybean, kefir, and fish are rich sources of bioactive peptides with hypolipidemic and hypocholesterolemic effects, which helps to reduce lipid levels in the blood through lipase inhibition and improvement of cholesterol ratios.[181] Moreover, researchers identified bioactive peptides from fermented goat milk that effectively lowered cholesterol levels in hypercholesterolemic rats.[182] Recently, there have been identified new peptides in fermented milk with antioxidant properties, which increased the expression of antioxidant enzymes, demonstrating their potential health benefits.[183] In the same way, peptides with antioxidant properties, capable of scavenging free radicals and chelating metal ions, have been isolated from fermented meat products.[184] Additionally, peptides exhibiting antitumoral activities and suppressing the metastasis and angiogenesis of fibrosarcoma HT1080 cells, lung cancer cells, and prostate PC3 cells *in vitro* have been also identified.[185,186]

3.2.6 Phenolic Compounds

The antioxidant properties of fermented foods are primarily attributed to the release of antioxidant molecules during fermentation, including phenolic compounds, bioactive peptides, γ-aminobutyric acid (GABA), and isoflavones. These substances exhibit scavenger activity against free radicals, thereby preventing oxidative damage.[187] The generation of free radicals can trigger a cascade of reactions, such as lipid degradation, inflammatory processes, and DNA damage, ultimately leading to cell damage and death.[188] Consequently, the balance between antioxidant components and free radicals plays a crucial role in determining the level of oxidative stress.

Phenolic compounds are a key contributor to the antioxidant properties of fermented foods. The activity of β-glucosidase in fermenting strains results in the conversion of complex or conjugated phenolic compounds into simpler, more bioavailable active compounds.[189,190] For example, cereal-based fermented products often exhibit an increase in total phenolic content following the fermentation process, as reported in quinoa and buckwheat fermented by *Pediococcus pentosaceous* and *L. paracasei.*[190] Similarly, other studies have confirmed that *L. plantarum* and *F. rossiae* strains could enhance the antioxidant properties of cereal-based fermented foods.[191]

The fermentation of fruit and plant-derived foods also leads to an increase in phenolic content. *L. plantarum* ATCC 14917 enhanced the total phenolic content and antioxidant properties of pomegranate juice compared to unfermented juice.[189] Several studies have reported the ability of various lactobacilli and *Bifidobacterium* strains to increase phenolic acids and, consequently, the antioxidant properties of different fruit-based beverages.[192,193] These studies often measure the *in vitro* antioxidant capacity of fermented foods using free-radical scavenging assays and they correlate higher scavenging capacity with a higher concentration of phenolics. Furthermore, the protection offered by fermented foods against oxidative stress in various cell lines has been confirmed *in vitro*. Thus, it has been demonstrated that *kombucha* (a fermented black tea) could restore the cell viability of mouse fibroblasts exposed to H_2O_2-induced oxidative stress, both in curative and preventive assays.[194] Fermented soybean foods have also exhibited increased antioxidant activity and protection against DNA damage following the fermentation process, and these effects were attributed to the rise in total phenolic and flavonoid concentrations.[195,196] During the fermentation of soybean products like *tempeh* (a traditional Indonesian food), *natto* (traditional Japanese food fermented with *B. subtilis*), or *miso* (a thick paste produced by fermentation with *Aspergillus oryzae*), there is a reduction in the levels of conjugated isoflavones, such as malonyl glycosides, and an increase in unconjugated bioactive isoflavones like daidzein and genistein.[197] These isoflavones have demonstrated effective antioxidant properties by increasing superoxide dismutase activity and reducing lipid peroxidation *in vitro*.[198] These compounds are more rapidly and efficiently absorbed in the gut than their glucosides, contributing to the prevention of chronic diseases.[199] Similar trends have been observed during the fermentation of *kombucha* tea, where the concentration of polyphenols, such as theaflavin, and catechins significantly increases, enhancing the antioxidant properties of the beverages.[194]

3.2.7 Conjugated Linoleic Acid

Conjugated linoleic acid (CLA) has gained significant attention due to its valuable properties and its ability to provide protection against various diseases, including atherosclerosis, obesity, chronic inflammatory conditions, and cancer. Many LAB possess linoleate isomerase activity, which leads to the conversion of linoleic acid (naturally present in milk) to CLA. Notably, dairy products serve as the primary source of CLA, accounting for approximately 70% of its daily intake.[200]

However, the typical daily intake of CLA (70–430 mg per day) is far from the recommended dosage needed to achieve its beneficial effects, which typically range from 3 to 4 grams of CLA per day.[200] Considering this, there has been a proposal to enhance CLA levels in fermented dairy products through milk fermentation with CLA-producing bacteria. Among them LAB producers can be categorized into two groups: those primarily producing the isomer C18:2 *cis*-9, *trans*-11, and those producing C18:2 *trans*-10, *cis*-12.[201] Various bacterial species, including *Bifidobacterium* spp., *Enterococcus* spp., *Lactococcus* spp. and *Streptococcus* spp. have been identified as capable of producing CLA. The existence of potential probiotic *S. cerevisiae* strains with the ability to generate high levels of CLA during dough fermentations has been recently reported.[202] Overall, the development of products rich in CLA has the potential to enhance the antioxidant, anti-obesity, anti-inflammatory, and atheroprotective properties of both dairy and non-dairy fermented products.[203]

3.2.8 γ-Aminobutyric Acid (GABA)

GABA, a non-protein amino acid, plays a pivotal role as the main inhibitory neurotransmitter within the central nervous system.[204] Beyond its neurological function, GABA also plays other physiological roles, including the induction of anti-hypertensive, diuretic and tranquilizing effects, as well as contributing to diabetes prevention.[205]

Researchers have explored the ability of LAB to synthesize GABA, with the goal of producing functional fermented foods. The key enzyme in this process is glutamate decarboxylase, which catalyzes the conversion of L-glutamate into GABA through a single-step α-decarboxylation reaction. The capacity of LAB strains to produce GABA during fermentation represents a promising strategy to enhance the functionality and nutritional properties of fermented foods. GABA-fortified foods have demonstrated efficacy in combating sleeplessness, autonomic disorders, and depression.[206,207] In another study, sourdough bread enriched with amaranth flour and prepared by fermentation with

L. brevis A7 and *Companilactobacillus farciminis* A11 exhibited a higher antioxidant capacity due to an increased content of GABA.[208] Similarly, various strains, including *L. brevis* CECT 8183, *L. plantarum* C48, *L. plantarum* CRL1905, and *L. mesenteroides* CRL1907, have been shown to produce bread with elevated GABA levels.[209,210] Dairy products also offer an interesting vehicle for GABA enrichment, with several research studies demonstrating promising results in the development of GABA-enriched fermented milks and yogurts.[179,211]

3.2.9 Fibrinolytic Enzymes

Fibrinolytic enzymes are considered potent thrombolytic agents to treat and prevent CVD. These sicknesses are a major cause of deaths worldwide and according to recent studies the number of deaths will increase from 18.6 million in 2019 to 23.3 million people by 2030.[212] Fibrinolytic enzymes could prevent CVD that fail in hemostasis and cause the formation of blood clots in the blood vessels, leading to thrombosis. Currently, commonly used drugs for antithrombotic clinical intervention include anti-coagulants (heparin and warfarin), thrombolytic agents, or platelet aggregation inhibitors (aspirin).[213,214] However, several side effects of these drugs may increase health complications such as cerebral hemorrhage,[215,216] therefore safer and more effective approaches to prevent thrombus formation are required. Fibrinolytic enzymes of microbial origin released during the fermentation process have become a research focus in recent years as they may be useful for thrombolytic therapy. Among others, traditional Asian fermented foods, mainly fermented soy products such as *natto, douche* (Chinese fermented black beans), and cheonggukjang (traditional fermented Korean food),[217,218] are a good source of a variety of fibrinolytic enzymes.

The main microorganisms involved in the fermentation of these foods and responsible for fibrinolytic enzyme production belong to the genus *Bacillus* (*B. subtilis, B. licheniformis,* and *B. velezensis*). Numerous fibrinolytic enzymes have been identified within the *Bacillus* genus, each exhibiting unique characteristics such as molecular weight and substrate specificity. These enzymes are a cost-effective strategy for direct thrombolysis, offering potential advancements in clot-dissolving treatments. Among these the nattokinase enzyme stands out. Nattokinase is a protease with potent fibrinolytic activity, which is known for its ability to combat blood clots.[219] The fibrinolytic activity and the potential to reduce the aggregation of platelets have been confirmed in different *in vitro* and *in vivo* studies.[220–222] Moreover, nattokinase has demonstrated antithrombotic and antihypertensive properties in small clinical trials.[223,224] In one of the studies, which involved 74 hypertensive individuals, nattokinase supplementation led to a significant reduction in both systolic and diastolic blood pressure after 8 weeks of analysis.[223] In a single-dose study, involving 12 young healthy males, nattokinase ingestion was associated with an enhancement of fibrinolysis and anti-coagulant factors.[224]

3.3 Perspectives

Fermented foods serve as a rich source of various strains of microorganisms, such as LAB, and bioactive compounds, many of which possess novel functional properties. Consequently, fermented foods hold significant promise as a delivery system for probiotic microorganisms and the supplementation of bioactive compounds. While in recent years there has been substantial research involving fermented foods, it is crucial to conduct comprehensive clinical trials to quantify the consistent effects of fermented foods in preventing and alleviating various health disorders. To gain a deeper understanding of the impact of consuming fermented foods on human health, future studies should extend their intervention periods to assess long-term effects. Furthermore, research studies that correlate the consumption of potentially functional fermented foods with an impact on the microbiota and gut–brain axis are of utmost importance.

Abbreviations

ACE	Angiotensin-1-converting enzyme
CLA	Conjugated linoleic acid
CVD	Cardiovascular diseases
EFSA	European Food Administration
EPS	Exopolysaccharide(s)
FAD	Flavin adenine dinucleotide
FDA	Food and Drug Administration (USA)
FMN	Flavin mononucleotide
GABA	γ-Aminobutyric acid
GIT	Gastrointestinal tract
GRAS	Generally recognized as safe
HeLAB LAB	Heterofermentative LAB
HePS	Heteropolysaccharide(s)
HoLAB	Homofermentative LAB
HoPS	Homopolysaccharide(s)
IFN-γ	Interferon gamma
IL-4	Interleukine 4
IL-6	Interleukine 6
IL-8	Interleukine 8
IL-10	Interleukine 10
IL-12	Interleukine 12
LAB	Lactic acid bacteria
NDC	Non-digestible carbohydrates
NF-κB	Nuclear factor kappa light chain enhancer of activated B cells
QPS	Qualified presumption of safety
RAA	Renin-angiotensin-aldosterone system
SFCA	Short-chain fatty acids
WT	Wild type

Acknowledgements

We thank Dr Stephen Elson for critical reading of the manuscript.

References

1. I. Diez-Ozaeta and O. J. Astiazaran, *Food Res. Int.*, 2022, **156**, 111133.
2. I. Siró, E. Kápolna, B. Kápolna and A. Lugasi, *Appetite*, 2008, **51**(3), 456.
3. J. P. Tamang, *Ethnic fermented foods and beverages of India: Science history and culture*, Springer Singapore, Singapore, 2020, DOI: 10.1007/978-981-15-1486-9.
4. Y. Wang, C. Zhang, F. Liu, Z. Jin and X. Xia, *Crit. Rev. Food Sci. Nutr.*, 2023, **63**(22), 5841.
5. F. Leroy and L. De Vuyst, *Trends Food Sci. Technol.*, 2004, **15**(2), 67.
6. M. L. Marco, D. Heeney, S. Binda, C. J. Cifelli, P. D. Cotter, B. Foligné, M. Gänzle, R. Kort, G. Pasin, A. Pihlanto, E. J. Smid and R. Hutkins, *Curr. Opin. Biotechnol*, 2017, **44**, 94.
7. H. Shi, F. An, H. Lin, M. Li, J. Wu and R. Wu, *Front. Microbiol.*, 2022, **14**(13), 1044820.
8. V. Capozzi, P. Russo, M. T. Dueñas, P. López and G. Spano, *Appl. Microbiol. Biotechnol.*, 2012, **96**(6), 1383.
9. A. Olanbiwoninu, A. Greppi, T. Awotundun, E. A. Adebayo, G. Spano, D. Mora and P. Russo, *Food Biosci.*, 2023, **55**, 102996.
10. P. Li, Q. Gu, L. Yang, Y. Yu and Y. Wang, *Food Chem.*, 2017, **234**, 495.
11. F. Saubade, Y. M. Hemery, J. P. Guyot and C. Humblot, *Crit. Rev. Food Sci. Nutr.*, 2017, **57**(18), 3894.
12. F. Saubade, Y. M. Hemery, I. Rochette, J. P. Guyot and C. Humblot, *Int. J. Food Microbiol.*, 2018, **266**, 79.
13. C. Misci, E. Taskin, M. Dall'Asta, M. C. Fontanella, F. Bandini, S. Imathiu, D. Sila, T. Bertuzzi, P. S. Cocconcelli and E. Puglisi, *Food Microbiol.*, 2021, **99**, 103820.
14. F. A. Oguntoyinbo, V. Fusco, G. S. Cho, J. Kabisch, H. Neve, W. Bockelmann, M. Huch, L. Frommherz, B. Trierweiler, B. Becker, N. Benomar, A. Gálvez, H. Abriouel, W. H. Holzapfel and C. M. A. P. Franz, *Front. Microbiol.*, 2016, **7**, 981.
15. P. Russo, V. Capozzi, M. P. Arena, G. Spadaccino, M. T. Dueñas, P. López, D. Fiocco and G. Spano, *Appl. Microbiol. Biotechnol.*, 2014, **98**(8), 3691.
16. J. You, X. Pan, C. Yang, Y. Du, T. Osire, T. Yang, X. Zhang, M. Xu, G. Xu and Z. Rao, *Metab. Eng.*, 2021, **68**, 46.
17. P. Russo, N. De Simone, V. Capozzi, M. L. Mohedano, J. Á. Ruiz-Masó, G. del Solar, P. López and G. Spano, *Methods Mol. Biol.*, 2021, **2280**, 3.
18. I. Diez-Ozaeta, L. Martín-Loarte, M. L. Mohedano, M. Tamame, J. Á. Ruiz-Masó, G. del Solar, M. T. Dueñas and P. López, *Front. Microbiol.*, 2023, **14**, 1154130.
19. C. Burgess, M. O'Connell-Motherway, W. Sybesma, J. Hugenholtz and D. Van Sinderen, *Appl. Environ. Microbiol.*, 2004, **70**(10), 5769.

20. C. M. Burgess, E. J. Smid, G. Rutten and D. van Sinderen, *Microb. Cell Fact.*, 2006, **5**, 24.
21. A. Yépez, P. Russo, G. Spano, I. Khomenko, F. Biasioli, V. Capozzi and R. Aznar, *Food Microbiol.*, 2019, **77**, 61.
22. V. Capozzi, V. Menga, A. M. Digesù, P. De Vita, D. Van Sinderen, L. Cattivelli, C. Fares and G. Spano, *J. Agric. Food Chem.*, 2011, **59**(14), 8013.
23. P. Russo, M. L. V. de Chiara, V. Capozzi, M. P. Arena, M. L. Amodio, A. Rascón, M. T. Dueñas, P. López and G. Spano, *LWT*, 2016, **68**, 288.
24. Y. Zhu, K. Thakur, J. Y. Feng, J. S. Cai, J. G. Zhang, F. Hu, P. Russo, G. Spano and Z. J. Wei, *Appl. Microbiol. Biotechnol.*, 2020, **104**(13), 5759.
25. R. Wang, K. Thakur, J. Y. Feng, Y. Y. Zhu, F. Zhang, P. Russo, G. Spano, J. G. Zhang and Z. J. Wei, *Food Chem.*, 2022, **387**, 132947.
26. A. M. Hernández-Alcántara, R. Chiva, M. L. Mohedano, P. Russo, J. Á. Ruiz-Masó, G. del Solar, G. Spano, M. Tamame and P. López, *Front. Nutr.*, 2022, **9**, 978831.
27. I. Spacova, S. Ahannach, A. Breynaert, I. Erreygers, S. Wittouck, P. A. Bron, W. Van Beeck, T. Eilers, A. Alloul, N. Blansaer, S. E. Vlaeminck, N. Hermans and S. Lebeer, *Front. Nutr.*, 2022, **9**, 916607.
28. M. P. Arena, P. Russo, V. Capozzi, P. López, D. Fiocco and G. Spano, *Appl. Microbiol. Biotechnol.*, 2014, **98**(17), 7579.
29. M. L. Mohedano, S. Hernández-Recio, A. Yépez, T. Requena, M. C. Martínez-Cuesta, C. Peláez, P. Russo, J. G. LeBlanc, G. Spano, R. Aznar and P. López, *Front. Microbiol.*, 2019, **10**, 1748.
30. P. Russo, I. Iturria, M. L. Mohedano, G. Caggianiello, S. Rainieri, D. Fiocco, M. Angel Pardo, P. López and G. Spano, *Appl. Microbiol. Biotechnol.*, 2015, **99**(8), 3479.
31. R. Levit, G. Savoy de Giori, A. de Moreno de LeBlanc and J. G. LeBlanc, *Nutrition*, 2018, **54**, 165.
32. R. Levit, G. Savoy de Giori, A. de Moreno de LeBlanc and J. G. LeBlanc, *J. Appl. Microbiol.*, 2017, **124**(1), 232.
33. R. Levit, G. Savoy de Giori, A. de Moreno de LeBlanc and J. G. LeBlanc, *J. Appl. Microbiol.*, 2021, **130**(6), 2063.
34. M. Juarez Del Valle, J. E. Laiño, A. De Moreno De Leblanc, G. Savoy De Giori and J. G. Leblanc, *Br. J. Nutr.*, 2016, **116**(7), 1229.
35. D. Perez Visñuk, M. del M. Teran, G. Savoy de Giori, J. G. LeBlanc and A. de Moreno de LeBlanc, *Neurochem. Res.*, 2022, **47**(5), 1269.
36. L. Gao, G.-S. Cho, E. Brinks, R. Pichner and C. M. A. P. Franz, *Microbiol. Resour. Announce.*, 2023, **12**(6), e00144-23.
37. M. P. Taranto, J. L. Vera, J. Hugenholtz, G. F. De Valdez and F. Sesma, *J. Bacteriol.*, 2003, **185**(18), 5643.
38. B. Bhushan, S. K. Tomar and S. Mandal, *Appl. Microbiol. Biotechnol.*, 2016, **100**(15), 6791.
39. H. Morita, T. O. H. Hidehiro, S. Fukuda, H. Horikawa, K. Oshima, T. Suzuki, M. Murakami, S. Hisamatsu, Y. Kato, T. Takizawa, H. Fukuoka,

T. Yoshimura, K. Itoh, D. J. O'Sullivan, L. L. Mckay, H. Ohno, J. Kikuchi, T. Masaoka and M. Hattori, *DNA Res.*, 2008, **15**(3), 151.
40. M. De Angelis, F. Bottacini, B. Fosso, P. Kelleher, M. Calasso, R. Di Cagno, M. Ventura, E. Picardi, D. Van Sinderen and M. Gobbetti, *PLoS One*, 2014, **9**(9), e107232.
41. A. C. Torres, V. Vannini, J. Bonacina, G. Font, L. Saavedra and M. P. Taranto, *BMC Microbiol.*, 2016, **16**(1), 240.
42. A. C. Torres, V. Vannini, G. Font, L. Saavedra and M. P. Taranto, *Front. Microbiol.*, 2018, **9**, 2256.
43. Q. Gu, C. Zhang, D. Song, P. Li and X. Zhu, *Int. J. Food Microbiol.*, 2015, **206**, 56.
44. B. Bhushan, S. K. Tomar and A. Chauhan, *Appl. Microbiol. Biotechnol.*, 2017, **101**, 697.
45. M. Kumari, B. Bhushan, A. Kokkiligadda, V. Kumar, P. Behare and S. K. Tomar, *Curr. Res. Food Sci.*, 2021, **4**, 646.
46. P. Varmanen, P. Deptula, B. Chamlagain and V. Piironen, *Int. J. Food Microbiol.*, 2016, **228**, 33.
47. P. Deptula, B. Chamlagain, M. Edelmann, P. Sangsuwan, T. A. Nyman, K. Savijoki, V. Piironen and P. Varmanen, *Front. Microbiol.*, 2017, **8**, 368.
48. R. Coronas, G. Zara, A. Gallo, G. Rocchetti, M. Lapris, G. L. Petretto, S. Zara, F. Fancello and I. Mannazzu, *Front. Microbiol.*, 2023, **14**, 1223741.
49. Y. Wang, C. Xie, M. Pulkkinen, M. Edelmann, B. Chamlagain, R. Coda, M. Sandell, V. Piironen, N. H. Maina and K. Katina, *LWT*, 2022, **161**, 113407.
50. C. Xie, R. Coda, B. Chamlagain, M. Edelmann, P. Deptula, P. Varmanen, V. Piironen and K. Katina, *J. Cereal Sci.*, 2018, **81**, 133.
51. C. Xie, R. Coda, B. Chamlagain, M. Edelmann, P. Varmanen, V. Piironen and K. Katina, *LWT*, 2021, **137**, 110431.
52. H. Du, Y. Hu, K. Lu, T. Li, Y. Tian and Y. Hu, *LWT*, 2023, **187**, 115357.
53. M. Czarnowska-Kujawska and B. Paszczyk, *Molecules*, 2021, **26**(19), 6063.
54. S. C. Chandrasekar Rajendran, B. Chamlagain, S. Kariluoto, V. Piironen and P. E. J. Saris, *J. Appl. Microbiol.*, 2017, **122**(6), 1663.
55. O. Zahed, K. Khosravi-Darani, A. M. Mortazavian and A. Mohammadi, *Biocatal. Agric. Biotechnol.*, 2022, **39**, 102267.
56. M. Khalili, A. H. Rad, A. Y. Khosroushahi, H. Khosravi and S. Jafarzadeh, *Probiotics Antimicrob. Proteins*, 2020, **12**(2), 756.
57. J. Zhang, D. Cai, M. Yang, Y. Hao, Y. Zhu, Z. Chen, T. Aziz, A. Sarwar and Z. Yang, *Food Funct.*, 2020, (7), 6308.
58. W. Turpin, C. Humblot and J. P. Guyot, *Appl. Environ. Microbiol.*, 2011, **77**(24), 8722.
59. O. E. Fayemi, G. B. Akanni, S. S. Sobowale, A. Oelofse and E. M. Buys, *J. Food Compos. Anal.*, 2023, **115**, 104850.
60. A. Tamene, K. Baye, S. Kariluoto, M. Edelmann, F. Bationo, N. Leconte and C. Humblot, *Nutrients*, 2019, **11**(11), 2819.

61. A. Tamene, K. Baye and C. Humblot, *Heliyon*, 2022, **8**(5), e09526.
62. A. Greppi, F. Saubade, C. Botta, C. Humblot, J. P. Guyot and L. Cocolin, *Food Microbiol.*, 2017, **62**, 169.
63. S. H. Panda, S. Das, P. Bal, S. K. Panda, J. K. Goli and N. Mohanty, *Food Biosci.*, 2018, **21**, 100.
64. A. L. Mosso, M. E. Jimenez, G. Vignolo, J. G. LeBlanc and N. C. Samman, *Food Res. Int.*, 2018, **109**, 168.
65. A. L. Mosso, J. G. LeBlanc, C. Motta, I. Castanheira, P. Ribotta and N. Sammán, *LWT*, 2020, **127**, 109339.
66. S. Li, C. Chen, Y. Ji, J. Lin, X. Chen and B. Qi, *J. Cereal Sci.*, 2018, **84**, 83.
67. C. Misci, E. Taskin, F. Vaccari, M. Dall'Asta, F. Vezzulli, M. C. Fontanella, F. Bandini, S. Imathiu, D. Sila, T. Bertuzzi, P. S. Cocconcelli and E. Puglisi, *Int. J. Food Microbiol.*, 2022, **362**, 109445.
68. E. R. Aminigo, L. Metzger and P. S. Lehtola, *Int. J. Food Sci. Technol.*, 2009, **44**(3), 560.
69. M. del M. Teran, A. de Moreno de LeBlanc, G. Savoy de Giori and J. G. LeBlanc, *Appl. Microbiol. Biotechnol.*, 2021, **105**(5), 2097.
70. M. del M. Teran, D. Perez Visñuk, G. Savoy de Giori, A. de Moreno de LeBlanc and J. G. LeBlanc, *Food Funct.*, 2022, **13**, 8056.
71. P. Sanalibaba and G. A. Cakmak, *Appl. Microbiol.: Open Access*, 2016, **2**, 2.
72. P. T. Nguyen, T. T. Nguyen, D. C. Bui, P. T. Hong, Q. K. Hoang and H. T. Nguyen, *AIMS Microbiol.*, 2020, **6**(4), 451.
73. U. Tahir and A. Yasmin, *Int. J. Environ. Sci. Technol.*, 2019, **16**, 8081.
74. E. Brdarić, S. Soković Bajić, J. Đokić, S. Đurđić, P. Rúas-Madiedo, M. Stevanović, M. Tolinački, M. Dinić, J. Mutić, N. Golić and M. Živković, *Front. Microbiol.*, 2021, **12**, 759378.
75. N. Castro-Bravo, J. M. Wells, A. Margolles and P. Rúas-Madiedo, *Front. Microbiol.*, 2018, **9**, 2426.
76. P. Ruas-Madiedo, J. Hugenholtz and P. Zoon, *Int. Dairy J.*, 2002, **12**, 163.
77. Y. Abid, A. Casillo, H. Gharsallah, I. Joulak, R. Lanzetta, M. M. Corsaro, H. Attia and S. Azabou, *Int. J. Biol. Macromol.*, 2018, **108**, 719.
78. M. Nabot, M. Guérin, D. Sivakumar, F. Remize and C. Garcia, *Biology*, 2022, **11**(2), 171.
79. E. Zannini, D. M. Waters, A. Coffey and E. K. Arendt, *Appl. Microbiol. Biotechnol.*, 2016, **100**(3), 1121.
80. S. Tiwari, D. Kavitake, P. B. Devi and P. Halady Shetty, *Int. J. Biol. Macromol.*, 2021, **183**, 1585.
81. K. Zarour, M. G. Llamas, A. Prieto, P. Rúas-Madiedo, M. T. Dueñas, P. F. de Palencia, R. Aznar, M. Kihal and P. López, *Carbohydr. Polym.*, 2017, **174**, 646.
82. A. Pérez-Ramos, M. L. Mohedano, M. Pardo and P. López, *Front. Microbiol.*, 2018, **9**, 1684.
83. M. Nácher-Vázquez, N. Ballesteros, Á. Canales, S. Rodríguez Saint-Jean, S. I. Pérez-Prieto, A. Prieto, R. Aznar and P. López, *Carbohydr. Polym.*, 2015, **124**, 292.

84. S. Notararigo, E. Varela, A. Otal, M. Antolín, F. Guarner and P. López, *Int. J. Mol. Sci.*, 2022, **23**(3), 1527.
85. A. Afreen, Z. Ahmed, N. Khalid, I. Ferheen and I. Ahmed, *Appl. Microbiol. Biotechnol.*, 2023, **107**(4), 1189.
86. J. Wu, Y. Zhang, L. Ye and C. Wang, *Carbohydr. Polym.*, 2021, **253**, 117308.
87. M. Andrew and G. Jayaraman, *Carbohydr. Res.*, 2020, **487**, 107881.
88. C. Sabater, N. Molinero-García, N. Castro-Bravo, P. Diez-Echave, L. Hidalgo-García, S. Delgado, B. Sánchez, J. Gálvez, A. Margolles and P. Rúas-Madiedo, *Front. Microbiol.*, 2020, **11**, 601233.
89. T. Bhandary, C. Kurian, M. Muthu, A. Anand, T. Anand and K. A. Paari, *J. Pure Appl. Microbiol.*, 2023, **17**(1), 35.
90. Y. R. Saadat, A. Y. Khosroushahi and B. P. Gargari, *Carbohydr. Polym.*, 2019, **217**, 79.
91. U. Surayot, J. Wang, P. Seesuriyachan, A. Kuntiya, M. Tabarsa, Y. J. Lee, J. K. Kim, W. J. Park and S. G. You, *Int. J. Biol. Macromol.*, 2014, **68**, 233.
92. C. Hidalgo-Cantabrana, P. López, M. Gueimonde, C. G. de los Reyes-Gavilán, A. Suárez, A. Margolles and P. Rúas-Madiedo, *Probiotics Antimicrob. Proteins*, 2012, **4**, 227.
93. P. Rúas-Madiedo, in *Food Oligosaccharides: Production, Analysis and Bioactivity*, ed. F. J. Moreno and M. L. Sanz, John Wiley & Sons, Chichester, UK, 1st edn, 2014, vol. 8, p. 118.
94. H. Kitazawa, T. Harata, J. Uemura, T. Saito, T. Kaneko and T. Itoh, *Int. J. Food Microbiol.*, 1998, **40**(3), 169.
95. S. Makino, S. Ikegami, H. Kano, T. Sashihara, H. Sugano, H. Horiuchi, T. Saito and M. Oda, *J. Dairy Sci.*, 2006, **89**(8), 2873.
96. H. Nakata, Y. Imamura, S. Saha, R. E. Lobo, S. Kitahara, S. Araki, M. Tomokiyo, F. Namai, M. Hiramitsu, T. Inoue, K. Nishiyama, J. Villena and H. Kitazawa, *Foods*, 2023, **12**(12), 2374.
97. D. Li, J. Li, F. Zhao, G. Wang, Q. Qin and Y. Hao, *Food Chem.*, 2016, **197**, 367.
98. F. Vaningelgem, M. Zamfir, F. Mozzi, T. Adriany, M. Vancanneyt, J. Swings and L. De Vuyst, *Appl. Environ. Microbiol.*, 2004, **70**, 900.
99. G. Vinderola, G. Perdigón, J. Duarte, E. Farnworth and C. Matar, *Cytokine*, 2006, **36**(5–6), 254.
100. O. K. Kwon, K. S. Ahn, M. Y. Lee, S. Y. Kim, B. Y. Park, M. K. Kim, I. Y. Lee, S. R. Oh and H. K. Lee, *Arch. Pharmacal Res.*, 2008, **31**(12), 1590.
101. A. Pérez-Ramos, M. L. Mohedano, A. Puertas, A. Lamontanara, L. Orru, G. Spano, V. Capozzi, M. Teresa Dueñas and P. López, *Genome Announc.*, 2016, **4**(6), e01381-16.
102. M. L. Werning, S. Notararigo, M. Nacher, P. F. de Palencia, R. Aznar and P. Lopez, in *Food Additive*, ed. Y. El-Samragy, InTechOpen, London, UK, 1st edn, 2012, vol. 5, p. 83.

103. M. T. Dueñas-Chasco, M. A. Rodríguez-Carvajal, P. T. Mateo, G. Franco-Rodríguez, J. L. Espartero, A. Irastorza-Iribas and A. M. Gil-Serrano, *Carbohydr. Res.*, 1997, **303**(4), 453.
104. A. Pérez-Ramos, M. L. Mohedano, P. López, G. Spano, D. Fiocco, P. Russo and V. Capozzi, *Int. J. Mol. Sci.*, 2017, **18**(7), 1588.
105. S. Notararigo, M. de Las Casas-Engel, P. F. De Palencia, A. L. Corbí and P. López, *Carbohydr. Polym.*, 2014, **112**, 109.
106. S. Notararigo, E. Varela, A. Otal, I. Cristobo, M. Antolín, F. Guarner, A. Prieto and P. López, *Front. Microbiol.*, 2021, **12**, 621280.
107. M. Sharifi-Rad, N. V. Anil Kumar, P. Zucca, E. M. Varoni, L. Dini, E. Panzarini, J. Rajkovic, P. V. Tsouh Fokou, E. Azzini, I. Peluso, A. Prakash Mishra, M. Nigam, Y. El Rayess, M. El Beyrouthy, L. Polito, M. Iriti, N. Martins, M. Martorell, A. O. Docea, W. N. Setzer, D. Calina, W. C. Cho and J. Sharifi-Rad, *Front. Physiol.*, 2020, **11**, 694.
108. A. T. Adesulu-Dahunsi, A. I. Sanni and K. Jeyaram, *LWT*, 2018, **87**, 432.
109. E. A. Elmansy, E. M. Elkady, M. S. Asker, A. M. Abdou, N. A. Abdallah and S. K. Amer, *World J. Microbiol. Biotechnol.*, 2022, **38**, 245.
110. S. Li, Y. Zhao, L. Zhang, X. Zhang, L. Huang, D. Li, C. Niu, Z. Yang and Q. Wang, *Food Chem.*, 2012, **135**(3), 1914.
111. Y. Zhang, X. Chen, P. Hu, Q. Liao, Y. Luo, J. Li, D. Feng, J. Zhang, Z. Wu and H. Xu, *CYTA – J. Food*, 2021, **19**(1), 228.
112. I. S. Choi, S. H. Ko, M. E. Lee, H. M. Kim, J. E. Yang, S. G. Jeong, K. H. Lee, J. Y. Chang, J. C. Kim and H. W. Park, *ACS Omega*, 2021, **6**(12), 8171.
113. M. Saito, H. Sakagami and S. Fujisawa, *Anticancer Res.*, 2003, **23**(6C), 4693.
114. B. Bhat and B. K. Bajaj, *Bioact. Carbohydr. Diet. Fibre*, 2019, **19**, 100191.
115. H. Nakajima, Y. Suzuki and T. Hirota, *J. Food Sci.*, 1992, **57**(6), 1327.
116. H. Maeda, X. Zhu, K. Omura, S. Suzuki and S. Kitamura, *BioFactors*, 2004, **22**(1–4), 197.
117. H. Maeda, X. Zhu, S. Suzuki, K. Suzuki and S. Kitamura, *J. Agric. Food Chem.*, 2004, **52**(17), 5533.
118. L. Ai, H. Zhang, B. Guo, W. Chen, Z. Wu and Y. Wu, *Carbohydr. Polym.*, 2008, **74**(3), 353.
119. K. Wang, W. Li, X. Rui, X. Chen, M. Jiang and M. Dong, *Int. J. Biol. Macromol.*, 2014, **63**, 133.
120. X. Zhou, T. Hong, Q. Yu, S. Nie, D. Gong, T. Xiong and M. Xie, *Sci. Rep.*, 2017, **7**, 14247.
121. J. Wu, D. Yan, Y. Liu, X. Luo, Y. Li, C. Cao, M. Li, Q. Han, C. Wang, R. Wu and L. Zhang, *Front. Microbiol.*, 2021, **12**, 644226.
122. S. S. Choi, Y. Kim, K. S. Han, S. You, S. Oh and S. H. Kim, *Lett. Appl. Microbiol.*, 2006, **42**(5), 452.
123. V. Deepak, S. Ramachandran, R. M. Balahmar, S. R. K. Pandian, S. D. Sivasubramaniam, H. Nellaiah and K. Sundar, *In Vitro Cell. Dev. Biol.: Anim.*, 2016, **52**, 163.

124. H. Kawanabe-Matsuda, K. Takeda, M. Nakamura, S. Makino, T. Karasaki, K. Kakimi, M. Nishimukai, T. Ohno, J. Omi, K. Kano, A. Uwamizu, H. Yagita, I. G. Boneca, G. Eberl, J. Aoki, M. J. Smyth and K. Okumura, *Cancer Discovery*, 2022, **12**(5), 1336.
125. L. Pan, Y. Han and Z. Zhou, *J. Funct. Foods*, 2020, **67**, 103853.
126. F. Dal Bello, J. Walter, C. Hertel and W. P. Hammes, *Syst. Appl. Microbiol.*, 2001, **24**(2), 232.
127. M. Polak-Berecka, A. Waśko, D. Szwajgier and A. Choma, *Pol. J. Microbiol.*, 2013, **62**(2), 181.
128. D. Das, R. Baruah and A. Goyal, *Int. J. Biol. Macromol.*, 2014, **69**, 20.
129. Y. Zhu, J. M. Zhou, W. Liu, X. Pi, Q. Zhou, P. Li, T. Zhou and Q. Gu, *LWT*, 2021, **139**, 110524.
130. P. Ruas-Madiedo, M. Gueimonde, A. Margolles, C. G. De Los Reyes-Gavilán and S. Salminen, *J. Food Prot.*, 2006, **69**(8), 2011.
131. C. Chelakkot, J. Ghim and S. H. Ryu, *Exp. Mol. Med.*, 2018, **50**, 103.
132. Y. Chen, M. Zhang and F. Ren, *Molecules*, 2019, **24**, 513.
133. P. Singh, Y. G. Ban, L. Kashyap, A. Siraree and J. Singh, in *Sugar and Sugar Derivatives: Changing Consumer Preferences*, ed. N. Mohan and P. Singh, Springer Singapore, Singapore, 1st edn, 2020, vol. 4, p. 39.
134. J. Bold, in *Gluten-Related Disorders: Diagnostic Approaches, Treatment Pathways, and Future Perspectives*, ed. M. Rostami-Nejad, Academic Press, London, United Kingdom, 1st edn, 2022, vol. 3, p. 33.
135. P. Fickers, H. Cheng and C. S. K. Lin, *Microorganisms*, 2020, **8**, 574.
136. P. Liang, M. Cao, J. Li, Q. Wang and Z. Dai, *Biotechnol. Adv.*, 2023, **64**, 108105.
137. M. Grembecka, *Eur. Food Res. Technol.*, 2015, **241**(1), 1.
138. F. Lu, W. Xu, W. Zhang, C. Guang and W. Mu, *Appl. Microbiol. Biotechnol.*, 2019, **103**(16), 6473.
139. Y. R. Rastogi, R. Thakur, P. Thakur, A. Mittal, S. Chakrabarti, S. S. Siwal, V. K. Thakur, R. V. Saini and A. K. Saini, *Int. J. Food Microbiol.*, 2022, **371**, 109666.
140. M. E. Ortiz, J. Bleckwedel, R. R. Raya and F. Mozzi, *Appl. Microbiol. Biotechnol.*, 2013, **97**(11), 4713.
141. M. Carocho, P. Morales and I. C. F. R. Ferreira, *Food Chem. Toxicol.*, 2017, **107**(Pt A), 302.
142. N. Z. Msomi, O. L. Erukainure and M. S. Islam, *J. Food Drug Anal.*, 2021, **29**(1), 1.
143. J. L. Buttriss, *Nutr. Bull.*, 2020, **45**(3), 244.
144. M. M. Abedin, R. Chourasia, L. C. Phukon, P. Sarkar, R. C. Ray, S. P. Singh and A. K. Rai, *Crit. Rev. Food Sci. Nutr.*, 2023, **64**(29), 10730.
145. A. Lenhart and W. D. Chey, *Adv. Nutr.*, 2017, **8**(4), 587.
146. M. Wojtuś, S. Tomaszuk and K. Wąsik, *J. Educ., Health Sport*, 2022, **12**(12), 146.
147. M. Rakicka-Pustułka, J. Miedzianka, D. Jama, S. Kawalec, K. Liman, T. Janek, G. Skaradziński, W. Rymowicz and Z. Lazar, *Microb. Cell Fact.*, 2021, **20**(1), DOI: 10.1186/s12934-021-01686-0.

148. R. Duan, H. Li, H. Li, L. Tang, H. Zhou, X. Yang, Y. Yang and Z. Ding, *Appl. Biochem. Biotechnol.*, 2018, **186**(4), 990.
149. P. Gaspar, A. L. Carvalho, S. Vinga, H. Santos and A. R. Neves, *Biotechnol. Adv.*, 2013, **31**(6), 764.
150. A. W. Sahin, T. Rice, E. Zannini, C. Axel, A. Coffey, K. M. Lynch and E. K. Arendt, *Int. J. Food Microbiol.*, 2019, **302**, 80.
151. T. Rice, A. W. Sahin, K. M. Lynch, E. K. Arendt and A. Coffey, *Int. J. Food Microbiol.*, 2020, **16**, 321.
152. S. Jeske, E. Zannini, K. M. Lynch, A. Coffey and E. K. Arendt, *Int. J. Food Microbiol.*, 2018, **286**, 31.
153. M. Silveira and R. Jonas, *Appl. Microbiol. Biotechnol.*, 2002, **59**, 400.
154. F. Patra, S. K. Tomar and S. Arora, *J. Food Sci.*, 2009, **74**(1), R16.
155. S. Bhattacharya, in *Snack Foods, Processing and Technology*, ed. S. Bhattacharya, Academic Press, London, United Kingdom, 1st edn, 2022, vol. 9, p. 211.
156. Y. Wang, J. Wu, M. Lv, Z. Shao, M. Hungwe, J. Wang, X. Bai, J. Xie, Y. Wang and W. Geng, *Front. Bioeng. Biotechnol.*, 2021, **9**, 612285.
157. Y. P. Silva, A. Bernardi and R. L. Frozza, *Front. Endocrinol.*, 2020, **11**, 25.
158. W. Schuh, D. Mielenz and H. M. Jäck, *Adv. Immunol.*, 2020, **146**, 57.
159. N. E. Nilsson, K. Kotarsky, C. Owman and B. Olde, *Biochem. Biophys. Res. Commun.*, 2003, **303**(4), 1047.
160. P. Markowiak-Kopeć and K. Śliżewska, *Nutrients*, 2020, **12**(4), 1107.
161. S. Patel, S. Seshadri and S. Dalai, *Prog. Mol. Biol. Transl. Sci.*, 2022, **191**(1), 175.
162. R. Havenaar, *Benefic. Microbes*, 2011, **2**(2), 103.
163. J. G. LeBlanc, F. Chain, R. Martín, L. G. Bermúdez-Humarán, S. Courau and P. Langella, *Microb. Cell Fact.*, 2017, **16**, 79.
164. A. Meimandipour, M. Hair-Bejo, M. Shuhaimi, K. Azhar, A. F. Soleimani, B. Rasti and A. M. Yazid, *Br. Poult. Sci.*, 2010, **51**(1), 52.
165. K. Sivieri, M. L. V. Morales, M. A. T. Adorno, I. K. Sakamoto, S. M. I. Saad and E. A. Rossi, *BMC Gastroenterol.*, 2013, **13**, 100.
166. N. Salazar, M. Gueimonde, C. G. de los Reyes-Gavilán and P. Rúas-Madiedo, *Crit. Rev. Food Sci. Nutr.*, 2016, **56**(9), 1440.
167. E. J. Culp and A. L. Goodman, *Cell Host Microbe*, 2023, **31**(4), 485.
168. A. Nogal, A. M. Valdes and C. Menni, *Gut Microbes*, 2021, **13**(1), e1897212.
169. R. Saleri, P. Borghetti, F. Ravanetti, V. Cavalli, L. Ferrari, E. De Angelis, M. Andrani and P. Martelli, *Porcine Health Manag.*, 2022, **8**, 21.
170. X. F. Liu, J. H. Shao, Y. T. Liao, L. N. Wang, Y. Jia, P. J. Dong, Z. Z. Liu, D. D. He, C. Li and X. Zhang, *Front. Immunol.*, 2023, **14**, 1186892.
171. J. Darzi, G. S. Frost and M. D. Robertson, *Proc. Nutr. Soc.*, 2011, **70**(1), 119.
172. A. A. Challa and E. D. Lewandowski, *JACC Basic Transl. Sci.*, 2022, **7**(7), 730.
173. E. Dimidi, S. R. Cox, M. Rossi and K. Whelan, *Nutrients*, 2019, **11**(8), 1806.

174. M. Mechmeche, H. Ksontini, M. Hamdi and F. Kachouri, *Int. J. Pept. Res. Ther.*, 2019, **25**, 137.
175. N. Şanlier, B. B. Gökcen and A. C. Sezgin, *Crit. Rev. Food Sci. Nutr.*, 2019, **59**, 506.
176. A. P. Niarchos, T. G. Pickering, D. B. Case, P. Sullivan and J. H. Laragh, *Circ. Res.*, 1979, **45**(6), 829.
177. S. Rysz, J. Al-Saadi, A. Sjöström, M. Farm, F. Campoccia Jalde, M. Plattén, H. Eriksson, M. Klein, R. Vargas-Paris, S. Nyrén, G. Abdula, R. Ouellette, T. Granberg, M. Jonsson Fagerlund and J. Lundberg, *Nat. Commun.*, 2021, **12**, 2417.
178. L. Beltrán-Barrientos, A. Hernández-Mendoza, A. González-Córdova, H. Astiazarán-García, J. Esparza-Romero and B. Vallejo-Córdoba, *Nutrients*, 2018, **10**(3), 262.
179. F. Nejati, C. G. Rizzello, R. Di Cagno, M. Sheikh-Zeinoddin, A. Diviccaro, F. Minervini and M. Gobbetti, *LWT - Food Sci. Technol.*, 2013, **51**, 183.
180. M. Á. Rendon-Rosales, M. J. Torres-Llanez, A. F. González-Córdova, A. Hernández-Mendoza, M. A. Mazorra-Manzano and B. Vallejo-Cordoba, *Nutrients*, 2019, **11**(9), 2150.
181. Y. Tung, H. Chen, H. Wu, M. Ho, K. Chong and C. Chen, *Mol. Nutr. Food Res.*, 2017, **62**(3), 1700505.
182. C. Mahdi, H. Untari, M. C. Padaga and S. J. Raharjo, *Int. Food Res. J.*, 2018, **25**(1), 17.
183. F. Tonolo, F. Fiorese, L. Moretto, A. Folda, V. Scalcon, A. Grinzato, S. Ferro, G. Arrigoni, A. Bindoli, E. Feller, M. Bellamio, O. Marin and M. P. Rigobello, *Antioxidants*, 2020, **9**(2), 117.
184. D. Yu, M. Feng, J. Sun, X. Xu and G. Zhou, *Meat Sci.*, 2020, **160**, 107958.
185. F. Huang, Y. Jing, G. Ding and Z. Yang, *Mol. Med. Rep.*, 2017, **16**(4), 4222.
186. F. Gong, M.-F. Chen, J. Chen, C. Li, C. Zhou, P. Hong, S. Sun and Z.-J. Qian, *J. Agric. Food Chem.*, 2019, **67**(32), 8855.
187. S. R. Joshi, K. Biswas, in *Health Benefits of Fermented Foods and Beverages*, ed. J. P. Tamang, CRC Press, Boca Ratón, Florida, 1st edn, 2015, vol. 19, p. 504.
188. E. Ivanišová, K. Meňhartová, M. Terentjeva, Ľ. Harangozo, A. Kántor and M. Kačániová, *J. Food Sci. Technol.*, 2020, **57**, 1840.
189. I. Mantzourani, A. Terpou, A. Alexopoulos, A. Kimbaris, E. Bezirtzoglou, A. A. Koutinas and S. Plessas, *Appl. Biochem. Biotechnol.*, 2019, **188**, 1096.
190. G. Rocchetti, F. Miragoli, C. Zacconi, L. Lucini and A. Rebecchi, *Food Res. Int.*, 2019, **119**, 886.
191. A. Lorusso, M. Verni, M. Montemurro, R. Coda, M. Gobbetti and C. G. Rizzello, *LWT*, 2017, **78**, 215.
192. E. Bujna, N. A. Farkas, A. M. Tran, M. S. Dam and Q. D. Nguyen, *Food Sci. Biotechnol.*, 2017, **27**, 547.
193. X. Yang, J. Zhou, L. Fan, Z. Qin, Q. Chen and L. Zhao, *Food Sci. Biotechnol.*, 2018, **27**(6), 1719.

194. F. Gaggìa, L. Baffoni, M. Galiano, D. Nielsen, R. Jakobsen, J. Castro-Mejía, S. Bosi, F. Truzzi, F. Musumeci, G. Dinelli and D. Di Gioia, *Nutrients*, 2018, **11**, 1.
195. A. Ahmad, K. Ramasamy, A. B. A. Majeed and V. Mani, *Pharm. Biol.*, 2015, **53**, 758.
196. C. W. C. Borges, M. C. Carrão-Panizzi, J. M. G. Mandarino, J. B. da Silva, S. Benedetti and E. I. Ida, *Pesqui. Agropecu. Bras.*, 2016, **51**(3), 271.
197. M. Nozue, T. Shimazu, S. Sasazuki, H. Charvat, N. Mori, M. Mutoh, N. Sawada, M. Iwasaki, T. Yamaji, M. Inoue, Y. Kokubo, K. Yamagishi, H. Iso and S. Tsugane, *J. Nutr.*, 2017, **147**, 1749.
198. Y.-D. Xi, H.-L. Yu, W.-W. Ma, B.-J. Ding, J. Ding, L.-H. Yuan, J.-F. Feng and R. Xiao, *J. Bioenerg. Biomembr.*, 2011, **43**, 399.
199. V. Mukund, D. Mukund, V. Sharma, M. Mannarapu and A. Alam, *Crit. Rev. Oncol. Hematol.*, 2017, **119**, 13.
200. L. Gorissen, F. Leroy, L. De Vuyst, S. De Smet and K. Raes, *Crit. Rev. Food Sci. Nutr.*, 2015, **55**, 1561.
201. J. C. Andrade, K. Ascençao, P. Gullón, S. M. S. Henriques, J. M. S. Pinto, T. A. P. Rochas-Santos, A. C. Freitas and A. M. Gomes, *Int. J. Dairy Technol.*, 2012, **65**, 467.
202. M. Palla, G. Conte, A. Grassi, S. Esin, A. Serra, M. Mele, M. Giovannetti and M. Agnolucci, *Foods*, 2021, **10**(9), DOI: 10.3390/foods10092087.
203. M. Tiss, Z. Souiy, N. ben Abdeljelil, M. Njima, L. Achour and K. Hamden, *J. Funct. Foods*, 2020, **67**, 103869.
204. B. E. Jewett and B. Thapa, Physiology, NMDA Receptor, *StatPearls*, StatPearls Publishing, Treasure Island, FL, 2024.
205. Q. Wang, L. Ren, Y. Wan and G. J. Prud'homme, *J. Cell. Physiol.*, 2019, **234**, 14432.
206. S. Bhanwar, M. Bamnia, M. Ghosh and A. Ganguli, *Int. J. Food Sci. Nutr.*, 2013, **64**, 77.
207. T. Okada, T. Sugishita, T. Murakami, H. Murai, T. Saikusa, T. Horino, A. Onoda, O. Kajimoto, R. Takahashi and T. Takahashi, *Nippon Shokuhin Kagaku Kogaku Kaishi*, 2000, **47**, 596.
208. M. Venturi, V. Galli, N. Pini, S. Guerrini and L. Granchi, *Foods*, 2019, **8**(6), 218.
209. J. A. Curiel, R. Coda, I. Centomani, C. Summo, M. Gobbetti and C. G. Rizzello, *Int. J. Food Microbiol.*, 2015, **196**, 51.
210. L. Ruiz Rodríguez, E. Vera Pingitore, G. Rollan, P. S. Cocconcelli, C. Fontana, L. Saavedra, G. Vignolo and E. M. Hebert, *J. Appl. Microbiol.*, 2016, **120**(5), 1289.
211. L. Chen, J. Alcazar, T. Yang, Z. Lu and Y. Lu, *J. Dairy Sci.*, 2018, **101**(12), 10685.
212. G. A. Roth, *et al.*, *J. Am. Coll. Cardiol.*, 2020, **76**(25), 2982.
213. N. Barzkar, S. T. Jahromi and F. Vianello, *Mar. Drugs*, 2022, **20**(1), 46.
214. M. De Carlo, O. Schlager, L. Mazzolai, M. Brodmann, C. Espinola-Klein, D. Staub, V. Aboyans, H. Sillesen, S. Debus, M. Venermo, J. Belch,

M. Ferrari and R. De Caterina, *Eur. Heart J.: Cardiovasc. Pharmacother.*, 2023, **9**(3), 201.

215. N. J. Reish, P. Jamshidi, B. Stamm, M. E. Flanagan, E. Sugg, M. Tang, K. L. Donohue, M. McCord, C. Krumpelman, M.-M. Mesulam, R. Castellani and S. H.-Y. Chou, *N. Engl. J. Med.*, 2023, **388**, 478.
216. J. Tanaka, T. Koseki, K. Sekido, M. Kimata, Y. Ito and S. Yamada, *J. Pharm. Pharm. Sci.*, 2023, **26**, 11263.
217. Y. Weng, J. Yao, S. Sparks and K. Wang, *Int. J. Mol. Sci.*, 2017, **18**(3), 523.
218. H. Yang, L. Yang, X. Li, H. Li, Z. Tu and X. Wang, *J. Gen. Appl. Microbiol.*, 2020, **66**(3), 153.
219. H. Chen, J. Chen, F. Zhang, Y. Li, R. Wang, Q. Zheng, X. Zhang, J. Zeng, F. Xu and Y. Lin, *Front. Cardiovasc. Med.*, 2022, **9**, 964977.
220. J.-Y. Jang, T.-S. Kim, J. Cai, J. Kim, Y. Kim, K. Shin, K. S. Kim, S. K. Park, S.-P. Lee, E.-K. Choi, M. H. Rhee and Y.-B. Kim, *Lab. Anim. Res.*, 2013, **29**, 221.
221. J. Xu, M. Du, X. Yang, Q. Chen, H. Chen and D.-H. Lin, *Acta Haematol.*, 2014, **132**, 247.
222. C. Yatagai, M. Maruyama, T. Kawahara and H. Sumi, *Pathophysiol. Haemostasis Thromb.*, 2007, **36**, 227.
223. G. Jensen, M. Lenninger, M. P. Ero and K. Benson, *Integr. Blood Pressure Control*, 2016, **9**, 95.
224. Y. Kurosawa, S. Nirengi, T. Homma, K. Esaki, M. Ohta, J. F. Clark and T. Hamaoka, *Sci. Rep.*, 2015, **5**, 11601.

CHAPTER 4

Role of Omega-3s in Functional Foods

PINKU CHANDRA NATH,[a] AMIYA OJHA,[a] MINAXI SHARMA,[b] BASKARAN STEPHEN INBARAJ,[c] KANDI SRIDHAR,*[d] PRAKASH KUMAR NAIK,*[e] PRAVEEN KUMAR DIKKALA[f] AND VENKATA SUNITHA SESHAMAMBA BURLA[g]

[a] Department of Bio Engineering, National Institute of Technology Agartala, Jirania 799046, India; [b] Research Centre for Life Science and Healthcare, Nottingham Ningbo China Beacons of Excellence Research and Innovation Institute (CBI), University of Nottingham Ningbo China, Ningbo 315000, China; [c] Department of Food Science, Fu Jen Catholic University, New Taipei City 242062, Taiwan; [d] Department of Food Technology, Karpagam Academy of Higher Education (Deemed to be University), Coimbatore 641021, India; [e] Department of Food Engineering and Technology, Central Institute of Technology Kokrajhar, Kokrajhar 783370, India; [f] Department of Food Technology, Koneru Lakshmaiah Education Foundation Deemed to be University, Vaddeswaram 522502, India; [g] Dr. NTR College of Food Science and Technology, Acharya NG Ranga Agricultural University, Bapatla 502101, India
*Emails: sridhar4647@gmail.com; pk.nayak@cit.ac.in

4.1 Introduction

The acquisition of nutrients through daily food consumption is essential for the sustenance and optimal performance of bodily organs. In addition to their energy-providing properties, they possess the capacity to combat degenerative and chronic illnesses. The presence of bioactive compounds in

Food Chemistry, Function and Analysis No. 44
Functional Foods of the Future
Edited by Vijai Kumar Gupta, Minaxi Sharma, Smriti Gaur and Ramesh Chander Kuhad

Published by the Royal Society of Chemistry, www.rsc.org

cereals, fruits, vegetables, and nuts is a crucial factor in disease prevention and promoting individual health.[1] The significance of omega-3 (ω-3) polyunsaturated fatty acids (PUFAs) has gained considerable attention in contemporary times due to their diverse functions in promoting health and mitigating disease risk.[2] The group of omega-3 polyunsaturated fatty acids (ω-3 PUFAs) comprises α-linolenic acid (ALA; 18 : 3), eicosapentaenoic acid (EPA; 20 : 5), stearidonic acid (SDA; 18 : 4), docosahexaenoic acid (DHA; 22 : 6), and docosapentaenoic (DPA; 22 : 5), and are a type of dietary lipids that offer a diverse range of health advantages.[3] The oils that contain these specific fatty acids (FAs) are predominantly derived from select plant sources or undergo modification within plants, in addition to marine, algal, and unicellular sources. Long-chain ω-3 FAs, specifically DHA and EPA, are present in the lipid composition of the liver of lean fish, fatty fish, and the blubber of marine mammals.[4] While ω-3 PUFA supplements and concentrated fish oil ethyl esters (EEs) and acylglycerols have been on the market for some time, innovative and renewable energy sources of long chain FAs such as algal, single-cell, fungal oils have lately gained in popularity. EPA- and DHA-rich phospholipid (PL) and triacylglycerol (TAG) preparations of krill oil have both been commercially sold.[5] Desaturation of linoleic acid, which is performed by 15-desaturase, produces ALA. Due to a deficiency of the enzyme needed to generate FAs, ALA must be consumed because it is necessary for human metabolic processes but cannot be produced by the body.[6] ALA is a necessary dietary precursor of DHA and EPA, the liver is the primary location for this conversion (at a rate of roughly 8% for EPA and 1% for DHA).[7] The essential fatty acids DHA and EPA are known to have a significant impact on the prevention and treatment of various health issues, contributing to improved wellbeing. These fatty acids are believed to possess advantageous properties for avoiding health problems such as cardiovascular disease, cancer, hypertension, inflammation, and others.[8] The significance of ω-3 FAs lies in their role in the synthesis of lipoxins, including leukotrienes (LTs), prostaglandins (PGs), prostacyclin (PGI), thromboxane (TX), and eicosanoids, which are vital for vascular physiology.[9]

Initial reports and epidemiological investigations indicated that the conventional diet of Greenland, which is abundant in marine mammals and fish, resulted in a notable decrease in the prevalence of cardiovascular disease (CD) among both the Inuit population and Danish settlers, albeit to varying degrees.[10] Numerous studies have demonstrated that omega-3 PUFAs exert a significant impact on modifying the blood lipid profile and membrane lipid composition. Additionally, they have been found to affect the biosynthesis of eicosanoids, cell-signaling cycles, and gene expression, ultimately influencing overall health.[11] Furthermore, extensive research has been reported on the advantageous impact of omega-3 PUFAs on individuals suffering from a multitude of health conditions and ailments, including but not limited to cardiovascular disease, rheumatoid arthritis, cancer, depression, various mental illnesses, periodontal disease, age-related cognitive decline, and diabetes.[12,13] Extensive research has been conducted on the significance of omega-3 oils, specifically DHA oils, in the dietary intake of expectant and nursing mothers, as well as in the growth and development of the fetal and infant brain and visual system.[14]

Over time, the American diet has undergone a shift towards a high consumption of saturated fatty acids (SFA) and a low consumption of ω-3 FAs. The alteration in dietary patterns revolves around the consumption of fast food that is rich in SFA and low in essential ω-3 PUFAs in comparison to home-cooked meals. The two biologically significant dietary ω-3 FAs, DHA and EPA, are primarily derived from seafood sources, including fish and fish-oil supplements.[15] The insufficient consumption of DHA and EPA through diet is believed to be linked with heightened inflammatory mechanisms, suboptimal fetal growth, overall cardiovascular wellbeing, and susceptibility to the onset of Alzheimer's disease (AD).[16] The present chapter centers on the manifold advantages of supplementing diet with ω-3 PUFAs and EPA and DHA.

4.2 Sources of ω-3 Fatty Acids

The recommended daily intake of ω-3 fatty acids for maintaining good health is between 200 mg per day and 1 g per day. The administration of elevated levels of ω-3 fatty acids may potentially induce oxidative stress.[17,18]

4.2.1 Fish Sources

At present, the primary origins of the fatty acids DHA and EPA are fish and fish oils. The triglyceride structures of fish oils derived from the flesh of oily fish and the offal of white fish are rich in long-chain ω-3 polyunsaturated fatty acids and various other fatty acids.[19] The flesh of oily fish, including but not limited to mackerel, salmon, sardines, anchovies, and pilchards, is notably abundant in DHA and EPA.[20] Krill oil is a significant source of DHA and EPA, with a content of 34.2 g of long-chain ω-3 PUFAs per 100 g.[20] Nonetheless, akin to other sources derived from the marine ecosystem, the assurance of sustainable krill oil production remains uncertain. According to recent research, there appears to be a correlation between the abundance of krill and the phenomenon of global warming. The lack of winter sea ice in areas where krill reside could potentially lead to a decrease in the population of krill.[21,22]

4.2.2 Animal Sources

Meat and poultry are significant sources of ω-3 FAs. It is important to note that the ω-3 content per 100 grams of meat derived from animal sources is significantly lower compared to that present in oily fish and other types of fish.[23] Poultry meat is a source of DHA and EPA, albeit in modest quantities. Empirical data from dietary surveys suggest that poultry and meat can account for approximately 43.0% of the ω-3 consumption among non-vegetarians.[24] Previous research has investigated the fortification of animal-based products with DHA and EPA through the utilization of fish oil supplemented diets.[25] There exist apprehensions regarding the prolonged and potentially escalated utilization of fish oils in the diets of animals, as it may not be a viable, economical, or enduring strategy.

4.2.3 Vegetarian Sources

The primary source of ω-3 in the diet is currently ALA, which is present in the lipid components of leafy green plants and certain seeds. The primary plant-based origin of ALA is flaxseed oil, which is also referred to as linseed oil.[26,27] Cold pressing of flaxseed oil can yield up to 56% ALA content. Echium, perilla seed, walnuts, and blackcurrant seed oil are some examples of other plant-based foods that are rich in LC ω-3 PUFAs.[28] Echium oil is a type of vegetable oil that has not been genetically engineered and is extracted from the seeds of the *Echium plantagineum* plant.[29] The plant itself is not genetically changed. There are large quantities of a LC ω-3 FAs found in it, which is sometimes referred to as stearidonic acid (18 : 4 ω-3; SDA).[30] Up to 47.1% of echium seed oil is composed of LC ω-3 PUFAs, with around 36.5% ALA and 10.6% SDA making up this percentage (Table 4.1).[31] Other excellent sources of ω-3 include the oil extracted from the seeds of the *Perilla frutescens* plant, which contains as much as 54% ALA, and blackcurrant seed oil, which is abundant in both ω-3 and ω-6 FAs, with ALA and SDA serving as its primary supply of LC ω-3 PUFAs.[32,33]

4.3 Metabolism of ω-3 Fatty Acids

The majority of these substances are primarily consumed as part of the diet as triglycerides (TAG) and subsequently undergo the process of digestion within the small intestine. The dismantling of the TAG architecture and consequent release of fatty acids (FAs) facilitates their uptake and systemic distribution to anatomical sites where they are assimilated into their respective organic frameworks.[45] FAs can undergo three distinct metabolic pathways. These include: (a) esterification into cellular lipids such as phospholipids (PL), TAG, and cholesteryl esters (CE); (b) β-oxidation, which generates energy for adenosine triphosphate (ATP) synthesis; and (c) serving as an initiating structure for the elongation and desaturation processes through enzymatic reactions, ultimately leading to the production of long-chain PUFAs, *via* the metabolic pathway show in Figure 4.1.[46–48] Phospholipids (PL) formed from fatty acids (FAs) are the primary component of the cell membrane and the preservation of cell membrane homeostasis is achieved through the equilibrium of FAs within the membrane. Additionally, FAs may detach from PL to serve as signaling mediators. Long-chain PUFAs are mostly synthesized in the liver. There is a negligible contribution from other tissues to this process.[49] Three of the PUFAs synthesized from LA-GLA, dihomogammalinolenic acid (DGLA) and

Table 4.1 Different sources and composition of omega-3 fatty acid.[a]

Sources	EPA (%)	DHA (%)	DPA (%)	ALA (%)	LA (%)	Ref.
Fish						
Salmon	0.89	1.19	—	—	—	2, 34 and 35
Butterfish oil	5.1	10.8	2.4	—	—	
Herring	1.09	1.16	—	—	—	
Haddock oil	14.8	24.1	1.8	—	—	
Halibut oil	9.6	30.6	2.5	—	—	
Sardine	0.51	1.16	—	—	—	
Capelin oil	9.3	4.1	0.9	—	—	
Cod oil	0.02	0.62	—	—	—	
Trout oil	0.15	0.5	—	—	—	
Animal						
Bearded seal oil	9.2	13.5	4.1	—	—	36–39
Harbor seal oil	9.6	7.5	4.3	—	—	
Grey seal oil	5.7	7.2	4.4	—	—	
Ringed seal oil	10.04	26.11	14.23	—	—	
Harp seal oil	6.2	7.8	4.6	—	—	
Lobster	17.2	7.6	1.23	—	—	
Shrimp	15.6	11.4	0.75	—	—	
Crab	20.12	10.6	2.33	—	—	
Octopus	16.56	20.14	1.58	—	—	
Squid	13.8	16.8	1.25	—	—	
Blue mussel	19.56	13.4	—	—	—	
Surf clam	22.14	14.52	—	—	—	
Cockle	22.14	16.23	0.11	—	—	
Scallop	26.5	25.9	—	—	—	
Beef	0.001	0.001	—	0.026	—	
Lamb	—	0.007	—	0.065	—	
Pork	0.006	0.008	—	0.02	—	
Poultry	0.013	0.022	—	0.015	—	
Vegetables						
Broccoli	—	—	—	0.0001	0.00003	40–44
Sprouts	—	—	—	0.0001	0.00008	
Lettuce	—	—	—	0.0001	0.00005	
Chia	—	—	—	0.017	0.005	
Walnuts	—	—	—	0.006	0.034	
Hazelnuts	—	—	—	0.0001	0.005	
Almonds	—	—	—	0.0003	0.01	
Sunflower	—	—	—	0.0003	0.049	
Wheat germ	—	—	—	0.005	0.055	
Soybean	—	—	—	0.007	0.051	
Corn	—	—	—	0.0006	0.049	
Canola	—	—	—	0.009	0.018	
Flaxseed	—	—	—	0.053	0.014	
Safflower	—	—	—	0.0001	0.012	

[a] EPA, eicosapentaenoic acid; DHA, docosahexaenoic acid; DPA, docosapentaenoic acid; ALA, α-linolenic acid; LA, linoleic acid.

arachidonic acid, stand out as the most common and biologically important.[50] Among the family of physiologically active fatty acids, ALA is the most basic and least efficiently metabolized to the more potent EPA and DHA.[51]

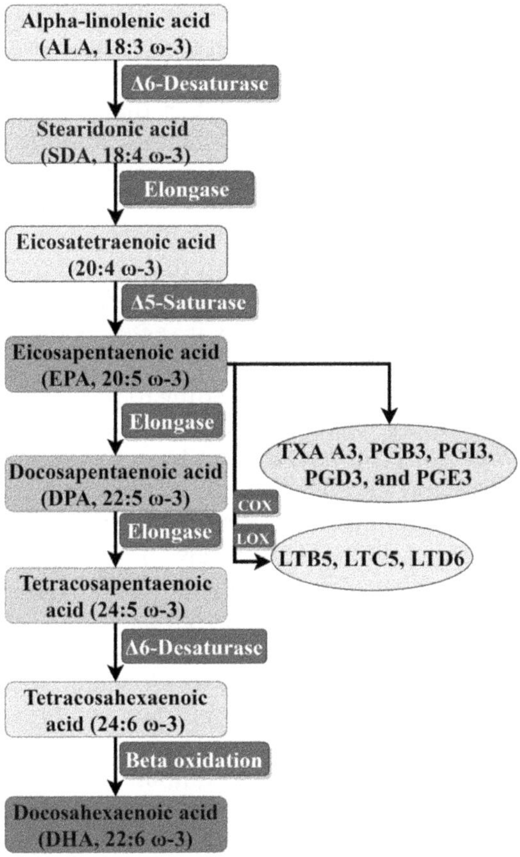

Figure 4.1 Metabolic pathway for the synthesis of ω-3 PUFA. COX, cyclooxygenase; LOX, lipoxygenase; LT, leukotriene; PG, prostaglandin.

Converting ALA to EPA uses the same enzymes as converting linoleic acid to arachidonic acid, which might lead to enzyme competition. In order to facilitate the effective digestion and absorption of FAs, the initial step involves the emulsification of fat through the secreting of bile into the small intestine. As per the scientific literature, triglycerides obtained from diverse food sources are ingested and metabolized in the small intestine, subsequently assimilated and conveyed to bodily tissues *via* plasma.[52] Mammals are incapable of biosynthesizing FAs with longer carbon chains. Hence, they depend on dietary sources to obtain the precursors linoleic acid and ALA to further synthesize other ω-3 and ω-6 PUFAs or as a last resort synthesize ω-9 PUFAs.[6] According to reports, the transformation of ALA to EPA and DHA is relatively inefficient in humans due to competition between ω-6 and ω-3 FAs for desaturase enzymes.[53] However, it has been suggested that DHA is necessary for proper brain function. However, due to genetic factors, mammals lack these enzymes, resulting in humans being incapable of converting oleic acid (OA) to ALA and LA.[54] The insufficiency of these particular enzymes renders the aforementioned polyunsaturated fatty acids indispensable. The concentration of EPA and DPA in an organism are primarily influenced by direct dietary intake.[55] This is due to the fact that a significant portion of dietary ALA undergoes β-oxidation within the mitochondria. The initial stage of long-chain PUFA synthesis involves desaturation, which is facilitated by the enzyme △-6 fatty acid desaturase (FADs).[56] This process entails the introduction of a double bond at the sixth carbon–carbon position from the carboxyl group of LA and ALA, resulting in the formation of GLA and SDA, respectively.[57] The enzyme △6-FADs plays a critical role in the synthesis of long-chain polyunsaturated fatty acids, as it serves as the limiting factor in both animals and humans.[58]

Subsequently, elongation takes place leading to the synthesis of DGLA and eicosatetraenoic acid (ETA).[58] The introduction of a double bond to the fifth carbon–carbon bond from the carboxy end by △5-FADS ultimately leads to the production of ARA and EPA. The process of desaturation occurring at the seventeenth carbon–carbon position leads to the conversion of GLA, DGLA, and ARA into SDA, ETA, and EPA, respectively.[2] The enzymes known as FADs play a pivotal role in the biosynthetic pathway of LC PUFAs, and are coded by the *FADs1* and *FADs2* genes situated on chromosome-11, also referred to as the *FADS* cluster.[59] The synthesis of DHA in a mammalian organism involves a three-step process, which includes two elongations and β-oxidation, starting from EPA.[60] The aforementioned biochemical process is commonly referred to as the Sprecher pathway in academic literature. The synthesis of DHA and EPA occurs most efficiently in human hepatoma cells when the ALA : LA ratio is maintained at 1 : 1. Based on the recorded transformation rates of 16.0% for EPA and 0.72% for DHA, it can be inferred that the supplementation of DHA is the most efficacious approach to enhance the levels of DHA in the body.[2] PUFAs such as DGLA, ARA, DHA, and EPA are synthesized biologically and stored in esterified form in phospholipids (PLs) or as neutral glycerides. These stored PUFAs can be released as unesterified free fatty acids by phospholipase A2 when required, and subsequently utilized by oxygenase enzymes to produce eicosanoids or other autacoids.[61] Cyclooxygenase-2 (COX-2) catalyzes the production of a set of prostaglandins (PG) derived from arachidonic acid (ARA).[62] On the other hand, lipoxygenase is responsible for the synthesis of a distinct set of leukotrienes (LT) (B4, C4, E4), which belong to series four.[62] In contrast, EPA undergoes metabolic processes that result in the production of series 3 PG (B3, D3, E3, I3) and series 5 LT (B5, C5, and D6), which exhibit notable anti-aggregative, anti-inflammatory, and vasodilatory properties.[63]

4.4 Absorption of ω-3 Fatty Acids

People who took ALA supplements had more ALA and EPA in their blood, but not more DHA. The plasma DHA pool is much bigger than the plasma EPA pool.[64] Because of this, it may take longer to notice a small amount of ALA transfer to the plasma DHA pool. Overall, ALA is good at increasing the amount of EPA and DPA ω-3 in the blood, but it has little effect on the amount of DHA. Changes in the gastrointestinal tract's environment can be caused by things like bile secretion, intestinal pH, and bacteria. This probably makes it harder for the body to absorb important fatty acids. Even though it is still not clear how the presence of other substances in the gut affects how ALA is absorbed and broken down, many epidemiological studies have shown that taking ALA with cholesterol supplements makes it easier for the body to absorb ALA.[65] This may be because it encourages the formation of micelles and the binding of gut contents to the brush edge of the intestine, which is how absorption happens. According to Shahidi *et al.* (2018), the hydrolysis of both DHA and EPA occurs through pancreatic carboxylic acid ester lipase, resulting in the formation of free fatty acids.[66] Initially, enterocytes are responsible for the absorption of the aforementioned substances. The process of FA uptake by enterocytes involves the participation of a membrane-bound FA binding protein. The presence of FA transport proteins is a significant determinant of the absorption of FAs. Subsequently, the enterocytes are assimilated into chylomicrons, which are eventually discharged into the lymphatic system.[67] The jejunum and ileum are responsible for facilitating efficient intraluminal absorption, thereby limiting the excretion of ingested fatty acids in fecal matter.

4.5 Bioavailability of Omega-3 Fatty Acids

A relative phrase, bioavailability can relate to both the rate of absorption and the amount of the material ingested. The rate at which the drug is absorbed from the digestive tract and travels to the portal system might be interpreted as the speed.[68] Only a small portion of the chemical is absorbed in the digestive system, depending on a variety of circumstances. The narrower definition of bioavailability is the amount of absorption and the rate of transfer to the portal circulation. Fatty acids can exist in the human body in various forms, including as free fatty acids or bound to glycerol to create TAG, diacylglycerol or monoacylglycerol.[69] Additionally, they can be incorporated into membrane phospholipids. LC PUFA typically occupy the second position in TAG molecules found in natural settings. The presence of DHA and EPA in cell membrane phospholipids can displace arachidonic acid from its position.[70] These FAs can be enzymatically released by phospholipase A2 and subsequently utilized for the biosynthesis of eicosanoids.[71]

The accessibility of LC ω-3 FAs is contingent upon a multitude of variables. Generally speaking, triacylglyceride-bound forms of LC ω-3 FAs are more efficiently utilized than ethyl esters.[72] Upon ingestion, the intracellular concentration of ALA is diminished through enzymatic activity and subsequently liberated into the extracellular milieu in the form of dihydrolipoic acid (DHLA). According to researchers, the β-oxidation pathway is primarily responsible for the complete bodily utilization of ALA.[73] It has been reported that ALA exhibits low bioavailability, primarily attributed to its high susceptibility to oxidation, resulting in the disappearance of approximately 84.9% of consumed ALA.[74] In contrast, it can be observed that DHA exhibits poor substrate characteristics for β-oxidation pathways, both peroxisomal and mitochondrial, but it displays comparatively higher bioavailability.[75] The bioavailability of FAs can be influenced by their position on the glycerol molecule within the triglyceride. The second position is considered the most preferred location for absorption, resulting in increased bioavailability. According to Eljack *et al.* (2020), the accumulation of DHA in the body is 86%, which is nearly three times as high as the accumulation of ALA at 30% of total intake.[76] Upon ALA ingestion, 59% of the compound was observed to be absent from the body, while 21.2% was accumulated as ALA.[77] A mere 0.4% was excreted, and the remaining portion of the total ALA was present as long-chain derivatives such as DHA, DPA, and EPA (Figure 4.2). Milk is considered to have a high bioavailability due to the dispersion of milk fat in very small micelles. According to various sources, the average diameter of fat globules in homogenized bovine milk is approximately (1–3 μm), which is believed to enhance the absorption and bioavailability of lipid-soluble compounds such as ω-3 FAs and vitamin E.[78]

LC ω-3 PUFAs are predominantly present in TAG and free FAs in fish and fish oils. Phospholipids constitute a significant proportion of the FAs present in krill oil, with phosphatidylcholine being the primary constituent, accounting for 30–65% of EPA and DHA.[79] Naturally occurring fish oils contain approximately 18% EPA and 12% DHA. As a result of the trans-esterification process, oil blends frequently exhibit significantly higher concentrations of both EPA and DHA. The aforementioned procedure pertains to the replacement of the glycerol backbone with ethanol, leading to the creation of ethyl esters (EE). These ethyl esters can subsequently be transformed into re-esterified TAG (rTAG) through enzymatic removal of ethanol, thereby liberating free FAs that can be reattached to the glycerol backbone with the aid of enzymes.[80] An alternative approach for augmenting the levels of DHA and EPA has been implemented in Epanova.[81] The process involves the substitution of glycerol with a hydrogen atom, leading to formation of the liberated fatty acid. The process of hydrolyzing the ester bond by pancreatic lipase is necessary for the release of free FAs from ethyl esters, the

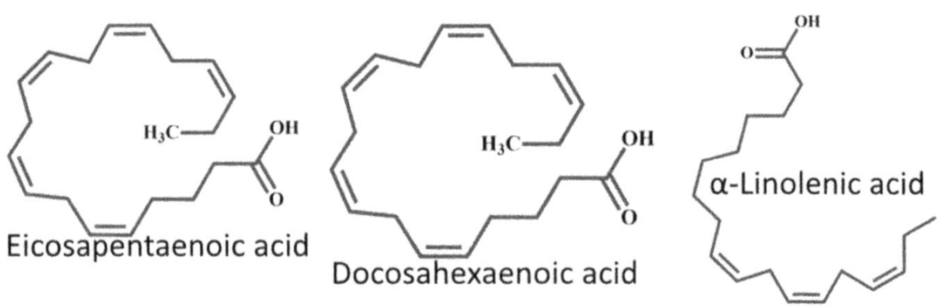

Figure 4.2 Structures of the three main omega-3 fatty acids.

primary component of Lovaza, which can subsequently be absorbed in the small intestine.[82] The hydrolysis step is unnecessary for carboxylic acids. It is noteworthy that EPA-EE and DHA-EE are absorbed without undergoing any alterations. Nonetheless, this particular form of omega-3 acid EE constitutes less than 1% of the overall EPA and DHA present in circulation subsequent to ingestion.[83]

4.5.1 Measuring Methods for Bioavailability of ω-3 Fatty Acids

The objective is to quantify the concentration of ω-3 FAs in various biological samples, including blood cells, plasma, lymph, and serum. The plasma's fatty acid (FA) composition is indicative of the short-to-medium-term availability of FAs in the diet, whereas the FA concentration in blood cells is a reliable indicator of their long-term bioavailability. Regarding LC ω-3 FAs, multiple markers can be utilized to detect the presence of DHA in a particular form.[84] However, only one marker, namely the plasma level of phospholipid EPA, is deemed effective in gauging the level of EPA. Normal dietary levels of EPA in erythrocytes may only be a modest dose-dependent indication of replacement of LC ω-3 PUFAs; the combined concentration of erythrocyte EPA and DHA appears to be a relatively effective indicator of long-term bioavailability.[26] Furthermore, this combined concentration is reflective of the content of LC ω-3 PUFAs in non-blood tissues, as will be discussed below. In Browning's study, EPA + DHA-phosphatidylcholine (in the case of sudden changes in intake) and platelet/mononuclear cell EPA + DHA (in the case of long-term consumption assessment) were considered biomarkers that best represent the intake of fish with high fat content in a typical UK population (1–4 servings a week).[85]

4.6 Functional Compounds in Marine Organisms

Algae contains a variety of functional components such as carotenoids, chlorophyll, DHA, EPA, and astaxanthin.[86] The aforementioned components offer numerous benefits for value-added food applications and are experiencing significant demand in the contemporary market. Due to the high production costs associated with nutraceutical products, scholars are investigating nanotechnology-based methods to enhance yield.[87] Marine invertebrates that are fit for consumption, including crabs, mollusks, and echinodermata, have the potential to provide biopolymers derived from organic sources, with a range of potential applications. Biopolymers, which consist of carbohydrate and proteins, have been utilized by humans for several decades as food and nutraceutical components to address diverse health conditions.[88] Biopolymers, including chitin, chitosan, fucosylated chondroitin sulfate, and glycosaminoglycans have emerged as promising therapeutic agents for a diverse range of medical conditions in the biomedical sector.[89] The aforementioned compounds exhibit various biological activities such as antibacterial, anti-cancer, wound healing, anti-inflammatory, hypoglycemic, and antimicrobial effects.[89] Additionally, they are utilized in animal nutrition. Despite the impact of detrimental human activities, the marine ecosystem serves as a significant reservoir of bio-active compounds. Seafood contains a diverse range of chemical compounds, such as carbohydrates, peptides, pigments, PUFAs, phenolics, enzymes, and other secondary metabolites, which have potential applications in the food industry.[90]

The aforementioned compounds have been observed across many types of organisms, including prokaryotes, microalgae, macroalgae, invertebrates such as crabs and sponges, and various vertebrates. The poor digestibility of the microalgae cell wall, owing to its cellulose structure, typically impedes protein absorption upon consumption.[91] Microalgae protein hydrolysates have enhanced digestibility, resulting in increased availability of peptides and amino acids, thereby rendering them more advantageous for utilization in food-related contexts. The significance of marine bio-peptides in human nutrition should be emphasized to improve their application in food products.[92]

4.7 Health Benefits of Omega-3 Fatty Acids

The consumption of a diet rich in omega-3 has garnered increased attention in recent times due to the demonstrated correlation between omega-3 intake and the prevention of various ailments such as cardiovascular diseases, certain types of cancer, inflammation, blood pressure, and childhood diseases.[26] This relationship has been established through numerous epidemiological studies. In order to maintain the positive impact on health attributed to omega-3 fatty acids, it is advisable to augment the intake of α-linolenic acid (ALA) through dietary means.[93]

4.7.1 Cardiovascular Diseases

The prevalence of heart diseases and associated mortality rates in the Western world can be attributed to the intake of a diet that is high in fat. Several research studies have been carried out to investigate the impact of ω-3 PUFAs on significant cardiovascular ailments. According to the American Heart Association (AHA), coronary heart disease (CHD) encompasses a number of physiological processes, including stable plaques of atherosclerosis progression (which occurs over years and results in angina), plaque instability (which occurs over weeks to months), acute plaque rupture (which occurs over seconds and results in acute coronary syndrome), thrombosis and coagulation (which occurs over minutes to hours and results in acute myocardial infarction), and ischemia-induced cardia.[94,95] According to Fodor *et al.* (2014), the assertion that Eskimos have a lower number of deaths from heart attacks compared to Danes living in the same area is inaccurate.[96] In fact, the Eskimos have much higher death rates from myocardial infarction than other people living in the same area. In a multicenter randomized controlled trial, Jones and colleagues (2014) discovered that the intake of a novel canola oil that is rich in DHA led to improvements in blood pressure, TAG, and high-density lipoprotein cholesterol (HDLPC).[97] These results indicate that the consumption of this oil may reduce the risk of coronary heart disease. Numerous academic studies and reviews, which have relied on clinical trials, have demonstrated that the consumption of ω-3 PUFAs may confer a survival advantage in the context of cardiovascular disease, specifically by mitigating the risk of sudden heart death.[98] Cannon *et al.* studied over a thousand patients. Giving them 1 g per day of ω-3 FAs showed a big drop of 20%, 30%, and 45% in death, cardiovascular disease, and heart failure.[99] Omega-3 fatty acids work to stop cardiovascular disease by lowering the amount of triacylglycerol in the blood, activating platelets, and expressing vascular adhesion molecules.[100] The 2016 guidelines from the European Society of Cardiology (ESC) and European Atherosclerosis Society (EAS) regarding the prevention of cardiovascular disease have raised questions about the potential protective effects of ω-3 FAs.[101] Additionally, the guidelines on the management of dyslipidemia have emphasized the need for further evidence on the effectiveness of ω-3 FA supplements in preventing clinical outcomes before their prescription can be justified.[102]

The impact of ω-3 PUFAs on cardiovascular parameters such as blood pressure and heart rate has been linked to various mechanisms, including an elevated production rate of nitric oxide, mitigation of vasoconstrictive responses to norepinephrine and angiotensin II, augmentation of vasodilatory responses, and enhancement of arterial compliance.[103] The ω-3 Index, as it stands, includes EPA and DHA in erythrocytes, while excluding DPA. DPA can be detected in fish and various fish oils, alongside EPA and DHA. However, there is a limited availability of sources that are abundant in DPA.[104]

According to a recent report by the American Heart Association (AHA), which relied on evidence from randomized controlled trials, the supplementation of ω-3 PUFA was found to be ineffective in lowering the risk of cardiovascular disease in people who have or are at risk of developing diabetes mellitus.[105] There is no reliable evidence that vitamin and mineral supplements prevent stroke in people who already have a high intake of those vitamins and minerals. According to their findings, the administration of ω-3 PUFA supplements may potentially lower the mortality rate associated with coronary cardiac disease in individuals with a history of the condition, potentially by mitigating the risk of sudden cardiac death induced by ischemia.[106] However, the treatment did not demonstrate a reduction in the occurrence of nonfatal myocardial infarction relapses. According to the US National Heart, Lung, and Blood Institute (NHLBI), it is recommended to increase the consumption of seafood as a means of obtaining ω-3PUFAs.[107]

4.7.2 Cancer Improvement

The impact of ω-3 PUFAs on multiple cancer types, such as breast, ovarian, prostate, colon, pancreatic, lung, colorectal, stomach, and skin, has been demonstrated.[108] Furthermore, the effectiveness and tolerability of chemotherapy have been enhanced by omega-3 PUFAs. Several biological ideas have also been put forward to explain how ω-3s prevent cancer. Larsson *et al.* (2004) indicated the following could be the case:[109] (a) reducing the production of arachidonic acid-derived eicosanoids, which changes the immune system's reaction to cancer cells and affects inflammation, cell growth, death, spreading, and new blood vessel growth; (b) impact on the activity of a TF (transcription factor) results in alterations in metabolism, cell growth, and differentiation; (c) modification of the metabolic activity of estrogen leads to a reduction in estrogen-induced cell growth; (d) adjustment of generated reactive oxygen species and free radicals; (e) regulation of insulin sensitivity and membrane fluidity (Figure 4.3). While several studies have looked at the possible links between dietary fat and prostate cancer, most have found no such link. By stimulating cancer cell apoptosis, EPA and DHA may be able to exhibit an anti-cancer effect, either on their own or in combination with standard treatments.[110] The risk of developing breast cancer was shown to be lower in women with higher concentrations of DHA, the most common ω-3 PUFA in erythrocyte membranes.

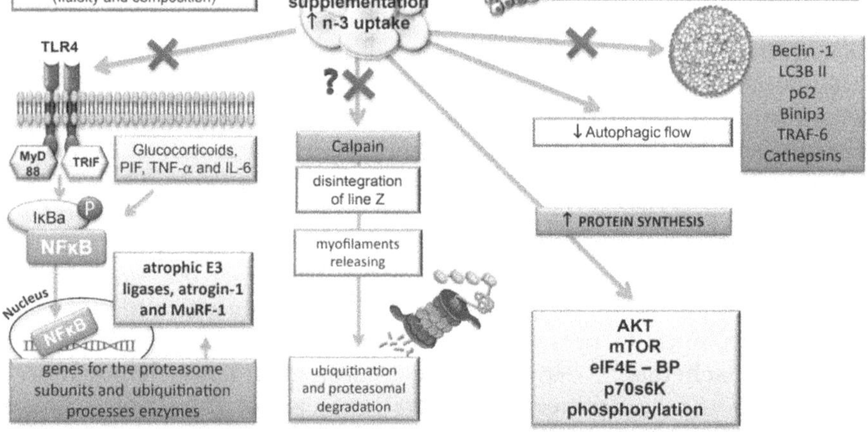

Figure 4.3 Omega-3 effects on cancer-producing cells. Reproduced from ref. 111 with permission from Elsevier, Copyright 2019.

In a recent study, it was shown that the combination of vitamin D_3 and phosphatidyl choline phospholipid complex greatly boosted the cell apoptosis in breast cancer cell lines.[112] Breast cancer patients benefit from dietary supplements that contain ω-3 FA, α-tocopherol, linolenic acid, fiber, and phytoestrogen.[113] Shahverdi and Niknam (2017) proposed that flaxseed has an anti-proliferative action in breast cancer patients and advised 26 g of flaxseed per day for pre-menopausal women.[114] A new comprehensive meta-analysis with 527 392 individuals and 16 178 breast cancer patients found a significant association between ω-3 consumption and breast cancer risk.[115] Supplementing with omega-3 FAs during radiation for neck and brain cancer improved patients' quality of life by allowing them to maintain their weight, thanks to an increase in protein and calories. Lung tumors caused by polycyclic aromatic hydrocarbons were similarly inhibited by omega-3 PUFAs in mice.[116] According to Finocchiaro *et al.* (2012), there was a notable rise in body weight alongside notable anti-oxidative and anti-inflammatory impacts of ω-3 PUFAs.[117] Evidence suggests that omega-3 PUFAs can improve nutritional status and immunological function in individuals after surgery on the gastrointestinal tract.

Furthermore, ω-3-rich menhaden oil was found to suppress human neuroblastoma tumor development by 40–70% when used to replace other dietary fats. Yan *et al.* (2021) suggested that the main reason omega-3 PUFAs stopped the growth of neuroblastoma tumors was because they stopped cell growth and caused cell death.[118] Since long-term use of ω-3s is thought to be safe for children, studies suggest using ω-3 PUFAs (DHA and EPA) to treat neuroblastoma. Silva *et al.* (2015) proposed that the administration of ω-3 PUFAs (ranging from 600 mg to 3.4 g) in conjunction with conventional cancer treatments, such as chemotherapy, radiotherapy, and chemo-radiotherapy, may have the potential to mitigate toxicity and enhance overall survival rate of individuals with cancer. Moreover, supplementation with ω-3 may have the potential to prevent neurotoxicity, in addition to its use as a cancer treatment.[119] In a separate investigation, it was observed that patients with advanced pancreatic and lung cancer who were administered ω-3 supplements (2 g EPA + 1 g DHA per day) experienced an increase in their plasma EPA level, while no significant alteration was observed in their plasma DHA levels.[120]

4.7.3 Maintaining Cholesterol

Studies including epidemiology and genetics have found evidence to support the hypothesis that hypercholesterolemia and cardiovascular illnesses are linked to one another. According to the findings of another study, a rise of one milligram per deciliter in the level of low-density lipoprotein (LDL) may raise the risk of cardiovascular diseases by two to three percent, and further elevations may be related to greater increases in risk.[121] The administration of prescription formulations containing DHA and EPA has been shown to enhance the lipid profile of individuals with normal lipid levels. Optimal amounts of ALA, DHA, and EPA have demonstrated significant efficacy in this regard. ALA exhibits lipid-lowering efficacy and anti-atherosclerotic characteristics, in addition to reducing plasma LDL level.[122] The cholesterol-reducing properties of α-linolenic acid (ALA) have been corroborated by numerous studies that have utilized walnuts, flaxseed, olive oil, and canola oil in subjects with either normal or elevated cholesterol levels. The study conducted an *in vivo* investigation where fish oil, perilla oil, and corn oil were administered in the diet. The findings indicated that the inclusion of dietary ALA resulted in a reduction of postprandial lipid levels, while simultaneously increasing the levels of DHA and EPA.[123]

4.7.4 Omega-3 Fatty Acids and Alzheimer's Disease (AD)

Alzheimer's disease (AD) is a debilitating condition that currently has limited treatment options and lacks a cure. One of the initial symptoms of the disease is memory loss, and it is a progressive condition that results in the patient's inability to perform self-care tasks and ultimately leads to mortality.[124] Presently, the estimated number of individuals affected by AD is 26.6 million, with projections indicating a surge to 106.2 million by the year 2050.[125] Docosahexaenoic acid (DHA) is abundantly present in phospholipids of neuron membranes and is crucial for the optimal functioning of the nervous system. This is the reason why it is believed to have a potential role in the pathogenesis of AD. A recent study discovered a correlation between a diet featuring elevated intake of ω-3 fatty acid-rich foods (such as nuts, fish, salad dressing, tomatoes, cruciferous vegetables, poultry, fruits, and green leafy vegetables) and a decreased risk of Alzheimer's disease.[126] Conversely, a diet with reduced intake of ω-3 fatty acid-poor foods (such as red meat, high-fat dairy products, organ meat, and butter) was also strongly linked to a lower risk of AD.[127] The activation of numerous inflammatory cells in the brain is considered a pivotal occurrence in AD. The potential dysfunction of neurons in the brain may be attributed to the release of interleukins (IL-1B, IL-6) and tumor necrosis factor (TNF) from microglia cells. In a particular study, patients diagnosed with AD who underwent EPA + DHA supplementation exhibited elevated levels of DHA and EPA in their plasma.[128] This increase in concentration was linked to a decrease in the activation of peripheral blood mononuclear cells to secrete inflammatory substances such IL-1B, IL-6, and granulocyte colony-stimulating factor.

PUFAs have been found to impact brain function through various mechanisms, including alterations to membrane fluidity, modulation of membrane-bound enzyme activity, changes in receptor number and affinity, modulation of ion channel function, influence on neurotransmitter production and activity, and regulation of signal transduction pathways that control neurotransmitter and neuronal growth factor activity.[129,130] Omega-3 PUFAs could cause a drop in the number of lymphocytes, the production of TNF, the activity of natural killer cells, and the production of interleukin 1 (IL-1) and interleukin 2 (IL-2) in people.[131] A deficiency in PUFAs during the early neonatal period has the potential to result in the development of Huntington's disease, schizophrenia, elevated blood pressure, and heightened appetite signaling in adulthood. Numerous academic studies have indicated that a lack of crucial fatty acids, specifically omega-3 polyunsaturated fatty acids, is a contributing factor to the development of attention deficit hyperactivity disorder.[132] According to a study by Yehuda *et al.* (2012), it was observed that the provision of ω-3 PUFAs resulted in a noteworthy enhancement in quality of life, concentration ability, sleep quality, and hemoglobin levels among children diagnosed with attention deficit hyperactivity disorder (ADHD) who were also experiencing iron deficiency and sleep disturbances.[133] In general, the consumption of fish and omega-3 PUFAs has demonstrated a beneficial impact on cognitive health among older adults who are in good health. However, the consumption of omega-3 PUFAs remains a topic of debate when considering individuals with Alzheimer's disease.[134]

The early stages of AD are often accompanied by neuropsychiatric symptoms, which have a tendency to intensify as the disease advances. A study involving 174 participants who were randomly assigned to either a placebo group or a group receiving treatment for mild to moderate Alzheimer's disease (MMSE score of 15) through daily intake of DHA (1.6 gm) and EPA (0.5 gm) revealed that after 6 months there was no significant difference in the decline of cognitive function between the two groups.[135,136] However, within a specific subset characterized by minimal cognitive impairment, a noteworthy decrease in the rate of decline in MMSE was noted in the group receiving DHA + EPA supplementation in comparison to the placebo group.[137] A further investigation was conducted to examine the effects of supplemental DHA on people with mild to severe AD. The study utilized the AD assessment scale-cognitive subscale, which assesses cognitive function using a 70-point scale across various domains, including attention, memory, language, praxis, and orientation.[138]

4.7.5 Omega-3 Fight Against Acne

The physiological mechanisms underlying acne are characterized by the interaction of four primary pathogenic factors. These factors include excessive sebum production, modified follicular keratinization, colonization of hair follicles by *Cutibacterium acnes* (also known as *Propionibacterium acnes*), and inflammation that involves innate and acquired immune responses.[139]

The Western diet is characterized by a relatively low proportion of ω-3 to ω-6 fatty acids in comparison to non-Westernized cuisine. The typical Western diet is characterized by a significantly higher proportion of ω-6 content compared to ω-3 content, with a ratio of nearly 10 : 1.[140] This disparity is primarily attributed to the limited consumption of fish, wild game, and wild plants. Furthermore, omega-3 fatty acids are commonly depleted or oxidized during the processes of food preparation and culinary techniques. A study conducted by Pessmier *et al.* (2021) studied the impact of dietary patterns on the occurrence of acne among the Korean population.[141]

The study revealed that individuals with acne exhibited a lower consumption of fish and a higher intake of junk food in comparison to their healthy counterparts. Likewise, a study conducted on an Italian population showed that the intake of fish provided a safeguarding impact in terms of the emergence of moderate and severe acne.[142] Milk and other dairy products, as well as high glycaemic index carbohydrates, have been identified as additional food types that can exacerbate acne.[143]

For several decades, the anti-inflammatory properties of ω-3 FAs have been acknowledged. These properties are responsible for the therapeutic benefits observed in the treatment of acne vulgaris, as inflammation is a crucial pathogenic factor in the development of acne (Figure 4.4). The role of inflammation in the advanced stages of acne, as evidenced by the presence of inflammatory lesions, has been widely acknowledged.[144] However, subsequent research has revealed that inflammation is among the initial occurrences in the pathogenesis of acne. The affirmation of the aforementioned statement is supported by the amplified infiltration of $CD3^+$ and $CD4^+$ T cells in the perifollicular and papillary dermis of unaffected skin in individuals with acne, as well as the efficacy of anti-inflammatory agents, including topical benzoyl peroxide, antibiotics, and dapsone in diminishing non-inflammatory lesions in acne patients.[145]

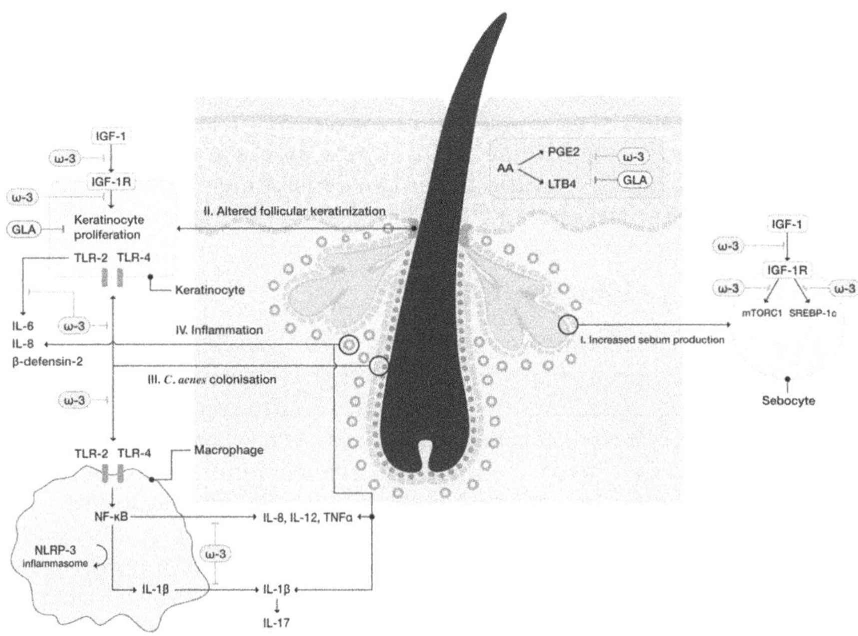

Figure 4.4 Mechanisms of fatty acid influence on pathogenetic factors in acne (Balić, *et al.*[2]). Reproduced from ref. 2, https://doi.org/10.3390/ijms21030741, under the terms of the CC BY 4.0 license, http://creativecommons.org/licenses/by/4.0/.

Omega-3 fatty acids have been found to impede the activation of the NLRP3 inflammasome within cells responsible for presenting antigens. The NLRP3 inflammasome is activated by the presence of *C. acnes*, which is released from the pilosebaceous follicle following disruption of the follicular epithelium and subsequently interacts with dermal macrophages.[146] Following this, the activation of caspase-1 occurs, resulting in the liberation of IL-1β derived from monocytes. The activation of Th17 is promoted by IL-1β, resulting in inflammation and keratinization that is mediated by IL-17.[147] The regulation of this process is mediated by both ROS and proteases. Snodgrass *et al.* (2019) have provided evidence that docosahexaenoic acid has an inhibitory effect on the expression of pro-interleukin-1β (pro-IL-1β) and the secretion of mature IL-1β in monocytes.[148]

Omega-3 FAs have been observed to compete with arachidonic acid (ARA) for incorporation into the phospholipids (PL) of the cellular membrane. This competition results in omega-3 fatty acids acting as a substrate for 5-lipoxygenase (5-LOX) and cyclooxygenase-2 (COX-2), ultimately leading to a decrease in the production of pro-inflammatory prostaglandin E2 (PGE2) and leukotriene B4 (LTB4).[149] Additionally, omega-3 FAs have been observed to impede mTORC1 signaling, resulting in the subsequent down-regulation of SREBP-1c pathways.[2]

4.7.6 Hypertension

Up to 20% of people have high blood pressure, which is a big cause of cardiovascular disease.[150] Omega-3 FAs may improve capillary function and lessen stiffness in the arteries and damage from atherosclerosis because they have antioxidant qualities. There is evidence that DHA and EPA are good for your health in many ways, such as fat metabolism, lowering systolic and diastolic blood pressure, heart rate, and cell function.[151] In order to manage blood lipids and hypertension, the American Heart Association recommended adhering to a healthy eating pattern that includes the consumption of 2 servings of fish per week.

4.7.7 ω-3s Reduce Inflammation

According to different studies, the anti-inflammatory properties of DHA are, in part, associated with two mechanisms: firstly, its impact on the reorganization of cell membranes, and secondly, its interaction with specific receptors. Specifically, DHA has been shown to regulate the expression of inflammatory genes through intracellular receptors (GPR120, PPARγ) *via* binding with these receptors.[152] Apart from conferring health advantages, ARA (arachidonic acid) and EPA also act as precursors for the synthesis of eicosanoids.[26] Numerous studies have demonstrated that ARA, which is obtained from ω-6 fatty acids, plays a significant role in regulating the inflammatory response. Conversely, ω-3 FAs have been found to impact the cells that are accountable for inflammation. EPA serves as a substrate for the enzymatic activities of lipoxygenase and cyclooxygenase, leading to the production of eicosanoids.[153] When combined with docosahexaenoic acid (DHA), EPA has been observed to inhibit the metabolism of arachidonic acid (ARA) in the cell membrane by reducing the expression of the *cox2* gene, which is responsible for inflammation. ω-3s exhibit the potential to compete for enzymes that facilitate the conversion of AA into eicosanoids such as PGs, TXs, and LTs, which are accountable for the occurrence of inflammation.[154] Eicosanoids that originate from EPA, such as PGE3 and LTB5, exhibit lower biological activity compared to eicosanoids derived from ARA (such as PGs, TXs, and LTs) which are known to cause inflammation.[155] Additionally, EPA-derived eicosanoids possess anti-inflammatory properties. ω-3 fatty acids have been found to displace ARA from the phospholipids of cell membranes, thereby reducing the inflammatory and pain-promoting properties that are typically associated with omega-6 fatty acids. Cadler *et al.* (2017) reported that DHA and EPA have been found to produce anti-inflammatory and inflammation-resolving mediators, which are commonly referred to as resolvins, protectins, and maresins.[156]

4.8 Adverse Effects of Omega-3 PUFA Supplementation

Omega-3 PUFA supplementation, especially long-term supraphysiological levels, might have negative side effects. The administration of other organic, seemingly beneficial substances has resulted in serious negative consequences. For instance, selenium consumption increased the risk of diabetes, and vitamin E supplementation led to a trend that was suggestive of an enhanced risk of prostate cancer.

Omega-3 fatty acid administration in intervention studies has not been associated with any severe side events at the levels used. Fishy belching, nausea, and loose stools are some of the most frequent side effects of fish-oil formulations, particularly at higher dose levels.[157] The efficacy of blinding procedures employed in certain studies investigating the potential of fish oils for treatment of mental health disorders has been called into question due to the aforementioned impacts. In a broader sense, it appears that ω-3 PUFAs do not exhibit significant clinical impacts on bleeding time.[17]

In relation to the adverse effects associated with ω-3 PUFAs, it is imperative to take into account the mode of consumption of these nutrients. The consumption of fish as well as the intake of supplements can present certain issues. The administration of natural ω-3 PUFAs requires careful consideration of various potential risks.[158] Fish and shellfish, which can contain methylmercury, dioxins, and polychlorinated biphenyls, are a significant natural source of these bioactives.[92] When eaten by a pregnant woman, these chemicals have the potential to damage the unborn child or raise the risk of certain malignancies.[159] In this regard, it has been discovered that eating fish once or twice a week is not troublesome.

Fish oil supplements have the potential to contain antioxidants and ω-3 PUFA oxidation products, which may result in unfavorable responses.[158] The susceptibility of omega-3 FAs to oxidative degradation is a well-established phenomenon. It has been observed that a considerable number of omega-3 fish-oil preparations, which are marketed in various countries, surpass the recommended safety thresholds for total oxidation as per the international voluntary guidelines. Some of the omega-3 fatty acid products that have been used in clinical trials may have had some of their oils altered, which could have affected the results of the studies.[160] Different levels of omega-3 oxidation also explain why taking omega-3 supplements and eating fish have different affects. Further investigation is warranted to explore the impact of oxidized oils on human health. It is imperative to investigate the potential negative outcomes associated with extended utilization of vitamin E, which is incorporated as an antioxidant in fish-oil supplements.[161] This is due to the fact that extensive trials have established a correlation between α-tocopherol supplementation and heightened occurrences of prostate cancer.

4.9 Conclusions

In recent years, there has been noteworthy advancement in the research pertaining to omega-3 fatty acids. Our knowledge of the sources and chemical structure of supplements has increased, leading to a better understanding of their significance in supplementation. Omega-3 fatty acids are primarily ingested by individuals through the consumption of fish, and to a lesser degree, through the intake of cereal products and meat. Certain entities, nonetheless, are not observable in the aforementioned commodities, or their existence remains unexamined. Furthermore, the distribution of specific acids within plants is contingent upon their respective locations, while in animals, it is influenced by their dietary habits. The consumption of essential fatty acids (EFA) as part of one's dietary patterns has been found to have a positive correlation with the reduction of morbidity and mortality related to cardiovascular disease (CVD). Additionally, such dietary patterns have been associated with potential anti-inflammatory and anticarcinogenic effects. Omega-3 fatty acids, including ALA, DHA, and EPA, are biologically active compounds. Following dietary intake of ALA, it undergoes conversion to the more biologically active DHA and EPA. Numerous studies have been conducted regarding the health benefits of omega-3s, however, there remains a dearth of information regarding the optimal dosage of omega-3s.

Conflicts of Interest

The authors report no declaration of interests.

References

1. T. G. Albuquerque, M. A. Nunes, S. M. Bessada, H. S. Costa, M. B. P. Oliveira, Biologically active and health promoting food components of nuts, oilseeds, fruits, vegetables, cereals, and legumes, in *Chemical analysis of food: Techniques and Applications*, Elsevier, 2020, pp. 609–656.
2. A. Balić, D. Vlašić, K. Žužul, B. Marinović and Z. Bukvić Mokos, *Int. J. Mol. Sci.*, 2020, **21**(3), 741.
3. C. Huang, L. Chiba and W. Bergen, *Livest. Sci.*, 2021, **243**, 104370.
4. D. Dave and W. J. Routray, *J. Cleaner Prod.*, 2018, **180**, 617–641.
5. J. Li, B. L. Pora, K. Dong and J. Hasjim, *Food Sci. Nutr.*, 2021, **9**(9), 5229–5243.
6. A. H. Mariamenatu and E. M. Abdu, *J. Lipids*, 2021, **2021**, 1–15.
7. A. H. Metherel, K. Rezaei, R. S. Lacombe and R. P. Bazinet, *Biochim. Biophys. Acta, Mol. Cell Biol. Lipids*, 2021, **1866**(8), 158942.
8. C. McReynolds, C. Morisseau, K. Wagner and B. Hammock, *Druggable lipid signaling pathways*, 2020, pp. 71–99
9. P. C. Calder, *Essays Biochem.*, 2020, **64**(3), 423–441.
10. P. Bjerregaard and C. V. Larsen, *Public Health Rev.*, 2018, **39**(1), 1–12.
11. A. K. Duttaroy and S. Basak, *Prostaglandins, Leukotrienes Essent. Fatty*, 2020, **155**, 102080.
12. F. Matrisciano and G. Pinna, Functional food and nutrition for the management of autism spectrum disorders and schizophrenia, in *Functional Foods and Mental Health*, Food Science Publisher, 2019, pp. 307–330.
13. O. Adi-Dako, D. Kumadoh, M. Fredua-Agyeman, M. A. Archer, F. Bentil and M. Okaijah, in *Nutraceuticals for Aging and Anti-Aging*, CRC Press, 2021. pp. 649–680.
14. J. Jin, Q. Jin, X. Wang and C. C. Akoh, *Engineering*, 2020, **6**(4), 424–431.
15. S. M. Azab, R. J. de Souza, K. K. Teo, S. S. Anand, N. C. Williams and J. Holzschuher, *et al.*, *J. Lipid Res.*, 2020, **61**(6), 933–944.
16. S. Basak, S. Vilasagaram and A. K. Duttaroy, *Prostaglandins, Leukotrienes Essent. Fatty Acids*, 2020, **158**, 102109.
17. I. Djuricic and P. C. Calder, *Nutrients*, 2021, **13**(7), 2421.
18. S. C. Gurumallu, T. Aqeel, A. Bhaskar, K. Chandramohan and R. Javaraiah, *Drug Chem. Toxicol.*, 2022, **45**(5), 2221–2232.
19. A. Cartoni Mancinelli, S. Mattioli, C. Twining, A. Dal Bosco, A. M. Donoghue and K. Arsi, *et al.*, *Nutrients*, 2022, **14**(9), 1969.
20. K. E. Lane and E. J. Derbyshire, *Crit. Rev. Food Sci. Nutr.*, 2018, **58**(1), 62–69.
21. M. Edwards, G. Beaugrand, L. Kléparski, P. Hélaouët and P. C. Reid, *Commun. Earth Environ.*, 2022, **3**(1), 162.
22. B. Meyer, A. Atkinson, K. S. Bernard, A. S. Brierley, R. Driscoll and S. L. Hill, *et al.*, *Commun. Earth Environ.*, 2020, **1**(1), 28.
23. T. Pérez-Palacios, J. Ruiz-Carrascal, J. C. Solomando and T. Antequera, *Food Rev. Int.*, 2019, **35**(5), 485–503.

24. E. Molina-Montes, E. Salamanca-Fernández, B. Garcia-Villanova and M. J. Sánchez, *Nutrients*, 2020, **12**(7), 2010.
25. S. Ghasemi Fard, F. Wang, A. J. Sinclair, G. Elliott and G. M. Turchini, *Crit. Rev. Food Sci. Nutr.*, 2019, **59**(11), 1684–1727.
26. R. K. Saini and Y.-S. Keum, *Life Sci.*, 2018, **203**, 255–267.
27. P. Prasad, P. Anjali and R. J. Sreedhar, *Crit. Rev. Food Sci. Nutr.*, 2021, **61**(10), 1725–1737.
28. G. Rizzo, L. Baroni and M. J. Lombardo, *Int. J. Environ. Res. Public Health*, 2023, **20**(3), 1683.
29. G. C. G. Carlini, G. G. Roschel, R. A. Ferrari, S. M. Alencar, H. C. Ota and T. F. F. da Silveira, *et al.*, *J. Food Sci.*, 2021, **86**(12), 5307–5317.
30. D. Vodolazska and C. J. Lauridsen, *J. Anim. Sci. Biotechnol.*, 2020, **11**, 1–18.
31. G. Vannice and H. J. Rasmussen, *J. Acad. Nutr.*, 2014, **114**(1), 136–153.
32. R. K. Saini, P. Prasad, R. V. Sreedhar, K. Akhilender Naidu, X. Shang and Y.-S. Keum, *Antioxidants*, 2021, **10**(10), 1627.
33. Q. V. Nguyen, B. S. Malau-Aduli, J. Cavalieri, P. D. Nichols and A. Malau-Aduli, *Nutrients*, 2019, **11**(4), 743.
34. F. Shahidi, Bailey's Industrial Oil and Fat Products, *Industrial and Nonedible Products from Oils and Fats*, John Wiley & Sons, 2005.
35. L. A. Copeman and C. C. Parrish, *J. Agric. Food Chem.*, 2004, **52**(15), 4872–4881.
36. M. T. Arts, R. G. Ackman and B. J. Holub, *Can. J. Fish. Aquat. Sci.*, 2001, **58**(1), 122–137.
37. S. M. Budge, S. J. Iverson, W. D. Bowen and R. G. Ackman, *Can. J. Fish. Aquat. Sci.*, 2002, **59**(5), 886–898.
38. J. J. Bourre, *J. Nutr., Health Aging*, 2005, **9**(4), 232–242.
39. G. Lenihan-Geels, K. S. Bishop and L. R. Ferguson, *Nutrients*, 2013, **5**(4), 1301–1315.
40. A. P. DeFilippis and L. S. Sperling, *Am. Heart J.*, 2006, **151**(3), 564–570.
41. J. C. Ruiz Ruiz, E. D. L. L. Ortiz Vazquez and M. R. Segura Campos, *Crit. Rev. Food Sci. Nutr.*, 2017, **57**(7), 1423–1434.
42. L. Cheng, L. Sanguansri, M. Hlaing, T. Singh, P. Shrestha and M. Augustin, *Food Chem.*, 2022, **370**, 131340.
43. W. Stonehouse, B. Klingner, R. Tso, P. S. Teo, N. S. Terefe and C. G. Forde, *Eur. J. Nutr.*, 2022, **61**(4), 2129–2141.
44. A. Pinar-Martí, S. Fernández-Barrés, F. Gignac, C. Persavento, A. Delgado and D. Romaguera, *et al.*, *Eur. Child Adolesc. Psychiatry*, 2022, 1–9.
45. R. Jamaledin, C. K. Yiu, E. N. Zare, L. N. Niu, R. Vecchione and G. Chen, *et al.*, *Adv. Mater.*, 2020, **32**(33), 2002129.
46. S. Zhong, L. Li, X. Shen, Q. Li, W. Xu and X. Wang, *et al.*, *Free Radicals Biol. Med.*, 2019, **144**, 266–278.
47. J. Dunn and M. H. Grider, Physiology, adenosine triphosphate, in *StatPearls*, StatPearls Publishing, Treasure Island (FL), 2024.
48. N. Koundouros and G. Poulogiannis, *Br. J. Cancer*, 2020, **122**(1), 4–22.

49. J. Choi, T. Yin, K. Shinozaki, J. W. Lampe, J. F. Stevens and L. B. Becker, *et al.*, *Mol. Cell. Biochem.*, 2018, **442**, 187–201.
50. R. Adili, M. Hawley and M. Holinstat, *Prostaglandins Other Lipid Mediators*, 2018, **139**, 10–18.
51. M. Pilecky, L. Závorka, M. T. Arts and M. J. Kainz, *Biol. Rev.*, 2021, **96**(5), 2127–2145.
52. C. Robert, L. Couëdelo, C. Vaysse and M.-C. J. B. Michalski, *Biochimie*, 2020, **169**, 121–132.
53. E. N. Ponnampalam, A. J. Sinclair and B. W. Holman, *Foods*, 2021, **10**(6), 1358.
54. N. Lakra, S. Mahmood, A. Marwal, N. Sudheep, K. J. P. Anwar and V. P. Human Health, *et al.*, *Plant Hum. Health*, 2019, 361–382.
55. Y. Fu, Y. Wang, H. Gao, D. Li, R. Jiang and L. Ge, *et al.*, *Mediators Inflammation*, 2021, **2021**(1), 8879227.
56. K.-K. Sam, P. Merosha, M. Janaranjani, I. Athirah and A. C. Shu-Chien, *Aquaculture*, 2021, **543**, 736942.
57. E. Baker, *Identifying and understanding anti-inflammatory effects of 18-carbon fatty acids from sustainable plant sources*, University of Southampton, 2018.
58. A. P. Simopoulos, *Am. J. Clin. Nutr.*, 1991, **54**(3), 438–463.
59. L. M. Reynolds, T. D. Howard, I. Ruczinski, K. Kanchan, M. C. Seeds and R. A. Mathias, *et al.*, *PLoS One*, 2018, **13**(3), e0194610.
60. A. K. Duttaroy, *Evidence-Based Nutrition and Clinical Evidence of Bioactive Foods in Human Health and Disease*, Academic Press, 2021.
61. W. W. Christie and J. L. Harwood, *Essays Biochem.*, 2020, **64**(3), 401–421.
62. A. J. Morgan, P. J. Kingsley, M. M. Mitchener, M. Altemus, T. A. Patrick and A. D. Gaulden, *et al.*, *ACS Chem. Neurosci.*, 2018, **9**(7), 1552–1559.
63. S. Davinelli, M. Intrieri, G. Corbi and G. Scapagnini, *Crit. Rev. Food Sci. Nutr.*, 2021, **61**(2), 259–274.
64. R. D. Rotarescu, K. Rezaei, D. M. Mutch and A. H. Metherel, *Essent. Fatty Acids*, 2022, **185**, 102494.
65. S. Punia, K. S. Sandhu, A. K. Siroha and S. B. Dhull, *PharmaNutrition*, 2019, **10**, 100162.
66. F. Shahidi and P. Ambigaipalan, *Annu. Rev. Food Sci.*, 2018, **9**, 345–381.
67. K. Hokkanen, A. Tirronen and S. Ylä-Herttuala, *Curr. Opin. Lipidol.*, 2019, **30**(5), 370–376.
68. S. D. Stamatis and J. Rose, *Mol. Pharmaceutics*, 2021, **19**(1), 213–226.
69. S. Senga, N. Kobayashi, K. Kawaguchi, A. Ando and H. Fujii, *Mol. Cell Biol. Lipids*, 2018, **1863**(9), 1057–1067.
70. R. Lordan, S. Redfern, A. Tsoupras and I. Zabetakis, *Food Funct.*, 2020, **11**(4), 2861–2885.
71. A. M. Astudillo, M. A. Balboa and J. Balsinde, *Biochim. Biophys. Acta, Mol. Cell Biol. Lipids*, 2019, **1864**(6), 772–783.

72. I. Oroz-Guinea, K. Zorn and U. T. Bornscheuer, *Eur. J. Lipid Sci. Technol.*, 2020, **122**(1), 1900115.
73. J. Wei, Z. Pengji, J. Zhang, T. Peng, J. Luo and F. Yang, *Water Res.*, 2023, **229**, 119397.
74. L. Dahabra, G. Broadberry, A. Le Gresley, M. Najlah and M. Khoder, *Molecules*, 2021, **26**(6), 1698.
75. T.-T. Zhang, J. Xu, Y.-M. Wang and C.-H. Xue, *Prog. Lipid Res.*, 2019, **75**, 100997.
76. B. Hassabelrasoul Eljack, N. Abdelwahab Ahmed, E. Ahmed Adam, N. Adam Abdelrahman and A. Abubakr Mohammed, *SUST J. Agric. Vet. Sci.*, 2020, 133–141.
77. Z. F. Bhat, J. D. Morton, S. Mason, A. E.-D. A. Bekhit and H. F. Bhat, *Crit. Rev. Food Sci. Nutr.*, 2019, **59**(8), 1294–1310.
78. A. Mishra, D. Pradhan, P. Biswasroy, B. Kar, G. Ghosh and G. Rath, *et al.*, *J. Drug Delivery Sci. Technol.*, 2021, **65**, 102693.
79. A. Colletti, G. Cravotto, V. Citi, A. Martelli, L. Testai and A. Cicero, *Mar. Drugs*, 2021, **19**(6), 306.
80. J. Park, Y.-S. Yoo, E. Shin, G. Han, K. Shin and D. H. Lim, *et al.*, *Br. J. Ophthalmol.*, 2021, **105**(11), 1504–1509.
81. E. D. Ahire, K. R. Surana, C. D. Patil, H. S. Shah, G. B. Sonwane, S. G. Talele, Role of omega-3 fatty acids in different neurodegenerative disorders, in *Applied Pharmaceutical Science and Microbiology*, Apple Academic Press, Applied Pharmaceutical Science and Microbiology, 2020, pp. 173–194.
82. K. Manusama, M. Balvers, M. Diepeveen-de Bruin, L. Headley, R. Bosi and M. Schwarm, *et al.*, *Prostaglandins, Leukotrienes Essent. Fatty Acids*, 2021, **164**, 102232.
83. M. K. Ahmmed, F. Ahmmed, H. Tian, A. Carne and A. E. Bekhit, *Compr. Rev. Food Sci. Food Saf.*, 2020, **19**(1), 64–123.
84. L. Han, S. Silvestre, O. Sayanova, R. P. Haslam and J. A. Napier, *Plant Biotechnol. J.*, 2022, **20**(9), 1833–1852.
85. L. M. Browning, C. G. Walker, A. P. Mander, A. L. West, J. Madden and J. M. Gambell, *et al.*, *Am. J. Clin. Nutr.*, 2012, **96**(4), 748–758.
86. S. Ramos-Romero, J. R. Torrella, T. Pagès, G. Viscor and J. Torres, *Nutrients*, 2021, **13**(2), 563.
87. F. Khan, P. Pandey and T. K. Upadhyay, *Agriculture*, 2022, **12**(10), 1672.
88. J. Baranwal, B. Barse, A. Fais, G. L. Delogu and A. Kumar, *Polymers*, 2022, **14**(5), 983.
89. M.-C. Wan, W. Qin, C. Lei, Q.-H. Li, M. Meng and M. Fang, *et al.*, *Bioact. Mater.*, 2021, **6**(12), 4255–4285.
90. S. Ghosh, T. Sarkar, S. Pati, Z. A. Kari, H. A. Edinur and R. Chakraborty, *Front. Mar. Sci.*, 2022, **9**, 76.
91. O. Córdova, J. Santis, G. Ruiz-Fillipi, M. E. Zuñiga, F. G. Fermoso and R. J. R. Chamy, *et al.*, *Renewable Sustainable Energy Rev.*, 2018, **82**, 2806–2813.
92. S. A. Ashraf, M. Adnan, M. Patel, A. J. Siddiqui, M. Sachidanandan and M. Snoussi, *et al.*, *Mar. Drugs*, 2020, **18**(5), 265.

93. N. Shrestha, S. L. Sleep, J. S. Cuffe, O. J. Holland, A. V. Perkins and S. Y. Yau, *et al.*, *Clin. Exp. Pharmacol. Physiol.*, 2020, **47**(5), 907–915.
94. Y. Hu, F. B. Hu and J. E. Manson, *J. Am. Heart Assoc.*, 2019, **8**(19), e013543.
95. B. Vogel, B. E. Claessen, S. V. Arnold, D. Chan, D. J. Cohen and E. Giannitsis, *et al.*, *Nat. Rev. Dis. Primers*, 2019, **5**(1), 39.
96. J. G. Fodor, E. Helis, N. Yazdekhasti and B. Vohnout, *Can. J. Cardiol.*, 2014, **30**(8), 864–868.
97. P. J. Jones, V. K. Senanayake, S. Pu, D. J. Jenkins, P. W. Connelly and B. Lamarche, *et al.*, *Am. J. Clin. Nutr.*, 2014, **100**(1), 88–97.
98. W.-H. Schunck, A. Konkel, R. Fischer and K.-H. Weylandt, *Pharmacol. Ther.*, 2018, **183**, 177–204.
99. C. P. Cannon, S. P. Curtis, G. A. FitzGerald, H. Krum, A. Kaur and J. A. Bolognese, *et al.*, *Lancet*, 2006, **368**(9549), 1771–1781.
100. J. Trebatická, A. Dukát, Z. Ďuračková and J. Muchová, *Physiol. Res.*, 2017, **66**(3), 363.
101. E. Sokoła-Wysoczańska, T. Wysoczański, J. Wagner, K. Czyż, R. Bodkowski and S. Lochyński, *et al.*, *Nutrients*, 2018, **10**(10), 1561.
102. T. Aung, J. Halsey, D. Kromhout, H. C. Gerstein, R. Marchioli and L. Tavazzi, *et al.*, *JAMA Cardiol.*, 2018, **3**(3), 225–233.
103. A. Daci, G. Özen, İ Uyar, E. Civelek, F. İ. A. Yildirim and D. K. Durman, *et al.*, *Prostaglandins Other Lipid Mediators*, 2017, **133**, 29–34.
104. G. Liu, R. A. Gibson, D. Callahan, X.-F. Guo, D. Li and A. J. Sinclair, *et al.*, *Food Funct.*, 2020, **11**(3), 2058–2066.
105. M. Simonetto, M. Infante, R. L. Sacco, T. Rundek and D. J. Della-Morte, *Nutrients*, 2019, **11**(10), 2279.
106. A. Elagizi, C. J. Lavie, K. Marshall, J. J. DiNicolantonio, J. H. O'Keefe and R. V. J. Milani, *Prog. Cardiovasc. Dis.*, 2018, **61**(1), 76–85.
107. J. P. Schuchardt, M. Cerrato, M. Ceseri, L. F. DeFina, G. E. Delgado and S. Gellert, *et al.*, *Prostaglandins, Leukotrienes Essent. Fatty Acids*, 2022, **179**, 102418.
108. I. S. Senatorov and N. H. J. Moniri, *Biochem. Pharmacol.*, 2018, **150**, 170–180.
109. S. C. Larsson, M. Kumlin, M. Ingelman-Sundberg and A. Wolk, *Am. J. Clin. Nutr.*, 2004, **79**(6), 935–945.
110. C.-H. Guo, S. Hsia, C.-H. Chung, Y.-C. Lin, M.-Y. Shih and P.-C. Chen, *et al.*, *J. Nutr. Biochem.*, 2021, **87**, 108504.
111. F. A. P. S. Malta, D. Estadella and D. C. Gonçalves, *J. Funct. Foods*, 2019, **54**, 1–12.
112. M. M. Agwa, M. M. Abu-Serie, D. A. Abdelmonsif, N. Moussa, H. Elsayed and S. N. Khattab, *et al.*, *Int. J. Pharm.*, 2021, **607**, 120965.
113. A. Jędrusek-Golińska, D. Górecka, M. Buchowski, K. Wieczorowska-Tobis, A. Gramza-Michałowska and K. Szymandera-Buszka, *et al.*, *Compr. Rev. Food Sci. Food Saf.*, 2020, **19**(2), 835–856.
114. E. Shahverdi and R. J. Niknam, *Focus Med. Sci. J.*, 2017, **3**(2), 1–7.
115. K. H. Lee, H. J. Seong, G. Kim, G. H. Jeong, J. Y. Kim and H. Park, *et al.*, *Adv. Nutr.*, 2020, **11**(5), 1134–1149.

116. S. Huerta-Yepez, A. Tirado-Rodriguez, M. R. Montecillo-Aguado, J. Yang, B. D. Hammock and O. Hankinson, *Sci. Rep.*, 2020, **10**(1), 1–17.
117. C. Finocchiaro, O. Segre, M. Fadda, T. Monge, M. Scigliano and M. Schena, *et al.*, *Br. J. Nutr.*, 2012, **108**(2), 327–333.
118. H.-F. Yan, T. Zou, Q.-Z. Tuo, S. Xu, H. Li and A. A. Belaidi, *et al.*, *Signal Transduction Targeted Ther.*, 2021, **6**(1), 49.
119. J. A. P. Silva, M. E. S. Fabre and D. Waitzberg, *Clin. Nutr.*, 2015, **34**, 359–366.
120. M. Newell, V. Mazurak, L. M. Postovit and C. J. Field, *Cancers*, 2021, **13**(6), 1206.
121. M. Farnier, M. Zeller, D. Masson and Y. Cottin, *Arch. Cardiovasc. Dis.*, 2021, **114**(2), 132–139.
122. P. Mourikis, S. Zako, L. Dannenberg, A. M. Nia, Y. Heinen and L. Busch, *et al.*, *Pharmacol. Ther.*, 2020, **213**, 107592.
123. V. Uppin, P. Acharya and R. R. Talahalli, *Modulatory potentials of n-3 polyunsaturated fatty acids in inflammatory diseases*, IntechOpen, London, UK, 2020.
124. Y.-J. Son, J. Choi and H.-J. Lee, *Int. J. Environ. Res. Public Health*, 2020, **17**(18), 6559.
125. H. Jin, S. Guan, R. Wang, X. Fang, H. Liu and Y. Wu, *et al.*, *J. Alzheimer's Dis.*, 2018, **65**(2), 433–442.
126. R. Schilling, Mediterranean Diet Reduces Heart Attacks and Strokes in Diabetes, 2020, Retrieved from https://www.linkedin.com/pulse/mediterranean-diet-reduces-heart-attacks-strokes-ray-schilling-md.
127. A. S. Manolis, T. A. Manolis, A. A. Manolis and H. J. Melita, *Curr. Vasc. Pharmacol.*, 2022, **20**(5), 383–408.
128. E. D. Miller, A. Dziedzic, J. Saluk-Bijak and M. J. Bijak, *Nutrients*, 2019, **11**(7), 1528.
129. M.-C. Hsu, Y.-S. Huang and W.-C. J. Ouyang, *Lipids Health Dis.*, 2020, **19**, 1–17.
130. K. M. Hixson, M. Cogswell, A. R. Brooks-Kayal and S. J. Russek, *BMC Genomics*, 2019, **20**(1), 1–16.
131. A. B. Rowaiye, O. A. Okpalefe, O. O. Adejoke, J. O. Ogidigo, O. H. Oladipo and A. C. Ogu, *et al.*, *J. Inflammation Res.*, 2021, **14**, 1487.
132. J. P.-C. Chang, K.-P. Su, V. Mondelli and C. M. Pariante, *Neuropsychopharmacology*, 2018, **43**(3), 534–545.
133. S. J. M. Yehuda, *Med. Hypotheses*, 2012, **79**(4), 456–461.
134. Y. Liao, B. Xie, H. Zhang, Q. He, L. Guo and M. Subramanieapillai, *et al.*, *Transl. Psychiatry*, 2019, **9**(1), 190.
135. R. Howard, O. Zubko, R. Bradley, E. Harper, L. Pank and J. O'brien, *et al.*, *JAMA Neurol.*, 2020, **77**(2), 164–174.
136. Y. Freund-Levi, H. Basun, T. Cederholm, G. Faxén-Irving, A. Garlind and M. Grut, *et al.*, *Int. J. Geriatric Psychiatry*, 2008, **23**(2), 161–169.
137. J. E. Lewis, J. Poles, D. P. Shaw, E. Karhu, S. A. Khan and A. E. Lyons, *et al.*, *J. Clin. Transl. Res.*, 2021, **7**(4), 575.
138. K. Rockwood, S. E. Howlett, D. Hoffman, R. Schindler and A. J. Mitnitski, *Alzheimer's Dementia*, 2017, **13**(10), 1098–1106.

139. I. Kurokawa, A. M. Layton and R. Ogawa, *Dermatol. Ther.*, 2021, **11**(4), 1129–1139.
140. A. Patel, S. S. Desai, V. K. Mane, J. Enman, U. Rova and P. Christakopoulos, *et al.*, *Trends Food Sci. Technol.*, 2022, DOI: 10.1016/j.tifs.2022.01.006.
141. B. De Pessemier, L. Grine, M. Debaere, A. Maes, B. Paetzold and C. Callewaert, *Microorganisms*, 2021, **9**(2), 353.
142. T. L. N. Tu, *Experiments in skin: Race and beauty in the shadows of Vietnam*, Duke University Press, 2021.
143. J. Meixiong, C. Ricco, C. Vasavda and B. K. Ho, *JAAD Int.*, 2022, 7, 95–112.
144. E. Firlej, W. Kowalska, K. Szymaszek, J. Roliński and J. J. Bartosińska, *J. Clin. Med.*, 2022, **11**(6), 1579.
145. A. Darjani, E. Aboutaleb, N. Alizadeh, R. Rafiei, K. Gharaee Nejad and S. Nabatchii, *et al.*, *Iran. J. Dermatol.*, 2022, **25**(2), 132–141.
146. C. C. Zouboulis, *Exp. Dermatol.*, 2020, **29**(9), 840–859.
147. C. Gupta, S. Das, V. Gaurav, P. K. Singh, G. Rai and S. Datt, *et al.*, *J. Med.*, 2022, 101331.
148. R. G. Snodgrass and B. J. Brüne, *Front. Pharmacol.*, 2019, **10**, 719.
149. Y. Pang, X. Liu, C. Zhao, X. Shi, J. Zhang and T. Zhou, *et al.*, *Free Radicals Biol. Med.*, 2022, **193**, 363–372.
150. J. R. Ayala Solares, D. Canoy, F. E. D. Raimondi, Y. Zhu, A. Hassaine and G. Salimi-Khorshidi, *et al.*, *J. Am. Heart Assoc.*, 2019, **8**(12), e012129.
151. G. Colussi, C. Catena, M. Novello, N. Bertin and L. A. Sechi, *Nutr., Metab. Cardiovasc. Dis.*, 2017, **27**(3), 191–200.
152. S. Punia , S. B. Dhull, A. K. Siroha, K. S. Sandhu and V. Chaudhary, Mechanism of Action of Essential Fatty Acids, in *Essential Fatty Acids*, CRC Press, Essential Fatty Acids, 2020. pp. 89–100.
153. D. Hathaway III, K. Pandav, M. Patel, A. Riva-Moscoso, B. M. Singh and A. Patel, *et al.*, *Infect. Chemother.*, 2020, **52**(4), 478.
154. P.-A. Regidor, A. Mueller, M. Sailer, F. Gonzalez Santos, J. M. Rizo and F. Moreno Egea, *Int. J. Mol. Sci.*, 2020, **22**(1), 384.
155. R. Zárate, N. el Jaber-Vazdekis, N. Tejera, J. A. Pérez and C. Rodríguez, *Molecules*, 2017, **6**, 1–19.
156. P. C. Calder, *Biochem. Soc. Trans.*, 2017, **45**(5), 1105–1115.
157. H. Watson, S. Mitra, F. C. Croden, M. Taylor, H. M. Wood and S. L. Perry, *et al.*, *Gut*, 2018, **67**(11), 1974–1983.
158. K. W. Lange, Y. Nakamura, A. M. Gosslau and S. Li, *J. Food Bioact.*, 2019, 7, 1–6.
159. Z.-R. Tang, X.-L. Xu, S.-L. Deng, Z.-X. Lian and K. Yu, *Int. J. Mol. Sci.*, 2020, **21**(4), 1519.
160. S. J. Nicholls, A. M. Lincoff, M. Garcia, D. Bash, C. M. Ballantyne and P. J. Barter, *et al.*, *JAMA*, 2020, **324**(22), 2268–2280.
161. S. Dong, X. Huang, J. Zhen, N. Van Halm-Lutterodt, J. Wang and C. Zhou, *et al.*, *Mol. Neurobiol.*, 2018, **55**, 9204–9219.

CHAPTER 5

Conjugated Linoleic Acids: Therapeutic Agents in Functional Food Formulations

GHOLAMREZA ABDI,*[a] RAVI KONDLE,[b] LOUDIYA KRISHNA NAYAK,[b] AB WAHEED WANI,[b] SHAKIR ALI,[b] HARJINDAR KAUR ZARINA,[b] VIMAL CHOUDHARY,[b] MUKUL JAIN,[c] MARYAM ABBASI TARIGHAT[d] AND CHARALAMPOS PROESTOS*[e]

[a] Department of Biotechnology, Persian Gulf Research Institute, Persian Gulf University, Bushehr, 7516, Iran; [b] Department of Horticulture, Lovely Professional University, Phagwara, Punjab – 144411, India; [c] Cell and Developmental Biology Lab, Centre of Research for Development, Parul University, Vadodara, Gujarat – 391760, India; [d] Faculty of Nano and Bio Science and Technology, Department of Chemistry, Bushehr, 75169, Iran; [e] Food Chemistry Laboratory, Department of Chemistry, National and Kapodistrian University of Athens, Panepistimiopolis Zographou, 15771 Athens, Greece

*Emails: abdi@pgu.ac.ir; harpro@chem.uoa.gr

5.1 Introduction

A collection of 18:2 free fatty acid positional and geometric isomers with two conjugated double bonds are referred to as conjugated linoleic acid (CLA) (Figure 5.1). The majority of frequently found fatty acids in nature have double bonds separated by methylene, such as *cis*-9 and *cis*-12 18:2 (linoleic acid).

Food Chemistry, Function and Analysis No. 44
Functional Foods of the Future
Edited by Vijai Kumar Gupta, Minaxi Sharma, Smriti Gaur and Ramesh Chander Kuhad

Published by the Royal Society of Chemistry, www.rsc.org

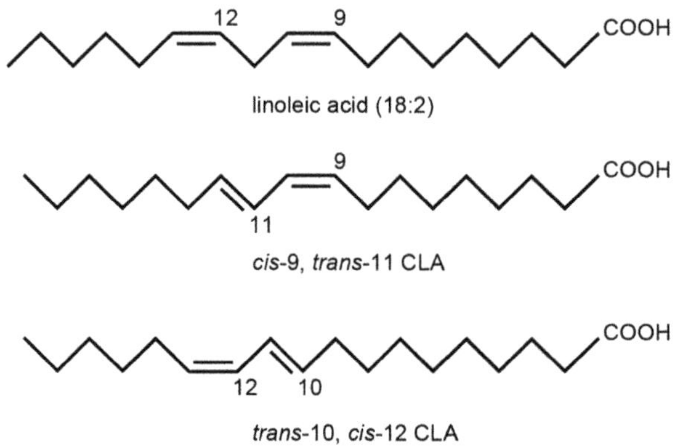

Figure 5.1 Structure of linoleic acid and its two main conjugated derivatives. CLA is normally generated by symbiotic bacteria in the stomach of ruminant animals.[1] The predominant isomer in dietary sources is *cis*-9, *trans*-11-CLA (*c*-9, *t*-11-CLA) (second panel), which constitutes up to ~90% of total CLA. 1,2 *trans*-10, *cis*-12-CLA (*t*-10, *c*-12-CLA) (bottom panel) is another common isomer that accounts for 1–10% of total CLA in dietary sources.[1,2]

Conjugated dienes, on the other hand, are considered to be minor constituents of dairy and ruminant lipids, as well as the byproducts of both autooxidation and partial hydrogenation of polyunsaturated fatty acids, and have been studied for more than 50 years. There is growing interested in CLA since studies on animals have suggested that it may have positive benefits on a number of illnesses, including cancer and type 2 diabetes, as well as on aspects of body composition including lean mass and bone composition. As a result, studies on a wide range of scientific subjects continues to flourish in this area. The goal of this review is to provide an overview of the present body of knowledge in this rapidly expanding field of study.

5.2 Occurrence in Natural Products

The meat and dairy products of ruminant animals are the most prevalent sources of natural CLA in the human diet. Although there are seasonal changes, the typical amount of CLA isomers in these materials is around 0.5% of the total fat. Of all the geometric configurations (*i.e. cis–cis*, *cis–trans*, *trans–cis* and *trans–trans*), the predominant isomer is the *cis*-9, *trans*-11 acid, accounting for 80–90% of the total CLA. Minor amounts of a large number of other isomers are also present, ranging from the 6,8 through to the 13,15 positional isomers. Natural CLA is assumed to be created mostly by the biohydrogenation of linoleic acid. The ruminant bacteria *Butyrivibrio fibrisolvens* uses linoleic acid isomerase to convert dietary linoleic acid into CLA, which is then used to create *trans*-11 18:1 vaccenic acid and 18:0 stearic acid. A second mechanism, involving the action of the D-9 desaturase enzyme on vaccenic acid within adipose and mammary tissues to produce the *cis*-9, *trans*-11 CLA isomer, has been proposed because the levels of CLA produced by *B. fibrisolvens* do not appear to be high enough to account for the amounts of CLA in ruminant tissues.

It is believed that a combination of double-bond migration and the activity of certain *cis* and *trans* isomerases in the rumen are responsible for the diversity of conjugated isomers seen in bovine milk and tissue. The amount of CLA in ruminant fats can be altered by changing the animals' diet. Animals raised on pasture had greater CLA levels than those fed commercial feed. An unsaturated fatty acid-enriched diet, however, can also raise the CLA levels in milk and meat. Amongst animals of different breeds and during various lactation phases, CLA output might vary greatly. This might be related to each animal's D-9 desaturase level. Fish containing CLA have been found, and the presence of CLA in farmed fish is likely a result of the CLA content of the vegetable oils used in fish feed. CLA does, however, elute on some gas chromatographic columns in the same area as the naturally occurring 18:4 isomers, raising concerns that it may be mistaken for other substances in fish tissues. As there are no known processes to account for CLA generation in marine foods, early claims of low amounts of CLA in trout, shrimp, and mussels may have been accurate. Due to the hydrogenation process, margarine, cooking oils, and other partly hydrogenated fats also contain trace levels of CLA (Figure 5.2).

The most common sources of linoleic acid or CLA are bovine milk, milk products, meat, and meat products. Linoleic acid is a complex group of positional and geometric isomers.[3] CLA is becoming more popular for treating fatal human ailments like cancer, atherosclerosis, and diabetes, similar to nutraceuticals, which are tiny lipids with supposed functional dietary status.[4] Despite several studies, it is difficult to pinpoint a single mechanism or molecular basis for CLA's *in vivo* activity pertinent to its alleged health advantages.[5] Nevertheless, the majority of CLA action at the molecular level relies on the conventional CLA-mediated activation of peroxisome proliferator-activated receptor (PPARs) and the ensuing "switching on and/or off" of the target genes to initiate a number of biochemical processes.[6] Figure 5.3 shows how one subtype of retinoic acid receptor (RXR), which must be activated by *cis*-9-retinoic acid to affect target gene transcription, forms a heterodimer with one of the well-studied PPARs (or subtypes) to bind to the peroxisome proliferator responsive element (PPRE) on the target gene during gene regulation.[7]

Only a relatively small amount of research has been done to integrate receptor-dependent activities, despite the fact that PPARs are engaged in the regulation of metabolic, immunological, and inflammatory processes.[8] The scientific knowledge of the biology of PPARs is mostly derived through trials using synthetic ligands to receptors because of the therapeutic importance of selective agonists. The observed biological effects of CLA are accounted for by a number of interconnected downstream molecular processes that occur in the *in vivo* environment in addition to the PPAR-mediated actions. Moreover, the transcription of a number of genes in the nucleus can be affected by polyunsaturated fatty acids (PUFAs) like CLA, in combination with

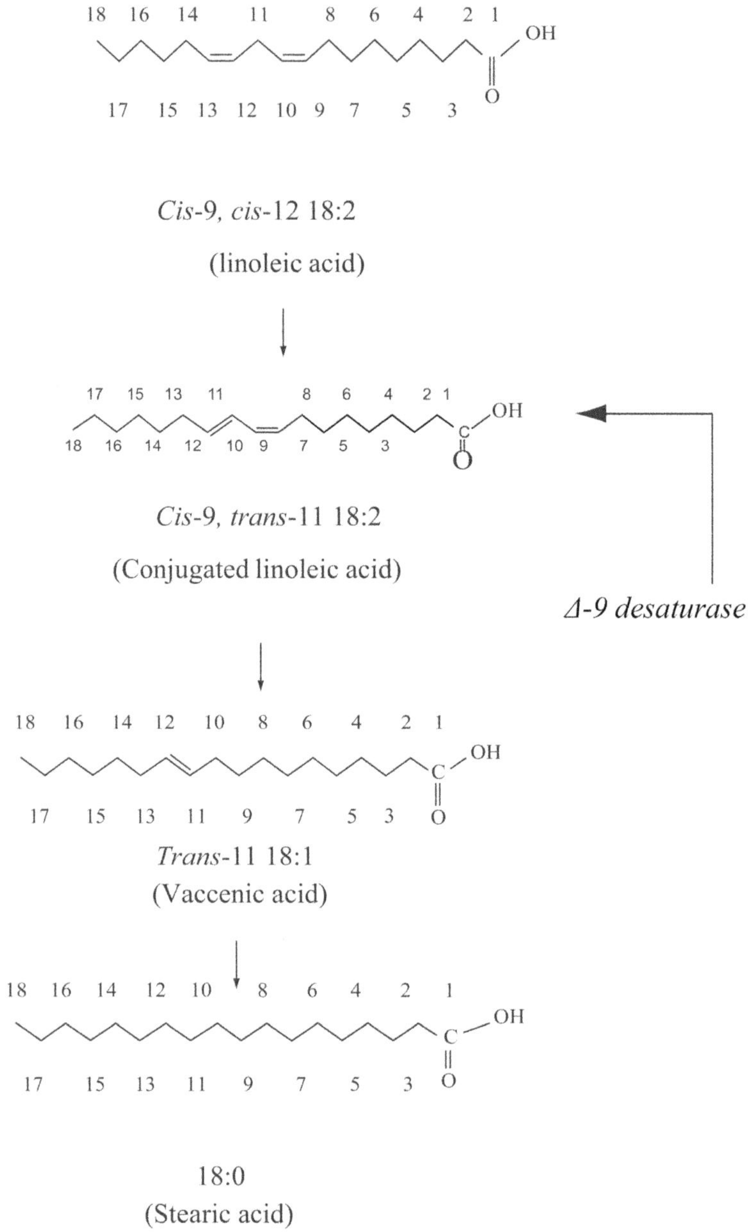

Figure 5.2 Linoleic acid is biohydrogenated to become CLA, both as a step in the production of stearic acid and by the desaturation of vaccenic acid.

other nuclear receptors and transcription factors. These comprise the transcription factors sterol-regulatory element binding protein (SREBP), liver receptor, nuclear factor-B (NFB), and hepatocyte nuclear factor (HNF)-4.[9] Another crucial pathway for adipocyte delipidation by CLA is opened through mitogen-activated protein kinase/extracellular signal-related kinase

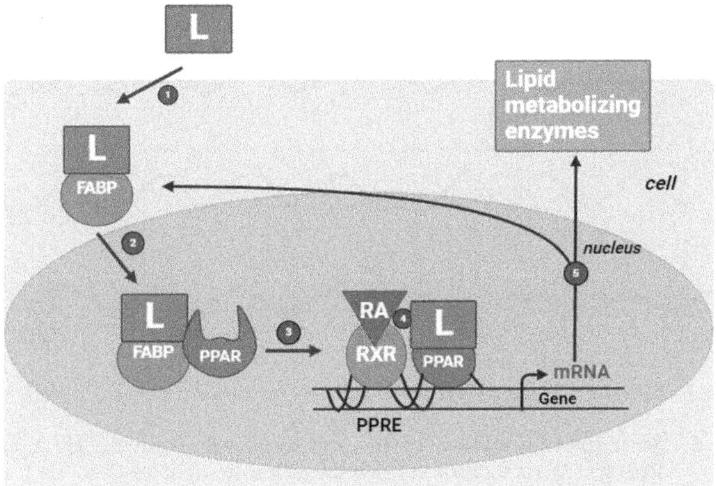

Figure 5.3 Proposed CLA-mediated signal transduction. (1) With the aid of particular membrane-bound fatty acid transporters, the dietary lipid nutrient L (here L = CLA) crosses the cell membrane and binds to tissue-specific fatty acid binding protein (FABP) in the cytosol; (2) L is transported to a particular subtype of the peroxisome proliferator-activated receptor (PPAR) when the L–FABP complex penetrates the nucleoplasm; (3) with the retinoic acid (RA)/retinoic acid receptor (RXR) subtype, the L–PPAR complex heterodimerizes; (4) the heterodimer attaches to the target gene's peroxisome proliferator response element (PPRE); and (5) the PPAR-RXR/PPRE complex recruits coactivators (e.g., mediator complex, histone acetylases) and RNA polymerase II, initiating transcription of the target gene (*e.g.*, genes involved in lipid metabolism).

(MEK/ERK) signalling *via* the autocrine/paracrine effects of interleukins-6 and 8.[10] Leptins, adiponectin, eicosanoids, vitamins, immunoglobulins, and PPAR-mediated gene expression are major mediators of the modulatory effects of CLAs on unique molecular signalling pathways, which also modify the properties of membrane proteins. However, clear objective evidence for these pathways is lacking, necessitating further research, particularly when hormones are involved.[11]

5.3 CLA Isomers

The scientific support for the beneficial health effects of CLA is mostly derived from animal and cell-culture models, with little to no direct human research. Similar to the anti-obesity benefits, the molecular mechanisms underpinning their anti-carcinogenic, anti-diabetogenic, anti-atherogenic and immunomodulatory effects, and even a few unfavorable effects, have not yet been fully uncovered.[12] Hence, it is necessary to clarify whether the biological activities attributed to CLAs are caused by a single isomer or a combination of isomers. The two most common isomers, *cis*-9, *trans*-11-CLA (9-CLA, rumenic acid), and *trans*-10, *cis*-12-CLA (10-CLA), which are largely

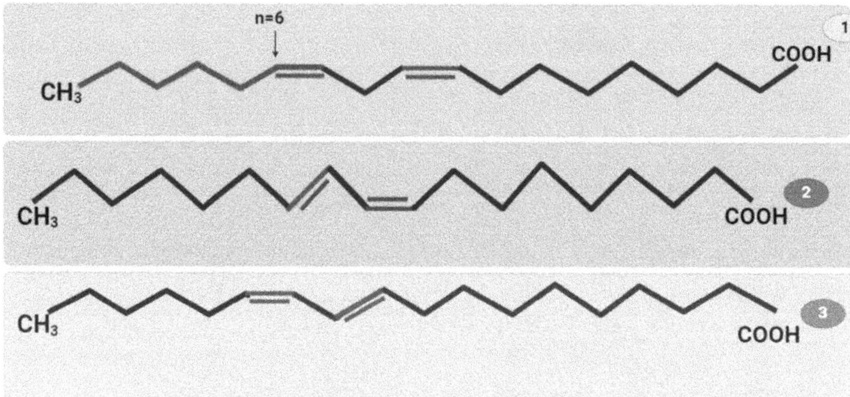

Figure 5.4 Structure of linoleic acid (LA) and its CLA derivatives. (1) LA (typical n-6 PUFA); (2) *cis*-9, *trans*-11-octadecadienoic acid (9-CLA, in fact an n-7 fatty acid); (3) *trans*-10, *cis*-12-octadecadienoic acid (10-CLA).

produced from linoleic acid, a typical n-6 fatty acid, are the subject of most research examining the biological activities of CLA (Figure 5.4). As a result, all CLA isomers are frequently incorrectly referred to as n-6 fatty acids, even though, for example, 9-CLA is an n-7 fatty acid. Results from research utilizing commercially available CLA, which is a combination of about equal amounts (40% each) of these two isomers, sparked interest in these isomers.[13] The majority of rodent (rat and mouse) research indicates that a CLA combination (9- and 10-CLA) may be more advantageous for the treatment of insulin resistance, in addition to the particular anti-obesity and hypocholesterolaemic benefits of 10-CLA. These investigations, however, demonstrate that CLA has no impact on insulin sensitivity or glucose metabolism in obese adults.[14] Thus, a single biochemical mechanism or the activity of a single isomer cannot account for the biological roles of CLA that have been documented. It seems that 9-CLA causes more widespread biological reactions than 10-CLA.[15] There is disagreement about the dietary supplementation of CLA in the form of free fatty acids or the triglyceride form, and additional research is necessary.[16]

In the rumen of cattle, deer, sheep, and goats, biotransformation of forage-derived unsaturated fatty acids like oleic acid and linoleic acid produces mostly 9-CLA, but in tiny amounts also 10-CLA. In mammalian cells, including those seen in humans, *trans*-11-vaccenic acid (*t*-VA), which is generated as a rumen biohydrogenation intermediate from both linoleic acid and linolenic acid, offers a significant alternative pathway for 9-CLA biosynthesis *via* 9-desaturation with stearoyl-CoA desaturase (SCD). With the use of food design and nutritional management of dairy cows, it is usually possible to achieve a several-fold increase in the concentration of CLA in milk fat by enhancing rumen outflow of *t*-VA and raising SCD activity.[17] In feed and diet, CLA is a low source of lipid nutrients; it is particularly abundant in dairy beef products. Milk contains more than 20 CLA isomers, although 9-CLA makes up about 75–90% of the total. The intramuscular CLA concentrations in meat and meat products from various animal species, the variables affecting these concentrations, the estimated daily requirements for humans, and the proportion of CLA that meat and meat products supply are all discussed here.[18] Also, our team has shown that people can synthesise 9-CLA from *t*-VA as an endogenous process.[19,20]

5.4 Types and Forms of CLNA Isomers

When applied to various induced oxidatively challenged models, the functionality of conjugated trienoic acid (CLNA) isomers has been proven to be extraordinarily successful (Figure 5.5). Dietary fatty acids play a significant role in shaping the membrane-lipid profile of cells in the systemic circulation. This, in turn, influences how these cells respond to various external mitogenic stimuli. In this situation, CLNA isomers have contributed fresh perspectives to our knowledge of the molecular effects of dietary fatty acids. Especially, it has been demonstrated that the geometric configuration of these conjugated n-5 fatty acids has a significant impact on their capacity to promote health.

To combat the negative effects of diabetes, it was discovered that the molecular expression of the proinflammatory transcription factor NF-B and other proinflammatory cytokines, such as IL-6 and TNF (tumour necrosis factor), were significantly downregulated. Controlled n-5 fatty acids in the diet had a positive impact on modulating nodes that transduce appropriate signals from the administered lipid environment to the level of gene expression against abnormal stress-related biochemical parameters, as shown by controlled blood glucose, antioxidative enzyme activities, and mitigated lipid profiles of the tissues belonging to CLNA-treated diabetes-afflicted rats.

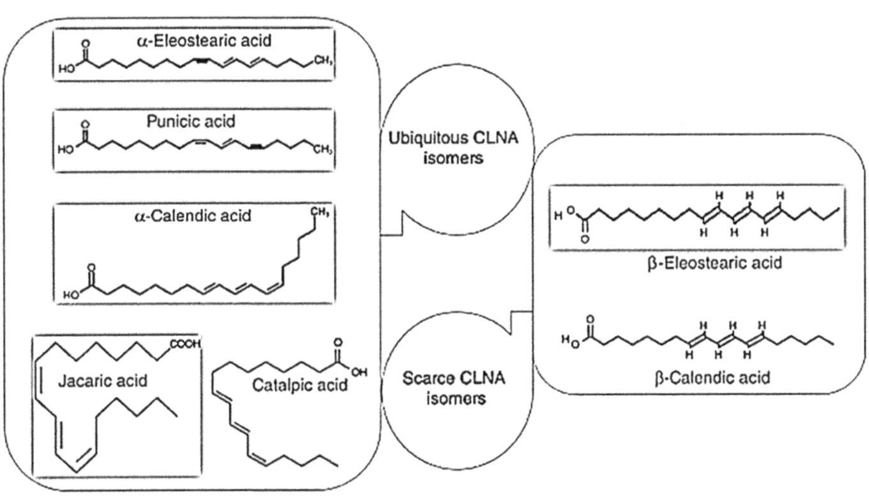

Figure 5.5 Structures of CLNA isomers.

Notwithstanding these findings, concrete data that might reflect unique regulatory mechanisms by which they support homeostasis and other pathological health-promoting situations are still absent. The ability of CLNA isomers to act as competitive inhibitors against COX-2 enzymes. Because of the signalling caused by oxidative stress, COX enzymes are regarded as nodal mediators of proteinoid production and inflammatory signalling. The study's findings imply that CLNA isomers may produce its antioxidative and anti-inflammatory effect by repressing the COX enzyme's activities while under stress. It has long been known that the CLNA isomers, at concentrations of 0.5% (w/v), exhibit strong hypocholesterolemic and hypolipidemic effects in a variety of species.

In the same research, doses that are higher or more than double this amount have also been demonstrated to have prooxidant effects. In general, replacing saturated fatty acids with unsaturated fat results in higher levels of HDL cholesterol and lower levels of total and VLDL cholesterol. HMG-CoA reductase, a rate-limiting enzyme in cholesterol biosynthesis, converts HMG-CoA into mevalonate during the production of cholesterol. By inhibiting the activities of these particular HMG-CoA reductase enzymes, CLNA isomers were able to limit the *in vivo* manufacture of cholesterols, as suggested by the research stated above. Moreover, the intestinal absorption of cholesterol might be successfully inhibited by CLNA isomers. Also, by raising plasma HDL levels, which then cleared cholesterol from the membranes, CLNA isomers helped maintain the homeostasis of cholesterol in cellular membranes. Both carcinogenesis and the control of cellular lipid metabolism are influenced by PPAR. The activation and facilitation of PPAR translocation in the nucleus, which has been demonstrated to inhibit cell proliferation and cause apoptosis in cancer cell lines, is positively influenced by CLNA isomers. Moreover, other transcription factors that are crucial for the development of apoptotic pathways were discovered to be elevated in the same study, including GADD45 and p53 expression. In a different study, it was shown that the CLNA isomer eleostearic acid has the innate potential to cause autophagy-induced cell death by producing reactive oxygen species (ROS) and inhibiting important apoptotic and autophagic signals, such as AKT/mTOR and ERK1/2, in human cervical cancer cells.

For improved and focused activity, a few attempts have also been undertaken to encapsulate these bioactive classes of conjugated octadecatrienoic fatty acids. The developed nanoparticle system has shown greater stability for the intended application in the presence of a natural bioactive fatty acid (lipoic acid) as opposed to a synthetic antioxidant (TBHT). In this study, we developed a novel nanoemulsion system enriched with a bioactive CLNA isomer, eleostearic acid (ESA), and assessed its health benefits compared to a conventional CLNA nanoemulsion in an acute alloxan-induced diabetic *in vivo* model. The results indicated that the standard emulsion was significantly less effective than the CLNA nanoemulsion in a nanotemplate for reducing oxidative enzyme parameters and promoting tissue repair. Apparently, the standard emulsion was considerably less effective in reducing the antioxidative enzyme parameters and remodeling damaged tissue than CLNA in a nanotemplate.

The results of this investigation might be attributable to the better bioavailability of CLNA in the designed nanoemulsion system than the standard emulsion system. In a mitogen (lipopolysaccharide)-induced oxidative-stressed human PBMC model, we discovered that nanoemulsion systems rich in ESA were capable of enhanced and targeted action in regulating intracellular ROS levels and safeguarding against induced cell death or necrosis, even under 100% bioavailability conditions.[21,22] The combined results of the two aforementioned studies show that formulations of CLNA made for more effective metabolism and targeted distribution have the potential for translational research with practical resolution of oxidative stress. In light of all the nutrigenomic research done thus far, it follows that the CLNA isomers, also known as octadecatrienoic n-5 conjugated fatty acids, have remarkable potential as essential nutraceuticals. To maximize the potential of these kinds of dietary PUFAs to enhance health in clinical settings, they must be included as designed and dosed supplements.

5.5 CLA Level Increases and Food Production

There are two main justifications for trying to raise CLA levels in milk and/or tissue in animals. First, CLA has an impact on body composition, for example, increasing lean meat and decreasing fat levels in pigs and cows. As public concern grows about potential associations between higher dietary fat levels and ailments like heart disease and obesity, this is becoming increasingly desired. The second reason is due to the fact that CLA itself may be advantageous in treating a number of disorders. Increased amounts of CLA in food are anticipated to have a preventive impact against illnesses like cancer, even though current dietary levels have been found to have little or no direct effect on a number of disorders.

The majority of experiments aiming to raise CLA levels have been done on cows. CLA levels can rise by up to five times when feed mixtures include unsaturated lipids like vegetable or marine oils. Nevertheless, increasing CLA typically results in a concomitant rise in other *trans* fatty acids, particularly *trans*-10-octadecenoic acid, which is harmful to both cows and humans. By providing a buffer, the levels of this fatty acid have been reduced. The immunological response of chicks fed CLA has also been found to be influenced, resulting in reduced tissue waste after immune stimulation, lowering the requirement for antibiotic treatment and enhancing the quality of the meat.

It has been hypothesised that the presence of CLA, which is found in human milk but not infant formula, contributes to the decreased prevalence of food allergies, immunological deficiencies, and diabetes in breast-fed children. Nevertheless, not enough substantial data sets have been evaluated for this. One strategy to increase CLA consumption is to enhance CLA in dietary components. Supplementation and the creation of functional foods are further techniques. CLA supplements are presently mostly marketed to people who are interested in bodybuilding and weight loss and are already accessible in the USA and Europe. These products have a similar composition to the commercial goods mentioned above and are either liquid or encapsulated CLA. A food item that has been modified to produce particular, advantageous biological effects is referred to as a functional food. Examples include stanol ester-containing margarines, which have been shown to decrease cholesterol in people. The typical customer may find these goods on the shelves of supermarkets. If it could be demonstrated that CLA has particular biological benefits in humans at acceptable (*i.e.* non-toxic) levels, it would be an excellent candidate for inclusion in a food product such as a dairy product.

5.5.1 Anti-obesity

The addition of CLA to the diet has the potential to change body composition due to its anti-obesogenic and hypolipidemic properties. Table 5.1 summarises the main biochemical effects of CLA linked with anti-obesity.[23] CLA dietary supplements have proven successful in lowering body fat percentages and raising protein percentages in the body.[24] Research on rodents, pigs, and cattle reveal that the effects of CLA on body composition may be partially attributable to increased lipolysis and decreased fat accumulation in adipocytes, presumably in conjunction with increased fatty acid oxidation in both muscle cells and adipocytes, as in rats.[25] It was discovered in a cross-species study (cattle breeds: Wagyu Limousin) that the *SCD-1* gene is a key participant in skeletal muscle fat metabolism, resulting in large amounts of monounsaturated fatty acids (MUFA) and CLA content, but with low amounts of saturated fatty acids. The SCD activity was shown to be suppressed in swine subcutaneous adipose tissue by a combination of CLA (9- and 10-CLA).[26]

Table 5.1 Primary biochemical effects of CLA on lipid metabolism.

Effect	Experimental proof	Ref.
Proliferation of preadipocytes	Suppressed proliferative	27
Differentiation of preadipocytes	In the absence of a PPAR ligand like CLA, human preadipocytes do not differentiate	28
Oxidation of fatty acids	Dietary CLA boosted the activity of carnitine palmitoyltransferase	29
Synthesis of lipids in adipose tissue	*De novo* lipogenesis is prevented by acetyl-CoA carboxylase and fatty acid synthase downregulation	30
Fat oxidation	Accelerated lipolysis and fat loss	31
Expenditure of energy	10-CLA-induced increase in oxygen uptake and energy expenditure	32
Desaturase of stearoyl-CoA	Inhibition caused by post-translational modification at the protein or activity level	33
Plasma leptin	Decrease in serum leptin, a hormone regulating fat level	34
Cell death	Adipocytes undergo apoptosis	27
Cancer-causing protein TNF	TNF expression was increased, and low fat level	23

Our team has earlier shown that dietary supplementation with *trans*-11- and *trans*-12-18:1 MUFA elevates 9-CLA in human peripheral blood mononuclear cell lipids (from 0.07 to 0.16%) without affecting immune function or inflammatory biomarkers.[19] In the same intervention we discovered that higher body fat accumulation was strongly correlated with higher SCD-1 isoform expression in control women, who had 25% more body fat mass than males.[19] It is interesting to note that after 42 days of intervention, GLUT 1 and 4 and SCD-1 were 10-fold down-regulated in women but not in males; this result can be attributable to endogenously produced 9-CLA.[35]

Moreover, we discovered a robust correlation between stored fats and gender-specific higher levels of leptin production. In comparison to their male counterparts, female participants had considerably greater plasma leptin and lower adiponectin concentrations. This demonstrates the hormonal modulation of genes.[36] SCD-1 becomes a crucial control point in the regulation of hepatic lipogenesis and lipid oxidation since it is the rate-limiting enzyme in the production of MUFA. Regardless of the model or species, 10-CLA consistently has inhibitory effects on SCD-1. In human breast cancer cells, human hepatocytes, and mouse adipocytes, treatment with mixed isomers, or more precisely 10-CLA, reduces either the activity or abundance of SCD-1.[10] SCD-1 therapeutic modification can be helpful in the treatment of obesity and metabolic syndrome, as Paton and Ntambi (2009) showed, as SCD-l is a critical enzyme in controlling hepatic lipogenesis and lipid oxidation. SCD-1 has become a therapeutic target in the management of obesity and metabolic syndrome as various hallmarks of metabolic syndrome and type 2 diabetes mellitus are linked to changes in intracellular lipid partitioning.[37]

Numerous studies have demonstrated a significant reduction in adipocyte lipid production. Evidence from both *in vivo* and *in vitro* studies suggests that, in addition to the major PPAR-mediated mechanism for CLA absorption, there may be other mechanisms for CLA-mediated fat reduction. It appears that CLA may function as a PPAR agonist or antagonist to regulate preadipocytes.[38] Intriguingly, human preadipocytes could not differentiate between the 9- and 10-CLA isomers and both equally decreased lipid accumulation when applied to pig preadipocytes, with the 10-CLA having a preferential impact. Furthermore, CLA may promote higher energy expenditure, improved lipolysis and fat oxidation, reduced preadipocyte differentiation and proliferation, and decreased lipogenesis.

Other putative pathways include decreased plasma leptin, increased adipocyte apoptosis, and elevated plasma tumour necrosis factor (TNF). The primary driver of the generalized mechanism is the modification of the transcription factor PPAR, also known as adipocyte determination and differentiation-dependent factor 1. This action would trigger a series of various metabolic pathways, including the suppression of SCD-1.[39] The antiproliferative effect of CLA in MCF-7 cells is evidence that delipidation through mitogen-activated protein kinase or extracellular signal-related kinase signalling occurs. The association between enhanced expression of TNF and a decrease in fat deposition may be caused by increased lipolysis and decreased lipoprotein lipase activities in addition to TNF's important mediating role in inflammatory reactions.[40] Studies made on animals and humans show that the consumption of CLA leads to loss of fat and total body weight, reduces the plasma concentrations of total and low-density lipoprotein (LDL)-cholesterol, and has an anti-inflammatory effect. Gene-by-diet interactions play an important role in the prevention of several diseases. To elaborate on this concept, we investigated the changes in gene expression in monocytes upon intervention with two *trans* fatty acids (*trans*-11 18:1 and *trans*-12 18:1) and endogenous synthesis of CLA from *trans*-11 18:1 (*t*-VA) as precursors in humans. Expression patterns of 20 candidate genes involved in glucose and lipid metabolism have been assessed, which were chosen based on the interaction of their proteins in fatty acid signalling and being partly regulated by PPAR and having a PPRE in their promoter. Interestingly, the expression patterns revealed prominent gender-based differences in gene expression.[20,35]

Researchers from the Ntambi group have shown that mice are protected against obesity when SCD-1 activity is lost. Reduced production of long-chain MUFAs like oleate, a favoured substrate for triglyceride synthesis, is probably to blame for this. Regardless of the model or species, 10-CLA consistently has inhibitory effects on SCD-1. In human breast cancer cells, human hepatocytes, and mouse adipocytes, treatment with mixed isomers, or more precisely 10-CLA, reduces either the activity or abundance of SCD-1.[37] Although though 10-CLA is known to be the antiadipogenic isomer, the effects of CLA on fat deposition in humans are less clear-cut and less substantial than those seen in animal studies. There is a need to determine if dietary CLA supplementation interacts with elements including calorie consumption and dietary fatty acid composition, particularly PUFA in people. Reduced caloric/food intake and increased energy expenditure, reduced preadipocyte differentiation and proliferation, reduced lipogenesis, and improved lipolysis and fat oxidation are some of the proposed antiobesity mechanisms of CLA.[32] According to Brown and McIntosh, 10-CLA partially inhibits adipocyte development by downregulating the expression of PPAR and its downstream targets, which are vital for glucose (*i.e.* GLUT4, ACC, SCD-1) and fatty acid (such as ACBP, A-FABP, LPL, and perilipin) metabolism. By the autocrine/paracrine effects of interleukins 6 and 8, 10-CLA lowered the triglyceride content of human adipocytes freshly differentiated from stromal vascular cells by triggering MEK/ERK activation. The 10-CLA may increase the expression and secretion of IL-6 and IL-8, which depends on the autocrine activation of MEK/ERK signalling in stromal vascular cells that are not adipocytes. MEK/ERK signalling in newly differentiated adipocytes is then impacted by paracrine actions, which results in insulin resistance and delipidation.[10]

Van Erk *et al.* (2008) found that CLA promoted higher expression of genes involved in lipid metabolism and decreased the activity of inflammatory genes in humans after consuming a specially formulated spread with higher levels of medium-chain triglycerides and polyunsaturated fatty acids (18:2 and 18:3).

A temporary change in diet activated genes involved in inflammatory processes in the adipose tissue of overweight patients. This method, which is an illustration of nutrigenomics research, enables the investigation of the impact of intricate combinations of functional dietary elements on adipose cells. These findings demonstrate the sensitivity of adipose tissue genes to dietary changes. The finding supports the idea that fat tissue plays a significant role in the development of diseases associated with obesity.[41] How food and nutrients consumed have an impact on the DNA is described by nutrigenomics (personalized nutrition). Although the aforementioned studies provide some insight into the possible advantages of CLA in humans, further research is necessary to establish the ideal CLA consumption amount, its short- and long-term effects, and any negative effects (if any) of each unique CLA isomer.[42] Before their widespread usage in humans can be advised, long-term randomised clinical studies with placebo controls must be conducted in large samples of patients to assess the effectiveness and safety of CLA isomers. Also, these experiments must be repeated in various labs, ideally in other continents and ethnicities with a range of dietary preferences.

5.5.2 Anti-carcinogenesis

By preventing tumour development and metastasis, CLA slows the spread of cancer. Fast-acting, CLA starts to suppress both cancerous and benign tumours right away. Whereas the 9-CLA isomer appears to alter arachidonic acid metabolism, the 10-CLA isomer appears to function predominantly through the regulation of apoptosis and cell cycle control.[15] The effects of CLA were seen on tumours in mice and rats that were chemically produced in their skin, stomach, breast, or colon. The strongest antiproliferative effects of CLA were seen *in vitro* on murine myeloid leukaemia (WEHI-3B JCS), human colorectal (HT-29, MIP-101) and prostate (PC-3) colorectal cells, as well as *in vivo* on breast and prostate malignancies. There are various and intricate cellular processes through which CLA modulates the development of cancer.[43] It may be accomplished by slowing down the metabolisms of prostaglandin (PG), vitamin A, lipid oxidation, and cell proliferation.[44] Furthermore, signal transduction may cause CLA to prevent cells from undergoing cellular transformation.[45] Additionally, CLA's capacity to disrupt the n-6 PUFA metabolic pathway for the manufacture of eicosanoids, including PGs, is thought to be largely responsible for their anticarcinogenic effects. Other key areas of ongoing study include altered phospholipid-associated fatty acid metabolism and eicosanoid (20-carbon derivatives such as PGs, thromboxanes, leukotrienes, and hydroxy eicosatetraenoic acids) synthesis. Eicosanoids control tissue differentiation, platelet aggregation, local and systemic immunology, and cell proliferation. Free CLAs could compete with other fatty acids for inclusion in phospholipids, which would change the subsequent generation of eicosanoids. Dietary CLA lowers PG-E2 and other arachidonic acid metabolism-derived leukotrienes (PGF2, leukotriene-B4, and leukotriene-C4).[46] By the inhibition of cyclooxygenase (COX) 1 and 2, the CLA-mediated decrease of arachidonate-derived eicosanoids may also occur. This comprises apoptosis (BCL-2) and cell cycle regulation (p21 (WAF/Cip1)) mechanisms as well as processes for the metabolism of arachidonic acid.

5.5.3 Cell Cycle and Apoptosis

By affecting cell replication, interfering with cell-cycle components, or boosting cell death by encouraging necrosis or/and apoptosis, CLAs may affect the formation of malignant cells. While apoptosis is a distinct energy-demanding process of programmed cell death characterised by DNA fragmentation, chromosome condensation, nuclear fragmentation, formation of apoptotic bodies, and inversion of phosphatidylserine in the plasma membrane, necrosis typically results from insult or toxicity reaction and causes inflammation.[47] It is anticipated that CLA might inhibit DNA synthesis and cell cycle proteins that control cell proliferation, and that CLA may promote increased apoptosis principally by inhibiting the expression of the antiapoptotic *bcl-2* gene.[48] This appears to be a function of the relative content of particular isomers and their capacity to elicit a p53 response that leads to cell growth arrest by inhibiting the expression of factors necessary for the G1 to S-phase transition, including cyclins D1 and E, or by the inhibition of the protein kinase activity that modulates the production of oxidant species, which significantly inhibits proliferation.[49]

5.5.4 Anti-atherosclerosis

The progressive condition known as atherosclerosis affects the medium and large arteries and is characterised by the buildup of lipids in inflammatory cells (leading to the creation of foam cells), cellular proliferation, platelet adhesion and aggregation, and calcium deposition.[50] In animal models of atherosclerosis, CLA is a strong anti-atherogenic dietary fatty acid that works through activating PPARs.[51] Through increases in apo A-I and apo A-II synthesis, one would anticipate (a) a decrease in atherogenic lipoprotein plasma levels, such as very low LDL cholesterols, and an increase in anti-atherogenic high-density lipoprotein cholesterol (HDL); (b) overexpression in HDL receptors capable of increasing cellular cholesterol efflux; and (c) a decrease in vascular inflammation by repressing nuclear NF-κB and apo A-.[52] Notwithstanding a few investigations, the aforementioned signalling pathways for drug delivery are not supported by any clear data using CLA isomers. It is interesting to note that rabbits given CLA had lower ratios of total cholesterol to HDL and LDL cholesterol to HDL cholesterol, as well as lower ratios of LDL cholesterol to HDL cholesterol.

5.5.5 Anti-diabetes

Type I diabetes can be brought on by insufficient insulin, type II diabetes by insulin resistance or both. CLA dietary supplements may help diabetics, particularly type II diabetics, better control their illness. Several studies overwhelmingly support the idea that the 10-CLA isomer may be the bioactive isomer of CLA that affects the changes in body weight seen in people with type-2 diabetes.[33] By suppressing the expression of PPAR and its downstream targets, which are essential for fatty acid and glucose metabolism, during the delipidation process, 10-CLA exerts a cascade of molecular activities that ultimately limits glucose and fatty acid absorption and metabolism.

Research examining the methods by which CLA functions at the cellular level reveals that the lipostat transcription factors, specifically PPAR, PPAR, SREBP1c, and LXR, are the key targets for CLA. Hence, it is likely that the effects of CLA on glucose metabolism are secondary effects mediated by elements under the control of these nuclear receptors, such as PPAR coactivator 1.[53] Alternatively, it could be brought on by intricate systems that control the expression of genes (such as uncoupling proteins) crucial for controlling adipogenesis, glucose and lipid metabolism, and potentially even whole-body thermogenesis. The sensitization of adiponectin, a recently identified hormone released by adipocytes and believed to improve insulin sensitivity, is another potential mechanism by which CLA may reduce hyperinsulinemia (in Zucker diabetic fatty rats).[54] It is crucial to understand the function of CLA in order to help with the management of type II diabetes and other associated disorders of insulin resistance. This requires identifying how CLA isomers affect glucose and lipid metabolism as well as indicators of insulin sensitivity.[55]

5.5.6 Immunomodulation

CLA modulates immunological function, as shown by *in vivo* animal models and *in vitro* investigations of immune cells. Contrary to publications using animal models, CLA feeding to young, healthy women did not change any of the immune status indicators examined, suggesting that short-term CLA supplementation in healthy volunteers is safe but does not improve their immune state.[56] According to Ringseis *et al.*, PPAR interaction between CLA and human vascular smooth muscle cells inhibits TNF-induced eicosanoid production.[57] According to reports, the active CLA isomers (9- and 10-CLAs) can both trigger innate immune responses and adaptive immunological responses.[58]

These results come from CLA's capacity to alter soluble immune mediators such eicosanoids, cytokines, and immunoglobulin synthesis.[59] Effects on human immunological function of two distinct combinations of 9- and 10-CLA glycerides [given daily for 12 weeks in the ratios of 50 : 50 (1.7 g) or 80 : 20 (1.6 g), respectively]. It was discovered that the 50 : 50 ratio helped to increase the levels of hepatitis B protective antibodies. *Via* TNF, CLA may change eicosanoid signalling, impacting a variety of biological processes, including the presentation of antigens. Changes in the properties of the membrane as well as changes in the function of membrane proteins that operate as ion channels, transporters, receptors, signal transducers, or enzymes might result from this.

The alternative theory is based on the fundamental PPAR-mediated signalling, which happens when immune system-related target genes are expressed. During instances of severe sickness, CLA can also offer protection against tissue degradation brought on by immunological activation.[60] This impact may be attributable to the regulation of NF-κB, which works against the catabolic and inflammatory effects brought on by cytokines, particularly TNF, as well as the inflammatory reactions brought on by lipopolysaccharide. The immunomodulatory properties of CLA may be used in cattle production as a functional food alternative to feed antibiotics, or as a way to enhance vaccination response and impart disease resistance.

In healthy human (male and female) volunteers, Song *et al.* (2005)[61] looked into the effects of dietary CLA supplementation (3 g per day; 50 : 50 mix of 9- and 10-CLA major isomers) on the immune system, plasma lipids, and glucose levels. It is interesting to note that when plasma IgE levels dropped, IgA and IgM levels rose. Also, taking CLA supplements resulted in lower levels of the pro-inflammatory cytokines TNF and IL-1 and higher levels of the anti-inflammatory cytokine IL-10. According to these results, delayed-type hypersensitivity response was reduced both during and after CLA treatment, although plasma glucose, lipids, and lymphocyte shape did not alter significantly. CLA (9- and 10-CLA mixture) supplementation modulated an increased release of markers (C-reactive protein, IL10, and IgM) related to inflammation and immune function in obese pre-menopausal Korean females, and this effect was much more subtle than those found in animals and a few other clinical studies. This observation was made very recently.[62]

5.5.7 Bone Formation

Animal studies have revealed that mixed CLA isomers can have a variety of impacts on bone osteosynthesis and resorption. The multiple isomers present in popular commercial preparations of CLA may be the cause of the varying effects of CLA on bone physiology; however, the effects of the two main isomers, 9- and 10-CLA, are unclear. Dietary CLA reduces prostaglandin-dependent bone resorption by preventing endosteal bone resorption, promoting endocortical bone formation, and altering the activity and expression of COX enzymes.[63] Moreover, CLA increases the rate of bone resorption in adult OVX rats and promotes bone formation and calcium absorption from the diet.[64]

It is theorized that CLA may be an effective method for preventing or reducing the symptoms of rheumatoid arthritis in people because it can influence inflammatory cytokines. Nevertheless, a CLA supplement of mixed isomers did not influence indicators of calcium or bone metabolism under the circumstances examined in this double-blind, placebo-controlled experiment in adult men. Eight weeks of CLA or placebo supplementation had no discernible effects on the levels of serum or urinary calcium or on the markers of bone formation and resorption (serum osteocalcin and bone-specific alkaline phosphatase, urinary pyridinoline and deoxypyridinoline, and serum C-telopeptide-related fraction of type 1 collagen degradation products). Several metabolic markers were present at similar baseline values in both subject groups. Contrary to the placebo, CLA administration caused total plasma lipids to include three times as much 9-CLA isomer.[65] In human osteoblast-like cells, the alkaline phosphatase activity in cell lysates is a hallmark of early osteoblast development. From 25 to 100 M, 9-CLA enhanced the size and quantity of mineralized bone nodules, but 10-CLA had no such impact.[66] A varied rise in alkaline phosphatase activity co-incided with the increase in mineralized bone nodule formation brought on by 9-CLA. Our findings support the isomer-specific effects of CLA on bone health by demonstrating that 9-CLA promotes the development of mineralized bone nodules in bone cells derived from humans.

5.6 Safety Concerns and Human Scenario

Although only the beneficial effects of CLA on health have been discussed here, adverse effects of CLA cannot be completely ruled out.[67] Studies on CLA supplementation in humans have occasionally produced contradictory and less compelling results. The stark differences across studies might be attributed to the often-antagonistic actions of distinct CLA isomers, which exhibit isomer-specific effects. Insulin resistance and increased liver and spleen weight are the main observations in mouse models. Another aspect is fatty liver hemorrhagic syndrome caused by CLA in birds.[68]

Data from human research suggest that 10-CLA might operate as a cancer promoter in colon carcinogenesis, presumably through pathways affecting NF-κB and cyclin D1. This is a significant issue because it mostly affects the gastrointestinal system.[69] While dietary CLA significantly enhanced the activity and mRNA levels of many lipogenic enzymes, such as hepatic 5- and 6-desaturases and SREBP-1 in the liver, 10-CLA significantly lowered body fat in mice, probably through higher energy expenditure. The physiological reaction of the animal to convert excess glucose to fatty acids for storage as triacylglycerols in the liver rather than adipose tissue may be represented by a significant increase in lipogenesis and buildup of triacylglycerol in the liver following CLA therapy. Hence, although perhaps being useful in avoiding fatty liver, it is possible that the antagonist of CLA-mediated stimulation of hepatic lipogenesis worsens hyperinsulinemia and glucose intolerance.[70]

Independent of changes in insulin concentrations, 10-CLA causes hyperproinsulinemia in obese males that is associated with decreased insulin sensitivity (hyperinsulinaemic-euglycemic clamp). The indiscriminate use of weight-loss supplements containing 10-CLA is warned against, since hyperproinsulinemia foretells diabetes and cardiovascular problems. In obese or diabetic patients, CLA isomers may have different effects from those in normal-weight or healthy subjects. Moreover, there may be changes based on gender and/or heredity, such as single nucleotide polymorphisms in associated genes.[71] Moreover, similar studies must be conducted in multiple labs with a focus on other age groups, cultural origins, or culinary preferences, as well as men and women.

The potential negative effects of CLA on the insulin balance have received less attention than the potential positive benefits of 10-CLA supplementation in reducing body fat mass. Contradictory to this is the observation of CLA-mediated hyperinsulinemia in several mouse investigations. Changes in plasma leptin levels may be connected to CLA-induced insulin resistance. Research on numerous animal models has demonstrated that CLA supplementation caused drops in plasma leptin levels. The use of mixed isomers, which may cancel out one another and result in no net change in adiposity, is to blame for the equivocal outcomes in human supplementation experiments. In addition, the dosages employed in human trials were significantly lower than those utilized in animal research. In adipocytes, 10-CLA reduces the expression of PPAR, which may encourage insulin resistance. The use of mixed CLA isomers or different experimental methods may be the cause of these perplexing results. In any case, the much-needed insight into possible human uses for CLA should come from isomer-specific dose-titrated clinical trials paired with mechanistic research in primary cell cultures. Dietary CLA has variable effects on different rat species and strains; consequently, when extrapolating rodent results to the human scenario, the variations in species sensitivity are crucial.

The effectiveness of CLA as a functional diet has recently been the subject of more than one hundred clinical studies, according to the literature. A quick analysis of the results of these trials reveals that CLA's overwhelmingly favourable effects encourage optimism. Clinical investigations show that CLA has a number of important effects on humans, which are included in Table 5.2. It appears that 9- and 10-CLA have different biological effects, with 10-CLA having greater negative outcomes. Yet, as affected in immunological function, a 50 : 50 ratio of these isomers may have a greater impact. In the human body, CLA typically makes up 0.1% of the total fatty acid content.[72] According to studies, an adult would get optimal biological benefits with daily consumption of roughly 2–3 g for 6 to 12 months. These effects would also be long-lasting.[73] The best way to consume CLA is as triacylglycerol, and both the 9- and 10-CLA isomers are absorbed into chylomicrons.

Recent research by Kelley *et al.* (2008) in mice showed that some negative effects associated with CLA, such as insulin resistance and non-alcoholic fatty liver disease, may actually be caused by a deficiency of n-3 PUFA.[91] These researchers also showed that these negative effects can be reversed by concurrently increasing intake of linolenic acid, an n-3 PUFA, and flax seed oil, a rich natural source of this fatty acid. Taking into account the adverse occurrences and safety worries reported, the results showed that healthy, overweight persons who used CLA supplements for 24 months saw lower body weight and body fat mass, as well as an increase in circulating

Table 5.2 Significant biological impacts of CLA on humans.

Therapy	Effects	Ref.
Conjugated linoleic acid with creatine	In older people, CLA and creatine supplementation improved mitochondrial function by lowering sarcopenia and oxidative stress	63
Omega-3 fatty acids and CLA	Boosts adiponectin and fat-free mass while preventing the accumulation of abdominal fat	28
9-Conjugated linoleic acid	No negative impact on coronary artery disease	72
Oryzanol and CLA	Body fat and blood pressure were both lower	74
9 & 10-CLA	Increased fat oxidation and energy use during sleep	29
10-Conjugated linoleic acid	Breast tissue of nursing mothers has an anti-lipogenic action	75
Conjugated linoleic acid	Lessened weight gain brought on by psychoactive drug use	76
Conjugated linoleic acid	Favorable impact on serum insulin, but no change in body composition or appetite-related energy expenditure	77
Conjugated linoleic acid	Elevated PPAR, hormone-sensitive lipase, and increased resting metabolic rate	40
Conjugated linoleic acid	No impact on insulin sensitivity or glucose metabolism in the obese population	14
trans-11/*trans*-12 fatty acids (18:1)	Gene expression dependent on gender	20
Conjugated linoleic acid	Insulin resistance brought on by leptin and adiponection inhibition	78
CLA and Ca	Reduced intracellular concentration of ionised free calcium in peripheral blood cells and pregnancy-induced hypertension	79
CLA	Did not alter diabetic or cardiovascular disease risk factors in a positive way	80
Vaccenic acid and CLA	In healthy young males, no difference in blood pressure or arterial flexibility	81
CLA	Decreased body weight, and lipid and carbohydrate metabolism	82
CLA and other PUFA	Potential contribution to avoiding renal cancer	83
CLA	Reduced colorectal carcinoma	84
CLA	Enhanced immunological function	61
CLA	Enhanced C-reactive protein	85
CLA	Suppressed rheumatoid arthritis	86
CLA	Lipid peroxidation	74
CLA	Not associated with breast cancer	87
10-Conjugated linoleic acid	Produced hyperproinsulinemia, which indicates a higher risk of diabetes and cardiovascular disease	88
CLA	Effect of protection on the likelihood of breast cancer spreading	89
10-Conjugated linoleic acid	Increased indicators of oxidative stress and inflammation in obese males	88
CLA	Positive impact on cardioprotective effect	90
CLA	Modification of atherosclerosis-related risk factors	71

lipoprotein, thrombocytes, and aspartate aminotransferase. The level of fasting blood sugar remained unchanged. LDL cholesterol and plasma total cholesterol were decreased, however, HDL cholesterol and triglycerides remained constant. Compared to the first 12 months of the trial, the reported adverse impact rate significantly reduced during the course of the 2-year study. Our findings show that the majority of the harmful effects that have been documented are linked to short-term human research.[92] It appears that adding more fish oil to mice's meals, which is a strong source of extremely long-chain fatty acids, helped to mitigate several of the physiological side effects such as hyperinsulinemia and fatty liver. Our findings suggest that a combination of all n-3, n-6 (CLA), and n-9 fatty acids in a sensible ratio for humans would be a superior solution to prevent the documented negative effects of CLA.

5.7 Conclusion

It is challenging to determine if CLA offers a 100% safe functional meal because of the variability of both *in vitro* and *in vivo* evidence on the efficacy of CLA. Despite their expense, obese persons are more inclined to eat more of these minor fatty nutrients. Although there is very little human clinical research, thus far it indicates that CLA is good for human health. To produce clear proof, more concentrated international clinical studies involving probands and patients from all continents are needed. The different functional properties of CLA isomers and the fact that most clinical trials employ a crude combination of CLA (predominately 9- and 10-CLAs) are crucial additional factors. However, the alleged detrimental consequences, such as fatty liver and spleen and the development of colon cancer, have not yet been proven beyond a reasonable doubt. Further convincing evidence is also required to identify additional molecular mechanisms and intricate signalling cascades induced by dietary CLA in addition to PPAR-mediated activation. Although most human research is based on blood, blood cells, milk, or biopsy materials – all of which would likely create variances in the overall data collected – strictly controlled investigations as carried out in animals or culture models may not be maintained in clinical trials. So, before drawing the conclusion that CLA is a foolproof functional food for humans, conclusive studies focusing on factors like the type of CLA isomer administered, variables measured, mode of administration (*e.g.* as free fatty acid or its triglyceride, liquid or solid), gender, age, and ethnic background remain to be undertaken. As a consequence, it is encouraging to note that new research highlights CLA and PUFA as the strongest combination for reducing the negative impacts thus far.

References

1. K. W. J. Wahle, S. D. Heys and D. Rotondo, Conjugated linoleic acids: are they beneficial or detrimental to health?, *Prog. Lipid Res.*, 2004, **43**, 553–587.

2. A. S. Salsinha, L. L. Pimentel, A. L. Fontes, A. M. Gomes and L. M. Rodríguez-Alcalá, Microbial production of conjugated linoleic acid and conjugated linolenic acid relies on a multienzymatic system, *Microbiol. Mol. Biol. Rev.*, 2018, **82**, 1–21.
3. S. Benjamin, T. Hanhoff, T. Börchers and F. Spener, An improved molecular test system for the screening of human PPAR transactivation by conjugated linoleic acid isomers and their precursor fatty acids, *Eur. J. Lipid Sci. Technol.*, 2005, **107**(10), 706–715.
4. Y. Park, K. Albright, J. Storkson, W. Liu and M. Pariza, Conjugated linoleic acid (CLA) prevents body fat accumulation and weight gain in an animal model, *J. Food Sci.*, 2007, **72**(8), S612–S617.
5. Z. Jiang, J. J. Michal, D. J. Tobey, T. F. Daniels, D. C. Rule and M. D. MacNeil, Significant associations of stearoyl-CoA desaturase (SCD1) gene with fat deposition and composition in skeletal muscle, *Int. J. Biol. Sci.*, 2008, **4**(6), 345.
6. B. Desvergne and W. Wahli, Peroxisome proliferator-activated receptors: nuclear control of metabolism, *Endocr. Rev.*, 1999, **20**(5), 649–688.
7. C. Schachtrup, T. Emmler, B. Bleck, A. Sandqvist and F. Spener, Functional analysis of peroxisome-proliferator-responsive element motifs in genes of fatty acid-binding proteins, *Biochem. J.*, 2004, **382**(1), 239–245.
8. J. Bassaganya-Riera, A. Guri, J. King and R. Hontecillas, Peroxisome proliferator-activated receptors: the nutritionally controlled molecular networks that integrate inflammation, immunity and metabolism, *Curr. Nutr. Food Sci.*, 2005, **1**(2), 179–187.
9. H. Sampath and J. M. Ntambi, Polyunsaturated fatty acid regulation of genes of lipid metabolism, *Annu. Rev. Nutr.*, 2005, **25**(1), 317–340.
10. J. M. Brown and M. K. McIntosh, Conjugated linoleic acid in humans: regulation of adiposity and insulin sensitivity, *J. Nutr.*, 2003, **133**(10), 3041–3046.
11. G. Agatha, A. Voigt, E. Kauf and F. Zintl, Conjugated linoleic acid modulation of cell membrane in leukemia cells, *Cancer Lett.*, 2004, **209**(1), 87–103.
12. J. K. Kramer, C. Cruz-Hernandez, Z. Deng, J. Zhou, G. Jahreis and M. E. Dugan, Analysis of conjugated linoleic acid and *trans* 18:1 isomers in synthetic and animal products, *Am. J. Clin. Nutr.*, 2004, **79**(6), 1137S–1145S.
13. C. G. Taylor and P. Zahradka, Dietary conjugated linoleic acid and insulin sensitivity and resistance in rodent models, *Am. J. Clin. Nutr.*, 2004, **79**(6), 1164S–1168S.
14. C. Syvertsen, J. Halse, H. O. Høivik, J. Gaullier, M. Nurminiemi and K. Kristiansen, *et al.*, The effect of 6 months supplementation with conjugated linoleic acid on insulin resistance in overweight and obese, *Int. J. Obes.*, 2007, **31**(7), 1148–1154.
15. J. J. Ochoa, A. J. Farquharson, I. Grant, L. Moffat, S. D. Heys and K. W. Wahle, Conjugated linoleic acids (CLAs) decrease prostate cancer cell proliferation: different molecular mechanisms for *cis*-9, *trans*-11 and *trans*-10, *cis*-12 isomers, *Carcinogenesis*, 2004, **25**(7), 1185–1191.

16. J.-I. Lee, Y.-J. Ha, J.-R. Lee, Y.-K. Joo, S.-J. Kwack and C.-H. Do, Effects of conjugated linoleic acid-triglyceride additives on quality characteristics of pressed ham, *Food Sci. Anim. Resour.*, 2007, **27**(1), 8–15.
17. A. L. Lock and D. E. Bauman, Modifying milk fat composition of dairy cows to enhance fatty acids beneficial to human health, *Lipids*, 2004, **39**(12), 1197–1206.
18. A. Schmid, M. Collomb, R. Sieber and G. Bee, Conjugated linoleic acid in meat and meat products: A review, *Meat Sci.*, 2006, **73**(1), 29–41.
19. K. Kuhnt, J. Kraft, H. Vogelsang, K. Eder, J. Kratzsch and G. Jahreis, Dietary supplementation with *trans*-11-and *trans*-12-18:1 increases *cis*-9, *trans*-11-conjugated linoleic acid in human immune cells, but without effects on biomarkers of immune function and inflammation, *Br. J. Nutr.*, 2007, **97**(6), 1196–1205.
20. K. Kuhnt, The Δ9-desaturation [Delta9-desaturation] of dietary *trans* octadeconoic acids (*trans*11 and *trans* 12 18:1) and the clinical relevance in humans, Jena, Univ., Diss., 2007, 2007.
21. D. Paul, T. K. Dey, A. Chakraborty and P. Dhar, Promising functional lipids for therapeutic applications, in *Role of Materials Science in Food Bioengineering*, Elsevier, 2018, pp. 413–449.
22. D. Paul, T. K. Dey, S. Mukherjee, M. Ghosh and P. Dhar, Comparative prophylactic effects of α-eleostearic acid rich nano and conventional emulsions in induced diabetic rats, *J. Food Sci. Technol.*, 2014, **51**, 1724–1736.
23. M. Rydén, E. Arvidsson, L. Blomqvist, L. Perbeck, A. Dicker and P. Arner, Targets for TNF-α-induced lipolysis in human adipocytes, *Biochem. Biophys. Res. Commun.*, 2004, **318**(1), 168–175.
24. A. Akahoshi, K. Koba, F. Ichinose, M. Kaneko, A. Shimoda and K. Nonaka, *et al.*, Dietary protein modulates the effect of CLA on lipid metabolism in rats, *Lipids*, 2004, **39**(1), 25–30.
25. M. Azain, Role of fatty acids in adipocyte growth and development., *J. Anim. Sci.*, 2004, **82**(3), 916–924.
26. S. Smith, T. Hively, G. Cortese, J. Han, K. Chung and P. Castenada, *et al.*, Conjugated linoleic acid depresses the Δ9 desaturase index and stearoyl coenzyme A desaturase enzyme activity in porcine subcutaneous adipose tissue, *J. Anim. Sci.*, 2002, **80**(8), 2110–2115.
27. M. Evans, C. Geigerman, J. Cook, L. Curtis, B. Kuebler and M. McIntosh, Conjugated linoleic acid suppresses triglyceride accumulation and induces apoptosis in 3T3-L1 preadipocytes, *Lipids*, 2000, **35**, 899–910.
28. R. L. McNeel, E. B. Smith and H. J. Mersmann, Isomers of conjugated linoleic acid modulate human preadipocyte differentiation, *In Vitro Cell. Dev. Biol.: Anim.*, 2003, **39**(8), 375–382.
29. R. N. Close, D. A. Schoeller, A. C. Watras and E. H. Nora, Conjugated linoleic acid supplementation alters the 6-mo change in fat oxidation during sleep, *Am. J. Clin. Nutr.*, 2007, **86**(3), 797–804.
30. C.-E. Lin, I.-J. Fang, Y.-J. Deng, W.-S. Liao, H.-T. Cheng and W.-P. Huang, Capillary electrophoretic studies on the migration behavior of cationic

solutes and the influence of interactions of cationic solutes with sodium dodecyl sulfate on the formation of micelles and critical micelle concentration, *J. Chromatogr. A*, 2004, **1051**(1–2), 85–94.
31. E. Ostrowska, R. F. Cross, M. Muralitharan, D. Bauman and F. Dunshea, Effects of dietary fat and conjugated linoleic acid on plasma metabolite concentrations and metabolic responses to homeostatic signals in pigs, *Br. J. Nutr.*, 2002, **88**(6), 625–634.
32. Y.-M. Wang, S. M. Rahman, K. Nagao, S.-Y. Han, Y. Buang and J.-Y. Cha, *et al.*, Conjugated linoleic acid reduces hepatic microsomal triacylglycerol transfer protein activity and hepatic triacylglycerol mass in obese rats, *J. Oleo Sci.*, 2003, **52**(3), 129–134.
33. M. A. Belury, A. Mahon and S. Banni, The conjugated linoleic acid (CLA) isomer, t10c12-CLA, is inversely associated with changes in body weight and serum leptin in subjects with type 2 diabetes mellitus, *J. Nutr.*, 2003, **133**(1), 257S–260S.
34. S. M. Rahman, Y.-M. Wang, H. Yotsumoto, J.-Y. Cha, S.-Y. Han and S. Inoue, *et al.*, Effects of conjugated linoleic acid on serum leptin concentration, body-fat accumulation, and β-oxidation of fatty acid in OLETF rats, *Nutrition*, 2001, **17**(5), 385–390.
35. K. Kuhnt, S. Flotho, S. Benjamin, T. Boerchers, R. Schubert and G. Jahreis, *et al.*, Gene expression after dietary intervention with *trans* fatty acids (*trans*-11/*trans*-12 18:1) in humans, *Eur. J. Lipid Sci. Technol.*, 2009, **111**(5), 442–450.
36. P. Cohen and J. M. Friedman, Leptin and the control of metabolism: role for stearoyl-CoA desaturase-1 (SCD-1), *J. Nutr.*, 2004, **134**(9), 2455S–2463S.
37. C. M. Paton and J. M. Ntambi, Biochemical and physiological function of stearoyl-CoA desaturase, *Am. J. Physiol.: Endocrinol. Metab.*, 2009, **297**(1), E28–E37.
38. H. J. Mersmann, Mechanisms for conjugated linoleic acid-mediated reduction in fat deposition, *J. Anim. Sci.*, 2002, **80**(E-suppl_2), E126–E134.
39. A. J. Lengi and B. A. Corl, Identification and characterization of a novel bovine stearoyl-CoA desaturase isoform with homology to human SCD5, *Lipids*, 2007, **42**, 499–508.
40. J.-A. Nazare, A. B. de La Perriere, F. Bonnet, M. Desage, J. Peyrat and C. Maitrepierre, *et al.*, Daily intake of conjugated linoleic acid-enriched yoghurts: effects on energy metabolism and adipose tissue gene expression in healthy subjects, *Br. J. Nutr.*, 2007, **97**(2), 273–280.
41. M. J. van Erk, W. J. Pasman, H. M. Wortelboer, B. van Ommen and H. F. Hendriks, Short-term fatty acid intervention elicits differential gene expression responses in adipose tissue from lean and overweight men, *Genes Nutr.*, 2008, **3**, 127–137.
42. A. M. Haro, R. Artacho and C. Cabrera-Vique, Linoleic conjugated acid: current interest in human nutrition, *Med Clin*, 2006, **127**(13), 508–515.
43. O.-L. Lui, N.-K. Mak and K.-N. Leung, Conjugated linoleic acid induces monocytic differentiation of murine myeloid leukemia cells, *Int. J. Oncol.*, 2005, **27**(6), 1737–1743.

44. N. Attar-Bashi, R. Weisinger, D. Begg, D. Li and A. Sinclair, Failure of conjugated linoleic acid supplementation to enhance biosynthesis of docosahexaenoic acid from α-linolenic acid in healthy human volunteers, *Prostaglandins, Leukotrienes Essent. Fatty Acids*, 2007, **76**(3), 121–130.
45. S. A. Khan and J. P. V. Heuvel, Reviews: current topicsrole of nuclear receptors in the regulation of gene expression by dietary fatty acids, *J. Nutr. Biochem.*, 2003, **14**(10), 554–567.
46. C. Bocca, F. Bozzo, G. Martinasso, R. A. Canuto and A. Miglietta, Involvement of PPARα in the growth inhibitory effect of arachidonic acid on breast cancer cells, *Br. J. Nutr.*, 2008, **100**(4), 739–750.
47. C. J. Field and P. D. Schley, Evidence for potential mechanisms for the effect of conjugated linoleic acid on tumor metabolism and immune function: lessons from n-3 fatty acids, *Am. J. Clin. Nutr.*, 2004, **79**(6), 1190S–1198S.
48. Z. Dong, X. Tian, Y. Chen, J. Hou, Y. Guo and J. Sun, *et al.*, A highly selective fluorescent chemosensor for Hg^{2+} based on rhodamine B and its application as a molecular logic gate, *Dyes Pigm.*, 2013, **97**(2), 324–329.
49. M. Q. Kemp, B. D. Jeffy and D. F. Romagnolo, Conjugated linoleic acid inhibits cell proliferation through a p53-dependent mechanism: effects on the expression of G1-restriction points in breast and colon cancer cells, *J. Nutr.*, 2003, **133**(11), 3670–3677.
50. S. Toomey, H. Roche, D. Fitzgerald and O. Belton, Regression of pre-established atherosclerosis in the apoE−/− mouse by conjugated linoleic acid, *Biochem. Soc. Trans.*, 2003, **31**(5), 1075–1079.
51. S. Weldon, S. Mitchell, D. Kelleher, M. J. Gibney and H. M. Roche, Conjugated linoleic acid and atherosclerosis: no effect on molecular markers of cholesterol homeostasis in THP-1 macrophages, *Atherosclerosis*, 2004, **174**(2), 261–273.
52. J. Fruchart, B. Staels and P. Duriez, New concepts on the mechanism of action of fibrates and therapeutic prospectives in atherosclerosis, *Bull. Acad. Natl. Med.*, 2001, **185**(1), 63–74; discussion.
53. A. Hammarstedt, P.-A. Jansson, C. Wesslau, X. Yang and U. Smith, Reduced expression of PGC-1 and insulin-signaling molecules in adipose tissue is associated with insulin resistance, *Biochem. Biophys. Res. Commun.*, 2003, **301**(2), 578–582.
54. K. Nagao, N. Inoue, Y.-M. Wang and T. Yanagita, Conjugated linoleic acid enhances plasma adiponectin level and alleviates hyperinsulinemia and hypertension in Zucker diabetic fatty (fa/fa) rats, *Biochem. Biophys. Res. Commun.*, 2003, **310**(2), 562–566.
55. M. L. Diaz, B. A. Watkins, Y. Li, R. A. Anderson and W. W. Campbell, Chromium picolinate and conjugated linoleic acid do not synergistically influence diet-and exercise-induced changes in body composition and health indexes in overweight women, *J. Nutr. Biochem.*, 2008, **19**(1), 61–68.
56. D. Kelley and K. Erickson, Modulation of body composition and immune cell functions by conjugated linoleic acid in humans and animal models: benefits *vs.* risks, *Lipids*, 2003, **38**(4), 377–386.

57. R. Ringseis, A. Müller, C. Herter, S. Gahler, H. Steinhart and K. Eder, CLA isomers inhibit TNFα-induced eicosanoid release from human vascular smooth muscle cells *via* a PPARγ ligand-like action, *Biochim. Biophys. Acta, Gen. Subj.*, 2006, **1760**(2), 290–300.
58. X. He, H. Zhang, X. Yang, S. Zhang, Q. Dai and W. Xiao, *et al.*, Modulation of immune function by conjugated linoleic acid in chickens, *Food Agric. Immunol.*, 2007, **18**(3–4), 169–178.
59. J. D. Ramakers, J. Plat, J.-L. Sébédio and R. P. Mensink, Effects of the individual isomers *cis*-9, *trans*-11 *vs. trans*-10, *cis*-12 of conjugated linoleic acid (CLA) on inflammation parameters in moderately overweight subjects with LDL-phenotype B, *Lipids*, 2005, **40**(9), 909–918.
60. H. Weiler, S. Austin, S. Fitzpatrick-Wong, E. Nitschmann, N. Bankovic-Calic and R. Mollard, *et al.*, Conjugated linoleic acid reduces parathyroid hormone in health and in polycystic kidney disease in rats, *Am. J. Clin. Nutr.*, 2004, **79**(6), 1186S–1189S.
61. H. Song, I. Grant, D. Rotondo, I. Mohede, N. Sattar and S. Heys, *et al.*, Effect of CLA supplementation on immune function in young healthy volunteers, *Eur. J. Clin. Nutr.*, 2005, **59**(4), 508–517.
62. H. K. Kwak, O. H. Kim, H. Jung and J. H. Kim, Effects of conjugated linoleic acid supplementation on inflammatory mediators and immunoglobulins in overweight Korean females, *FASEB J.*, 2009, **23**, 563.21.
63. M. A. Tarnopolsky and A. Safdar, The potential benefits of creatine and conjugated linoleic acid as adjuncts to resistance training in older adults, *Appl. Physiol., Nutr., Metab.*, 2008, **33**(1), 213–227.
64. O. Kelly and K. Cashman, The effect of conjugated linoleic acid on calcium absorption and bone metabolism and composition in adult ovariectomised rats, *Prostaglandins, Leukotrienes Essent. Fatty Acids*, 2004, **71**(5), 295–301.
65. L. Doyle, C. Jewell, A. Mullen, A. Nugent, H. Roche and K. Cashman, Effect of dietary supplementation with conjugated linoleic acid on markers of calcium and bone metabolism in healthy adult men, *Eur. J. Clin. Nutr.*, 2005, **59**(3), 432–440.
66. I. Platt, L. G. Rao and A. El-Sohemy, Isomer-specific effects of conjugated linoleic acid on mineralized bone nodule formation from human osteoblast-likecells, *Exp. Biol. Med.*, 2007, **232**(2), 246–252.
67. M. W. Pariza, Perspective on the safety and effectiveness of conjugated linoleic acid, *Am. J. Clin. Nutr.*, 2004, **79**(6), 1132S–1136S.
68. G. Cherian and M. P. Goeger, Hepatic lipid characteristics and histopathology of laying hens fed CLA or n-3 fatty acids, *Lipids*, 2004, **39**(1), 31–36.
69. J. Rajakangas, M. Mutanen, S. Basu and I. Salminen, Adenoma growth stimulation by the *trans*-10, *cis*-12 isomer of conjugated linoleic acid (CLA) is associated with changes in mucosal NF-κB and cyclin D1 protein levels in the Min mouse, *J. Nutr.*, 2003, **133**(6), 1943–1948.
70. T. Ide, Interaction of fish oil and conjugated linoleic acid in affecting hepatic activity of lipogenic enzymes and gene expression in liver and adipose tissue, *Diabetes*, 2005, **54**(2), 412–423.

71. Y. K. Nakamura, N. Flintoff-Dye and S. T. Omaye, Conjugated linoleic acid modulation of risk factors associated with atherosclerosis, *Nutr. Metab.*, 2008, **5**, 1–20.
72. S. N. Zlatanos, K. Laskaridis and A. Sagredos, Conjugated linoleic acid content of human plasma, *Lipids Health Dis.*, 2008, **7**, 1–6.
73. L. D. Whigham, A. C. Watras and D. A. Schoeller, Efficacy of conjugated linoleic acid for reducing fat mass: a meta-analysis in humans, *Am. J. Clin. Nutr.*, 2007, **85**(5), 1203–1211.
74. J.-H. Kim, O.-H. Kim, Y.-L. Ha and J.-O. Kim, Supplementation of conjugated linoleic acid with γ-oryzanol for 12 weeks effectively reduces body fat in healthy overweight Korean women, *Prev. Nutr. Food Sci.*, 2008, **13**(3), 146–156.
75. A. Hasin, J. M. Griinari, J. E. Williams, A. M. Shahin, M. A. McGuire and M. K. McGuire, Consumption of *c* 9, *t* 11–18:2 or *t* 10, *c* 12–18:2 Enriched Dietary Supplements Does Not Influence Milk Macronutrients in Healthy, Lactating Women, *Lipids*, 2007, **42**, 835–843.
76. M. A. Katzman, L. Jacobs, M. Marcus, M. Vermani and A. C. Logan, Weight gain and psychiatric treatment: is there as role for green tea and conjugated linoleic acid?, *Lipids Health Dis.*, 2007, **6**, 1–4.
77. E. V. Lambert, J. H. Goedecke, K. Bluett, K. Heggie, A. Claassen and D. E. Rae, *et al.*, Conjugated linoleic acid *versus* high-oleic acid sunflower oil: effects on energy metabolism, glucose tolerance, blood lipids, appetite and body composition in regularly exercising individuals, *Br. J. Nutr.*, 2007, **97**(5), 1001–1011.
78. T. Pérez-Ruiz, C. Martínez-Lozano and M. D. García-Martínez, Simultaneous determination of thiamine and its phosphate esters by a liquid chromatographic method based on post-column photolysis and chemiluminescence detection, *J. Pharm. Biomed. Anal.*, 2009, **50**(3), 315–319.
79. J. A. Herrera, M. Aréalo-Herrera, A. Shahabuddin, G. Ersheng, S. Herrera and R. G. Garcia, *et al.*, Calcium and conjugated linoleic acid reduces pregnancy-induced hypertension and decreases intracellular calcium in lymphocytes, *Am. J. Hypertens.*, 2006, **19**(4), 381–387.
80. E. Naumann, Y. A. Carpentier, A. Saebo, T. S. Lassel, J.-M. Chardigny and J.-L. Sébédio, *et al.*, *Cis*-9, *trans*-11 and *trans*-10, *cis*-12 conjugated linoleic acid (CLA) do not affect the plasma lipoprotein profile in moderately overweight subjects with LDL phenotype B, *Atherosclerosis*, 2006, **188**(1), 167–174.
81. M. Raff, T. Tholstrup, K. Sejrsen, E. M. Straarup and N. Wiinberg, Diets rich in conjugated linoleic acid and vaccenic acid have no effect on blood pressure and isobaric arterial elasticity in healthy young men, *J. Nutr.*, 2006, **136**(4), 992–997.
82. D. O. Genis, G. Bilge, B. Sezer, S. Durna and I. H. Boyaci, Identification of cow, buffalo, goat and ewe milk species in fermented dairy products using synchronous fluorescence spectroscopy, *Food Chem.*, 2019, **284**, 60–66.

83. K. Hoffmann, J. Blaudszun, C. Brunken, W.-W. Höpker, R. Tauber and H. Steinhart, Distribution of polyunsaturated fatty acids including conjugated linoleic acids in total and subcellular fractions from healthy and cancerous parts of human kidneys, *Lipids*, 2005, **40**(3), 309–315.
84. S. C. Larsson, L. Bergkvist and A. Wolk, High-fat dairy food and conjugated linoleic acid intakes in relation to colorectal cancer incidence in the Swedish Mammography Cohort, *Am. J. Clin. Nutr.*, 2005, **82**(4), 894–900.
85. A. Smedman, B. Vessby and S. Basu, Isomer-specific effects of conjugated linoleic acid on lipid peroxidation in humans: regulation by α-tocopherol and cyclo-oxygenase-2 inhibitor, *Clin. Sci.*, 2004, **106**(1), 67–73.
86. S. J. Hur and Y. Park, Effect of conjugated linoleic acid on bone formation and rheumatoid arthritis, *Eur. J. Pharmacol.*, 2007, **568**(1–3), 16–24.
87. S. E. McCann, C. Ip, M. M. Ip, M. K. McGuire, P. Muti and S. B. Edge, *et al.*, Dietary intake of conjugated linoleic acids and risk of premenopausal and postmenopausal breast cancer, Western New York Exposures and Breast Cancer Study (WEB Study), *Cancer Epidemiol., Biomarkers Prev.*, 2004, **13**(9), 1480–1484.
88. U. Riserus, B. Vessby, P. Arner and B. Zethelius, Supplementation with *trans* 10 *cis* 12-conjugated linoleic acid induces hyperproinsulinaemia in obese men: close association with impaired insulin sensitivity, *Diabetologia*, 2004, **47**, 1016–1019.
89. V. Chajès, F. Lavillonnière, V. Maillard, B. Giraudeau, M.-L. Jourdan and J.-L. Sébédio, *et al.*, Conjugated linoleic acid content in breast adipose tissue of breast cancer patients and the risk of metastasis, *Nutr. Cancer*, 2003, **45**(1), 17–23.
90. D. Luongo, P. Bergamo and M. Rossi, Effects of conjugated linoleic acid on growth and cytokine expression in Jurkat T cells, *Immunol. Lett.*, 2003, **90**(2–3), 195–201.
91. D. S. Kelley, M. Vemuri, Y. Adkins, S. H. S. Gill, D. Fedor and B. E. Mackey, Flaxseed oil prevents *trans*-10, *cis*-12-conjugated linoleic acid-induced insulin resistance in mice, *Br. J. Nutr.*, 2008, **101**(5), 701–708.
92. J.-M. Gaullier, J. Halse, K. Høye, K. Kristiansen, H. Fagertun and H. Vik, *et al.*, Supplementation with conjugated linoleic acid for 24 months is well tolerated by and reduces body fat mass in healthy, overweight humans, *J. Nutr.*, 2005, **135**(4), 778–784.

CHAPTER 6

Curcumin: Therapeutic Nutrient in Human Health and Recent Advancements in Food Formulations

GHOLAMREZA ABDI,*[a] MUKUL JAIN,[b,c] MOHD. TARIQ,[d] ABHISHEK BARNWAL,[d] NISHA PRASAD,[d] SUMEET TAYADE,[d] PANKAJ KUMAR,[d] SHIPRA CHOUDHARY,[e] LETSABIA EMELDA LETSIE,[d] AB WAHEED WANI,[f] MARYAM ABBASI TARIGHAT[g] AND CHARALAMPOS PROESTOS*[h]

[a] Department of Biotechnology, Persian Gulf Research Institute, Persian Gulf University, Bushehr, 75169, Iran; [b] Cell and Developmental Biology Lab, Centre of Research for Development, Parul University, Vadodara, Gujarat – 391760, India; [c] Department of Life Sciences, Parul Institute of Applied Sciences, Parul University, Vadodara, Gujarat – 391760, India; [d] Department of Environmental Sciences, Parul Institute of Applied Sciences, Parul University, Vadodara, Gujarat – 391760, India; [e] Department of Microbiology and Biotechnology, Meerut Institute of Engineering and Technology, Meerut, Uttar Pradesh – 250005, India; [f] Department of Horticulture, Lovely Professional University, Phagwara, Punjab – 144411, India; [g] Faculty of Nano and Bio Science and Technology, Department of Chemistry, Bushehr, 75169, Iran; [h] Food Chemistry Laboratory, Department of Chemistry, National and Kapodistrian University of Athens, Panepistimiopolis Zographou, 15771 Athens, Greece
*Emails: abdi@pgu.ac.ir; harpro@chem.uoa.gr

Food Chemistry, Function and Analysis No. 44
Functional Foods of the Future
Edited by Vijai Kumar Gupta, Minaxi Sharma, Smriti Gaur and Ramesh Chander Kuhad

Published by the Royal Society of Chemistry, www.rsc.org

6.1 Introduction

Curcumin (1,7-bis(4-hydroxy-3-methoxyphenyl)-1,6-heptadiene-3,5-dione; Figure 6.1), also denoted as diferuloylmethane,[1] is the main naturally occurring bright-yellow polyphenol found in the root and rhizome of *Curcuma longa* (Family: Zingiberaceae)[2] and other *Curcuma* spp. *C. longa* is perennial herb with wide distribution throughout the tropical and sub-tropical regions of the world.[3] Curcumin is a diarylheptanoid, belonging to the group of curcuminoids,[4] that is found in turmeric, a spice commonly used in Indian and Southeast Asian cuisine. It has been extensively studied for its potential health benefits, including anti-inflammatory, antioxidant, anti-cancer, anti-obesity, and antidiabetic properties (Figure 6.2). Wide investigation has suggested that curcumin has a range of potential health benefits, including in improving brain function and lowering the risk of heart disease.

Curcumin has emerged as a potential active constituent with its primary source from the turmeric roots. Turmeric root contains other bioactive compounds that may also contribute to its health benefits. Some of these other compounds include, demethoxycurcumin, bisdemethoxycurcumin, turmerones, atlantone, zingiberene, curcumene *etc.* These compounds, along with curcumin, are collectively referred to as curcuminoids. The medicinal properties and yellow color of the turmeric is due to these curcuminoids. They are

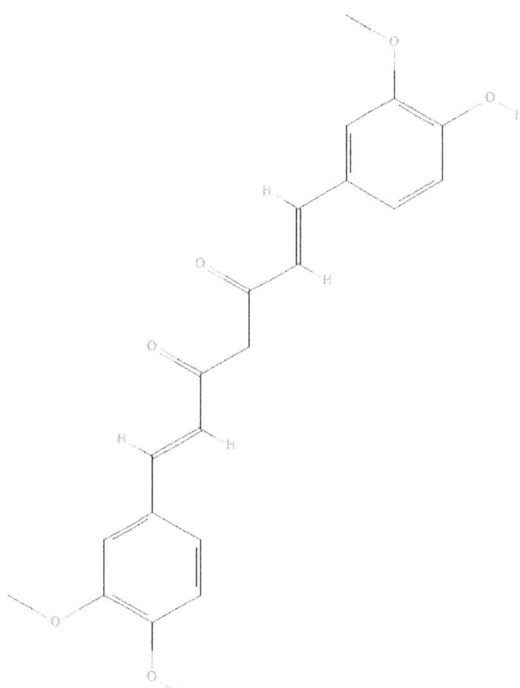

Figure 6.1 Structure of curcumin.

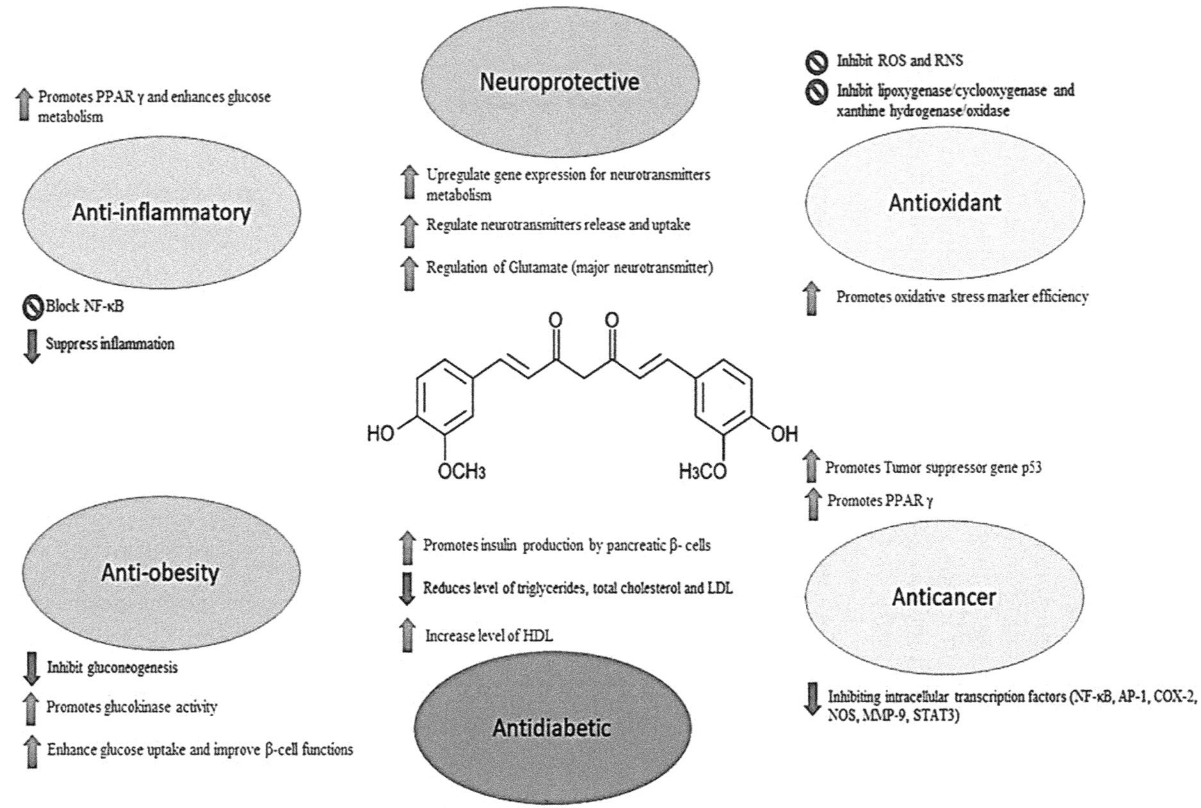

Figure 6.2 Detailed mechanism of curcuminoid therapeutic activities. (PPAR γ, peroxisome proliferator-activated receptor gamma; ROS, reactive oxygen species; RNS, reactive nitrogen species; L/HDL, low-/high-density lipoprotein; NK-κB, nuclear factor kappa B; AP, activator protein-1; COX2, cyclooxygenase-2; NOS, nitric oxide synthase; MMP9, matrix metalloproteinase-9; STAT3, signal transducer and activator of transcription 3).

Table 6.1 Major active constituents in different *Curcuma* spp.

Species	Country of origin	Bioactive component	Ref.
C. aeruginosa Roxb.	Malaysia	8,9-Dehydro-9-formyl-cycloisolongifolene	5
C. amada Roxb.	India	Myrcene	6
C. angustifolia Roxb.	India	Germacrone	7
C. angustifolia Roxb.	Thailand	β-Elemenone	8
C. aromatica Salisb.	China	Turmerone	9
C. aromatica Salisb.	Japan	Curcumol	10
C. aromatica Salisb.	Sri Lanka	Camphor	11
C. aromatica Salisb.	Bangladesh	Camphor	12
C. harmandii Gagnep.	Vietnam	1,8-Cineole	13
C. inodora Blatt.	Malaysia	Curzerenone	14
C. longa L.	Pakistan	Turmerone	15
C. longa L.	Brazil	Turmerone	16
C. longa L.	Nigeria	Turmerone	17
C. mangga Valeton & Zijp	Malaysia	Caryophyllene oxide	5
C. zanthorrhiza Roxb.	Indonesia	α-Curcumene	18
C. zedoaria (Christm.) Roscoe	Taiwan	Epicurzerene	19

believed to have a synergistic effect, which means that they work together to enhance the overall health benefits of turmeric. In addition to curcuminoids, turmeric root also contains volatile oils, including turmerone and zingiberene, which contribute to its characteristic aroma and flavor and may also have health-promoting properties. The various therapeutic activities of the curcuminoids with their mechanisms of action are depicted in Figure 6.2.

6.2 Species That Contain Curcumin

Curcuminoids are mainly extracted from the underground part of turmeric plant (*Curcuma longa*). Turmeric is widely used as an ingredient in South Asian cuisine and has a long history of use in traditional medicine as well. In addition to turmeric, other species of *Curcuma*, such as *Curcuma zedoaria* and *Curcuma aromatica*, also contain curcuminoids (listed in Table 6.1). However, turmeric is the most commonly studied and widely used source of curcuminoids.

Curcuminoids have also been found in other plant species outside of the *Curcuma* genus, including ginger (*Zingiber officinale*), galangal (*Alpinia galanga*), and black pepper (*Piper nigrum*). However, the levels of curcuminoids in these plants are generally lower than in turmeric. It is important to note that the concentration of curcuminoids in these plants can vary depending on factors such as the plant's age, growing conditions, and processing methods.

6.3 Chemical Structure and Properties

Curcumin, which is also know as diferuloylmethane ($C_{21}H_{20}O_6$), is a low molecular mass (368.37 g mol^{-1}) compound of polyphenol having physical properties such as a melting point of approximately 183 °C (Figure 6.3).

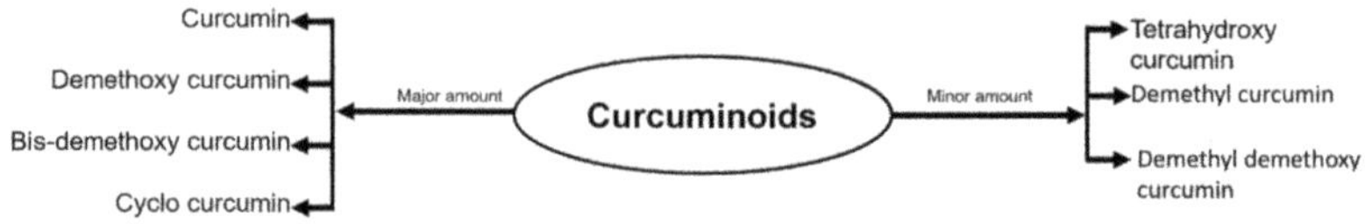

Figure 6.3 The various types of curcuminoid derivatives present in *Curcuma* plants, distinguishing into major and minor components.

The IUPAC name of curcumin is (1*E*,6*E*)-1,7-bis(4-hydroxy-3-methoxyphenyl)-1,6-heptane-3,5-dione, and the structure has two aryl rings with an *ortho*-methoxyphenolic group symmetrically linked to an α,β-unsaturated β-diketone moiety.[20]

Curcuminoids are a crude form, from which curcumin is the major bioactive analyte present, and by-products of other components like demethoxycurcumin and bisdemethoxycurcumin are extracted.[21] Curcumin has various health benefits that are mostly related to antioxidant and anti-inflammatory properties, which make it the most useful spice. Due to its rapid metabolism and its low oral absorbability, the clinical application of curcumin is still limited but there are various techniques like deprotonation of all the hydroxy substitutes of curcumin that can increase the solubility up to a 12–18%.[21] So, to find solutions to all these drawbacks, numerous approaches have been constructed such as structural modification, employment of an adjuvant that interferes with its glucuronidation, and nanoencapsulation.[21]

Various curcuminoids are present in the underground rhizome of turmeric (2.5–6%). Curcumin I, the most common curcuminoid, also known as diferuloylmethane (85%), is a yellow-orange powder extensively studied for medicinal properties (Table 6.2).[22] Curcumin has shown some potential benefits in preventing and treating a variety of health conditions including arthritis, diabetes, and certain types of cancer. Curcumin, a yellow-colored analyte extracted from the rhizome of the turmeric plant, has the name (1*E*,6*E*)-1,7bis(4-hydroxy-3methoxyphenyl)-1,6-heptane-3,5-dione, and can also be in oil and resin form (Figure 6.4).[23]

Curcumin II is another component, also known as demethoxycurcumin or monodemethoxycurcumin [(1*E*,4*Z*,6*E*)-5-hydroxy-7-(4-hydroxy-3-methoxyphenyl)-1-(4-hydroxyphenyl)-1,4,6-heptatrien-3-one)], is a significant active curcuminoid; it has anti-inflammatory qualities and can kill human cancer cells by inducing apoptosis.[24] This curcuminoid makes up 17%. It is β-diketone, enone and a diaryheptanoid, in which in comparison to curcumin one of the methoxy groups is replaced by hydrogen. Demethoxycurcumin is present in *Curcuma amada*, *Curcuma aeruginosa* and other organisms.[25]

Bisdemethoxycurcumin lacks two methoxy groups on the aromatic rings in comparison to curcumin. (Table 6.3).[20] The β-diketone bisdemethoxycurcumin is a methane molecule with two hydrogens replaced by 4-hydroxycinnamoyl groups. It functions as an α-amylase inhibitor. It is a β-diketone, a polyphenol, an enone and a diarylheptanoid. It is functionally

Table 6.2 Physical appearance, chemical composition and uses in daily life of different preparations of turmeric.[27]

Product	Appearance	Chemical constituents	Uses
Whole rhizome (dried)	Red-yellow, orange brown or pale yellow	Essential oils (1.5–5%) and curcuminoids (3–15%)	Medicinal purposes
Ground *C. longa*	Yellow or red-yellow	Processing, as well as exposure to light, can decrease curcuminoids and essential oils	Dietary supplement, condiment, medicine and dye
C. longa oil	Yellow to brown oil	Essential oils from the leaves and rhizomes contain monoterpenes and sesquiterpenes, respectively	Dietary supplement, medicine and spice
C.longa oleoresins	Reddish-brown or dark yellow viscous fluid	Curcuminoids (37–55%) and essential oils (25%)	Food, medicine and cosmetics
Curcumin	Yellow to orange-red crystalline powder	Curcumin and it's bisdemethoxy and demethoxy derivatives; the three primary curcuminoids may account for up to 90% of the total curcuminoids; oils and resins may make up a small percentage of the total composition	Therapeutic supplement and medicine

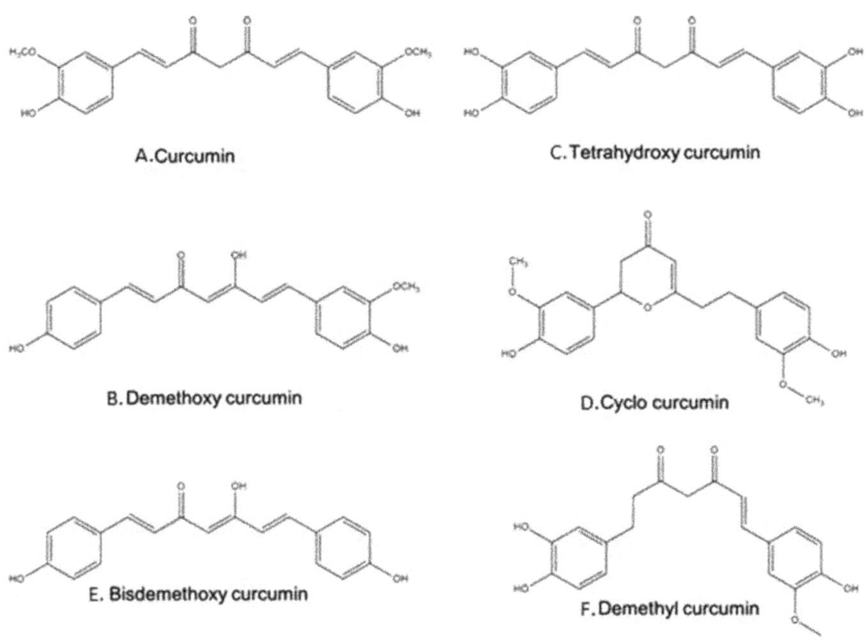

Figure 6.4 Chemical structures of various curcuminoids.

Table 6.3 A complete list of all curcuminoid derivatives with additional details on IUPAC name, chemical modification, formula, CAS number, source and references.

Derivative	IUPAC name	Chemical modification	Formula	CAS no.	Source	Ref.
Curcumin	(1*E*,6*E*),-1,7-Bis(4-hydroxy-3-methoxyphenyl)-1,6-heptadiene-3,5-dione	—	$C_{21}H_{20}O_6$	458-37-7	Curcuma aromata, Curcuma longa	ChEBI
Demethoxycurcumin	(1*E*,6*E*)-1-(4-Hydroxy-3-methoxyphenyl)-7-(4-hydroxyphenyl)-1,6-heptadiene-3,5-dione	One methoxy group is replaced by hydrogen	$C_{20}H_{18}O_5$	24939-17-1	Curcuma zedoria, Etlingera elatior	ChEBI
Bisdemethoxycurcumin	(1*E*,6*E*)-1,7-Bis(4-hydroxyphenyl)-1,6-heptadiene-3,5-dione	Two methoxy groups are replaced by hydrogen	$C_{19}H_{16}O_4$	52328-96-8	Curcuma longa	ChemSrc
Tetrahydroxycurcumin	(1*E*,6*E*)-1,7-Bis(3,4-dihydroxyphenyl)-1,6-heptadiene-3,5-dione	Two methoxy groups have been replaced by hydroxy groups	$C_{19}H_{16}O_6$	60831-46-1	Curcuma aromata, Curcuma longa	PubChem
Cyclocurcumin	2-(4-Hydroxy-3-methoxyphenyl)-6-[(*E*)-2-(4-hydroxy-3-methoxyphenyl)vinyl]-2,3-dihydro-4*H*-pyran-4-one	Central diketone is cyclised	$C_{21}H_{20}O_6$	153127-42-5	Curcuma longa	ChEBI
Demethylcurcumin	(1*E*,6*E*)-1-(3,4-Dihydroxyphenyl)-7-(4-hydroxy-3-methoxyphenyl)-1,6-heptadiene-3,5-dione	One methoxy group is replaced by a hydroxy groups	$C_{20}H_{18}O_6$	149732-51-4	Curcuma longa	ChemSrc
Demethyldemethoxycurcumin	(1*E*,6*E*)-1-(3,4-Dihydroxyphenyl)-7-(4-hydroxyphenyl)-1-6-heptadiene-3,5-dione	One methoxy group is replaced by hydrogen and one methoxy group is replaced by a hydroxy group	$C_{20}H_{16}O_5$		Curcuma longa	ChemSrc

related to 4-coumaric acid.[25] Compared to the others, bisdemethoxycurcumin is more resistant to alkaline degradation. It was found to be effective in sensitizing PC cells resistant against gemcitabine.[26]

6.3.1 Phenolic Compounds with Structural Similarities to Curcuminoids

Apart from *Curcuma longa* as a source of curcuminoids, there are other plants that produce phenolic compounds with structural similarities to curcuminoids (Figure 6.5).

From the rhizome of *Zingiber officinale,* three diarylheptanoids have been identified and isolated in the ethanol extract. The new compounds that were isolated possess a resemblance to curcuminoids except lacking unsaturation in the heptadienone linkage.[28]

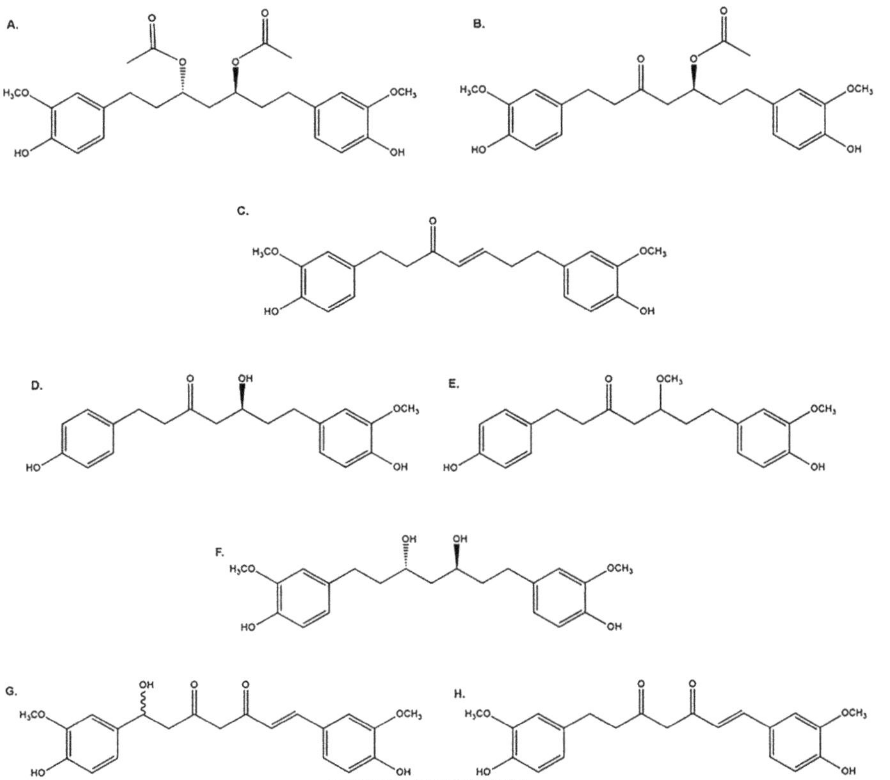

Figure 6.5 Phenolic compounds related to curcumin extracted from other plants. Diarylheptanoids isolated from *Zingiber officinale* (A, B and C), diarylheptanoids isolated from *Juglans mandshurica* (D and E) and diarylheptanoids isolated from *Curcuma xanthorrhiza* (F, G and H).

From the roots of *Juglans mandshurica maximowicz* (Juglandaceae), diarylheptanoids have been isolated which are extensively used as a folk medicine in the treatment of cancer in Korea.[29]

From the rhizomes of temu lawal, *Curcuma xanthorriza* (Zingiberaceae),[30] three diarylheptanoids have been isolated which include (3*S*,5*S*)-1,7-bis(4-hydroxy-3-methoxyphenyl)heptane-3,5-diol, (1*S*)-1-hydroxy-1,7-bis(4-hydroxy-3-methoxyphenyl)heptane-3,5-dione, dihydrocurcumin.

6.3.2 Chemical and Physical Properties of Curcumin

Curcumin (CUR) exhibits tautomerism in two forms, a keto form and an enol form. The keto form of CUR is mainly present in neutral or acidic media, whereas the enol form is mainly present in alkaline media.[31] Curcumin is a crystalline yellow-orange diferuloylmethane with a molecular weight of 368.39 $g\,mol^{-1}$ and a melting point of 183 °C. CUR is lipophilic in nature, having a $\log P$ value of ~3.0, due to which it tends to be soluble in various organic solvents like acetone, dimethyl sulfoxide, methanol, and ethanol. In an aqueous solution, solubility of CUR increases under alkaline conditions, but under both alkaline and neutral conditions curcumin degrades quickly. CUR shows maximum absorption in methanol at 430 nm, whereas in acetone it is 415–420 nm. CUR shows maximum absorbance at 467 nm under alkaline pH and is fully deprotonated. The pK_a value of CUR is 8.54 and at neutral pH it contains 3 labile protons, of which 2 are phenolic protons and 1 is enolic.[32] Curcumin (CUR) has several significant limitations, including low bioavailability, poor water solubility, limited oral absorption, rapid elimination from the body, and susceptibility to degradation under alkaline pH. Under acidic conditions, less than 20% of CUR is degraded after 1 hour; however, in alkaline environments, CUR undergoes rapid degradation, typically within approximately 30 minutes. Outside of biological systems, curcumin is degraded by solvolysis, oxidation, and photodegradation, which results in a five percent loss of the parent chemical when synthesized in clear glass as compared to amber glass.[33] It gets absorbed in a small proportion and is metabolized in the liver and plasma while the rest is eliminated from the body.[34] Curcumin is a versatile molecule that can be used to treat a variety of ailments, because it is known to be able to modulate growth factors, transcription factors, enzymes, kinase, antiapoptotic and proapoptotic proteins, and inflammatory cytokines.[35]

6.3.3 Pharmacokinetics

According to numerous studies of the pharmacokinetics, curcumin is largely metabolized in the liver. There, it goes through substantial reduction by alcohol dehydrogenase before being converted to glucuronate and sulphate. Moreover, Perkins and associates reported that individuals need 1.6 g of curcumin every day to obtain the intended outcome (Tables 6.4–6.6).[31] Molecules that enter the bloodstream are susceptible to rapid metabolic

Table 6.4 Serum and tissue levels of curcumin under different doses and administration routes in humans.

Species	Route	Dose	Plasma/tissue	Level	Ref.
Humans	Oral	$2\ g\,kg^{-1}$	Serum	$0.006 \pm 0.005\ \mu g\,mL^{-1}$	36
Humans	Oral	$4–8\ g\,kg^{-1}$	Serum	$0.4–3.6\ \mu mol\,mL^{-1}$	36
Humans	Oral	10 g	Serum	$50.5\ ng\,mL^{-1}$	36
Humans	Oral	12 g	Serum	$51.2\ ng\,mL^{-1}$	36
Humans	Oral	3.6 g	Plasma	$11.1 \pm 0.6\ nmol\,mL^{-1}$	36
Humans	Oral	0.4–3.6 g	Colorectum	$7–20\ nmol\,g^{-1}$	36

Table 6.5 Molecular properties of curcumin and derivatives.

Property	Curcumin	Demethoxy-curcumin	Bisdemethoxy-curcumin	Cyclocurcumin
Molecular weight	368.385	338.359	308.333	368.385
log *P*	3.3699	3.3613	3.3527	3.7428
Rotatable bonds	8	7	6	5
Acceptors	6	5	4	6
Donors	2	2	2	2

transformations in the liver, other organs, and the blood. Curcumin levels in serum of human participants could not be identified one to four hours after an ingested dose of 500–8000 mg. Excretion occurs through the biliary system and, to a smaller extent, the renal system.[33] The low solubility could result in reduced curcumin bioavailability. The glucuronides of tetrahydrocurcumin (THC) as well as hexahydrocurcumin (HHC), which interact with biliary salts, are the primary bile metabolite compounds of CUR in humans and rats. Dihydroferulic acid, along with trace amounts of ferulic acid, is a minor metabolite of curcumin in water (approximately $11\ ng\,mL^{-1}$ at pH 5.0). These compounds also contribute to the limited gastrointestinal absorption and the formation of biliary metabolites.[36]

6.4 Therapeutic Role of Curcumin

The dietary polyphenol curcumin, which is found in the root of the turmeric plant, is widely used as a spice and herb across the world. Because of its biopharmacological properties, curcumin is frequently referred to as the "spice of life".[37] For millennia curcumin has been also utilized historically for its usefulness as a medicine and as a wound healer, although in 1937 it was used for the first time as a medication to treat biliary illness.[35] Curcumin's chemistry, particularly its bisphenolic hydroxyl groups, double-conjugated bonds, unsaturated diketone, and methoxy groups, are what give it its numerous biological functions and medicinal properties (Figure 6.6).[37] Except for mice, other animals can have longer life expectancy because of curcumin's role in regulating protein homeostasis.[38] Like other phytochemicals,

Table 6.6 ADMET values of curcumin and derivatives.

ADMET	Model (unit)	Curcumin	Demethoxy-curcumin	Bisdemethoxy-curcumin	Cyclo-curcumin
Absorption	Water solubility ($\log \text{mol L}^{-1}$)	−4.01	−3.566	−3.38	−2.892
	Caco-2 permeability ($\log P_{app}$ in 10^{-6} cm s^{-1})	−0.093	1.023	0.957	1.653
	Intestinal absorption (human) (% absorbed)	−0.093	91.393	91.159	81.358
	Skin permeability ($\log K_p$)	−2.764	−2.768	−2.804	−2.735
	P-Glycoprotein substrate (categorical)	Yes	Yes	Yes	No
	P-Glycoprotein I inhibitor (categorical)	Yes	Yes	Yes	No
	P-Glycoprotein II inhibitor (categorical)	Yes	Yes	No	No
Distribution	V_{Dss} (human) ($\log \text{L kg}^{-1}$)	−0.215	−0.075	0.139	0.011
	Fraction unbound (human) (fu)	0	0	0.045	0.382
	BBB permeability (log BB)	−0.562	−0.337	−0.089	−0.242
	CNS permeability ($\log P_S$)	−2.99	−2.458	−2.271	−1.1442
Metabolism	CYP2D6 substrate (categorical)	No	No	No	No
	CYP3A4 substrate (categorical)	Yes	Yes	Yes	No
	CYP1A2 inhibitor (categorical)	Yes	Yes	Yes	Yes
	CYP2C19 inhibitor (categorical)	Yes	Yes	Yes	No
	CYP2C9 inhibitor (categorical)	No	Yes	Yes	No
	CYP2D6 inhibitor (categorical)	Yes	No	No	No
	CYP3A4 inhibitor (categorical)	Yes	Yes	Yes	No
Excretion	Total clearance ($\log \text{mL min}^{-1} \text{kg}^{-1}$)	−0.002	0.026	−0.008	−39.151
	Renal OCT2 substrate (categorical)	No	No	No	No

Table 6.6 (*Continued*)

ADMET	Model (unit)	Curcumin	Demethoxy-curcumin	Bisdemethoxy-curcumin	Cyclo-curcumin
Toxicity	AMES toxicity (categorical)	No	No	No	Yes
	Max. tolerated dose (human) (log mg per kg per day)	0.081	−0.12	−0.081	0.438
	hERG I inhibitor (categorical)	No	No	No	No
	Oral rat acute toxicity (LD_{50}) (mol kg^{-1})	1.833	1.973	2.09	2.482
	Oral rat chronic toxicity (LOAEL) (log mg per kg_bw per day)	2.228	2.236	1.616	11.508
	Hepatotoxicity (categorical)	No	No	No	No
	Skin sensitization (categorical)	No	No	No	No
	T. pyriformis toxicity (log μg L^{-1})	0.494	0.593	0.968	0.285
	Minnow toxicity (log mM)	−0.081	−0.426	0.069	7.417

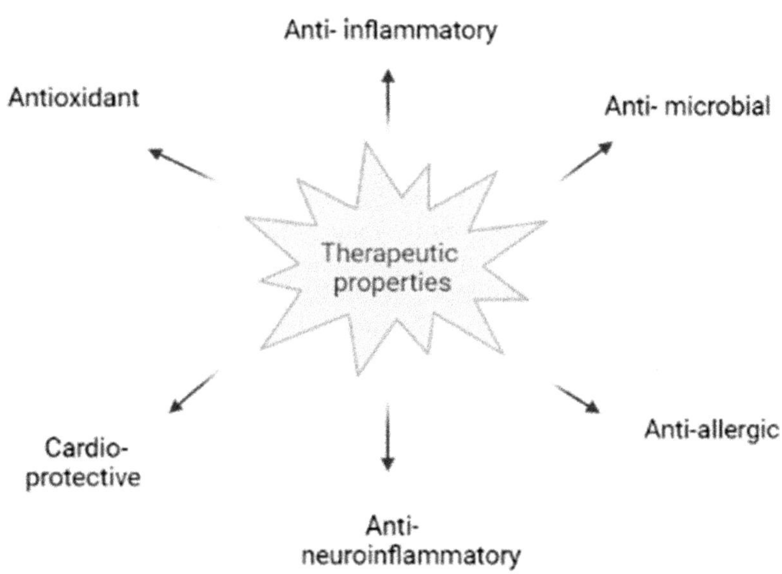

Figure 6.6 Therapeutic properties of curcumin.

curcumin affects cells in a variety of ways. In fact, because of this affinity to interact with numerous proteins, it can also trigger cellular reactions to environmental stimuli. Furthermore, curcumin can alter a cell's epigenetic state, and up- and down-regulates a number of miRNAs.[37] Diets that are predominantly drawn from natural sources have several cell protecting qualities, including antioxidant, anti-arthritis and anti-inflammatory effects and enhancing memory and cognitive capacities.[36] Due to its ability to affect a number of signalling molecules, contemporary medicine has shown that curcumin exhibits a wide range of biological and therapeutic properties, including antioxidant, neuroprotective, anti-inflammatory, and chemoprotective.[39] According to a recent study, curcumin dramatically decreased lead-induced brain damage in rats by boosting the levels of glutathione and antioxidant enzymes such as superoxide dismutase and catalase.[36] According to the Joint United Nations and the World Health Organization's Expert Committee on Food Additives (JECFA), curcumin is regarded as a safe ingredient and is hence appropriate for everyday dietary usage. Thus, there are currently a number of products containing free curcumin available.[39]

6.4.1 Significance of Curcumin in Neurodegenerative Disorders

Neurodegenerative diseases (NDs) are the chronic, incurable, irreversible, pathologies of CNS and PNS[40] that cause the neurons to degenerate slowly, resulting in loss of brain function and problems in movement, and represent a major threat to human health. Examples of neurodegenerative disorders are Alzheimer's disease, Huntington's disease, epilepsy, Parkinson's disease, amylotropic lateral sclerosis *etc.*[41] Since they are incurable, to address these horrifying diseases new and more effective therapeutic options are urgently needed. Several studies conducted both *in vivo* and *in vitro* have demonstrated that curcumin has a protective effect against several neurodegenerative disorders.[37] The onset and development of NDs are significantly influenced by a number of variables, including neuroinflammation, the buildup of misfolded amyloid proteins, and oxidative damage.[35] Nutraceuticals, or dietary supplements, may protect the CNS by protecting neurons from stress-related damage, decreasing neuroinflammation, and improving neurocognitive function, according to recent research.[37] Curcumin offers a number of advantages, including being a secure, affordable, and widely accessible polyphenol that can readily traverse the BBB, drawing interest from researchers to employ it for treating neurodegenerative diseases.[35] Curcumin has the ability to regulate cytokines, adhesion molecules, redox state protein kinases, and enzymes linked to inflammation (Figure 6.7). Many cells, including neurons, astrocytes, microglia, and several CNS regions such the hippocampus, mesencephalic region, cortical region, and spinal cord, have been found to be protected by curcumin. Curcumin was tested at various levels as an anti-inflammatory dietary additive in APPSw mice (an Alzheimer's-like model). According to the findings, a minimal dose

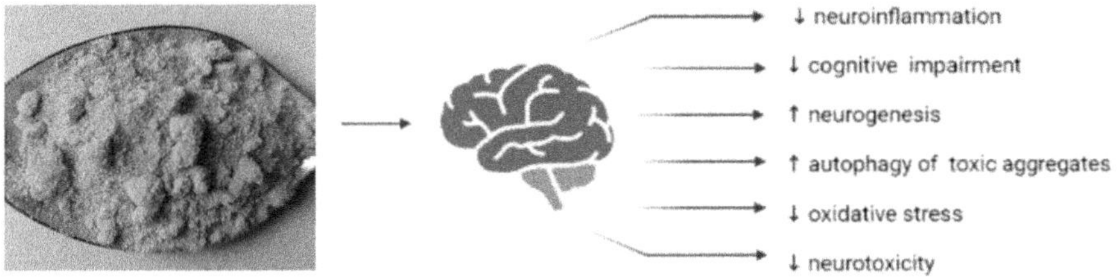

Figure 6.7 Effects of curcumin in the brain of neurodegenerative disease individuals.

of curcumin of around 160 ppm decreased GFAP, which is an astrocytic marker linked to inflammatory processes. Moreover, curcumin dramatically reduces GFAP mRNA level in hippocampus astrocytes, which improves spatial memory in a rat model of Alzheimer's disease.

Recent studies on amyloid-β (Aβ) and curcumin have shown that curcumin protects neurons from numerous toxic insults of ageing and Aβ in humans by preventing amyloid aggregation and crossing the BBB. Due to its hydrophobic nature and interactions with the aromatic rings of amyloid-β dimers, as well as the keto and enol groups, curcumin disrupts the molecular interactions required for the formation of amyloid fibrils and plaques. While short-term use of curcumin appears to be safe, the discrepancies observed between *in vivo* and *in vitro* results, as well as human clinical trial outcomes, may be attributed to various factors, including limited bioavailability and suboptimal study design (Table 6.7).[36] Curcumin's ability to inhibit amyloid plaques seems dose- as well as time-dependent. For instance, curcumin concentrations of 15–30 M work better for short trials (less than 24 h), while those of 5–15 M work better for prolonged trials (4–6 days). The rats given Aβ42 treatment displayed apparent cognitive abnormalities in behavioral activities. Curcumin doses (50, 100, and 200 $mg\,kg^{-1}$) given chronically (once daily for seven days) rather than once enhanced cognitive abilities in a dose-dependent manner. Curcumin may have anti-amyloid effects in AD, according to a number of lines of evidence: (1) curcumin inhibits Aβ aggregation and disaggregates to generate fibrillar Aβ 40, according to research with the cell line SHSY5Y *in vitro*; (2) numerous *in vivo* investigations have demonstrated that curcumin helps in promoting the disaggregation of pre-existing amyloid deposits, inhibits the creation of fresh amyloid deposits, and even helps to shrink the size of the pre-existing deposits; (3) *in vitro* studies have shown that curcumin as well as its derivatives suppress the production of fibrillar Aβ from Aβ monomer and also destabilize preformed fibrillar Aβ, suggesting that curcumin is protective against the toxicity of Aβ.[36]

6.4.2 Protective Effects on Cardiovascular Disease

Cardiovascular diseases (CVDs) are a major cause of death globally. There are various known factors which increases the risk of CVD development and

Table 6.7 Role of curcumin in various neurodegenerative diseases.

Disease	Model system/patients	Studies	Effects of curcumin	Ref.
Alzheimer's disease	Mice	Vertebrate/ *in vivo*	• Downregulates BACE1 • Aβ formation reduces • Expression of γ-secretase component, *i.e.* presenilin-2 • Expression of Aβ degrading enzymes increases	36 and 37
Alzheimer's disease	SHSY5Y cells	*In vitro*	• PS1 and GSK-3β levels decreases • Aβ production reduces • Inhibition of Aβ-induced tau hyperphosphorylation • Prevention of aggregation in NFTs • Mitochondrial biogenesis • Stimulates clearance of Aβ aggregates	36, 37 and 41
Alzheimer's disease	Microglia from AD patients	*In vitro*	• Aβ phagocytosis increases • Blocks ERK1/2 and p38 kinase signalling • Production of TNF-α, interleukin-1β and interleukin-6 reduced	35 and 37
Alzheimer's disease	Hippocampus and subventrical zone of adult rat	*In vitro*	• Modulates neurogenesis by activation of Wnt pathway • Remarkable increase in brain Ach, GSH, BCL2, paraoxenase levels	35 and 36
Alzheimer's disease	Adult female albino rat	Vertebrate/ *in vivo*	• GFAP mRNA decreases in hippocampal astrocytes • Improves spatial memory	36 and 42
Alzheimer's disease	AD rat	Vertebrate/ *in vivo*	• Enhances Aβ uptake by macrophages • Co-localization of Aβ with intracellular component • Improves cognitive decline	36 and 43
Alzheimer's disease	AD patients	Vertebrate/ *in vivo*	• Improves spatial memory and learning ability • Improves cerebral glucose uptake	36 and 44

Alzheimer's disease	APPswe/PS1dE9 double transgenic mice	Vertebrate/ *in vivo*	• Promotes neural stem cell differentiation • Clears or reduces the size of senile plaques	43 and 44
Alzheimer's disease	APPswe/PS1dE9 mice	Vertebrate/ *in vivo*	• Significantly lowered oxidized proteins and IL-1β, elevated in the brains of mice	44
Alzheimer's disease	Tg2576l line APPSw mouse model	*In vitro*	• Strongly inhibit BACE-1 activity	44
Parkinson's disease	PINK1 knockdown model of PD	*In vitro*	• Protective effect against mitochondrial dysfunction and cell death	37
Parkinson's disease	Rotenone mouse with PD	Vertebrate/ *in vivo*	• Improves α-synuclein levels and reduces Lewy bodies • Reduces carbonylated protein and nitrotyrosine-modified proteins	31 and 37
Parkinson's disease	PD mouse	Vertebrate/ *in vivo*	• Reduces oxidative stress • Improves motor performance • Inhibits NF-κB translocation and activation in astrocytes	36 and 37
Parkinson's disease	Swiss albino mice	Vertebrate/ *in vivo*	• Conserves mitochondrial enzyme complex • Improves motor behaviour	31
Parkinson's disease	SH-SY5Y	*In vitro*	• Improves mitochondrial membrane potential, reduces apoptosis • Attenuates neurotoxicity • Inhibits increases in ROS levels • Provides mitochondrial protection • Reduces p53-mediated apoptosis • Inhibits α-synuclein expression • Inhibition of dopamine neuron apoptosis • Improvement in movement disorder	31, 37 and 45
Parkinson's disease	MPTP-induced PD mouse	Vertebrate/ *in vivo*	• Improved abnormal motor behaviour • Reduced dopaminergic neuron degeneration • Enhances the survival of snpc neurons and striatal fibres • Exerts neuroprotective properties *via* 7-nachr-mediated mechanism	45

Table 6.7 *(Continued)*

Disease	Model system/patients	Studies	Effects of curcumin	Ref.
Parkinson's disease	6-OHDA-induced animal model of PD	Vertebrate/ *in vivo*	• Protects against oxidative stress-induced neurodegradation by stimulating Wnt/ catenin pathway • Reduces neuronal apoptosis	45
Parkinson's disease	Rotenone-induced and human α-synuclein-expressing *Drosophila*	Vertebrate/ *in vivo*	• Increases lifespan • Improves locomotor abilities Decreases oxidative stress and decreases DA neuron degeneration	45
Multiple sclerosis	U373-MG	Vertebrate/ *in vivo*	• Reduced the release of both IL-6 and MMP9 activity	37
Multiple sclerosis	MS patient	Clinical	• Remarkable decrease in expression of miRNAs including mir-145, mir-132, mir-16 • Reduction of inflammatory mediators	37
Epilepsy	PTZ-induced models of epilepsy, larval and adult zebrafish	Vertebrate/ *in vivo*	• Inhibits tonic-clonic seizures	37
Epilepsy	Male Wistar rats	Vertebrate/ *in vivo*	• Improvement in the onset of PTZ-induced seizures • Reduction in oxidative stress • Decrease in cognitive decline	37
Huntington's disease	Yeast	Invertebrate/ *in vivo*	• Inhibits formation of mHTT aggregation • Destabilizes preformed aggregates • Prevents recruitment of misfolded protein to perivascular zone	35

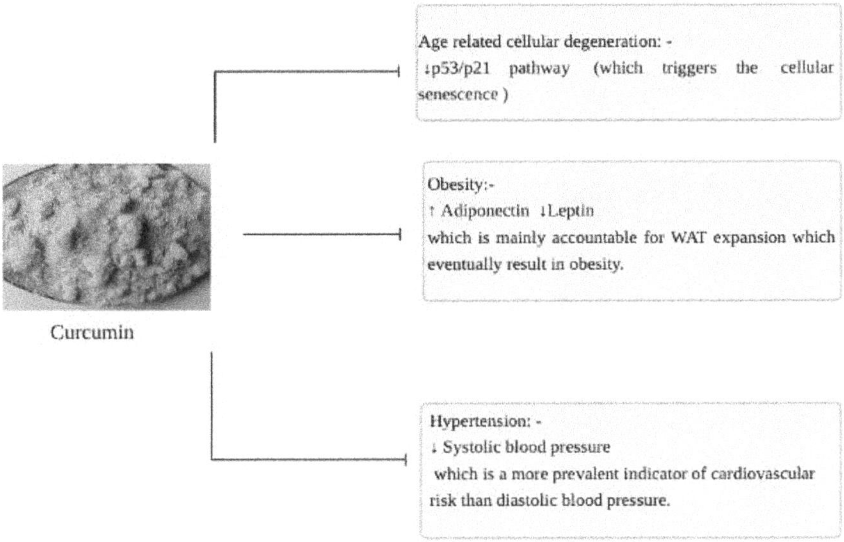

Figure 6.8 Effect of curcumin on risk factors of cardiovascular disease.

progression. Metabolic syndrome, diabetes, obesity, nonalcoholic fatty liver disease, hypertension and dyslipidemia are major risk factors that increase atherosclerosis, which many believe to be the main element in CVD pathologies. As a result, limiting the factors which contribute to CVD disease is mandatory to limiting the death and disability rates associated with this condition.[46] The ageing process is a prominent risk factor that cannot be modified because it cannot be inverted. Other factors, such as obesity, which is a major cause of heart attack and stroke, are modifiable. Both cellular senescence and oxidative stress-related senescence, which both result in vascular dysfunction, can be postponed and reduced by curcumin (Figure 6.8).[47] Along with the vascular wall and the heart tissues; brown adipose tissue (BAT), which was previously thought to be only responsible for thermogenesis, also plays a role. In addition to being thermogenic, BAT is involved in the management of glucose and lipid metabolism, as well as the monitoring of energy balance and body weight. The process by which white adipocytes, which constitute white adipose tissue (WAT), the chief organ of energy storage, are transformed into brown adipose tissue (BAT) is known as beiging. Besides ageing, obesity and the related WAT (white adipose tissue) expansion are primary factors for high risk of the progression of CVDs. Curcumin had a significant impact on glucose management, WAT expansion, and beiging. Curcumin reduces excess WAT and it also reduces the number of white adipocytes, which is an intriguing approach to lowering the risk of CVD development.[47]

CVDs affect the blood vessels and the heart in a variety of manners. Atherosclerosis is a complex disease characterized by chronic inflammation (Figure 6.9). A fibrous plaque develops on the artery wall as the disease

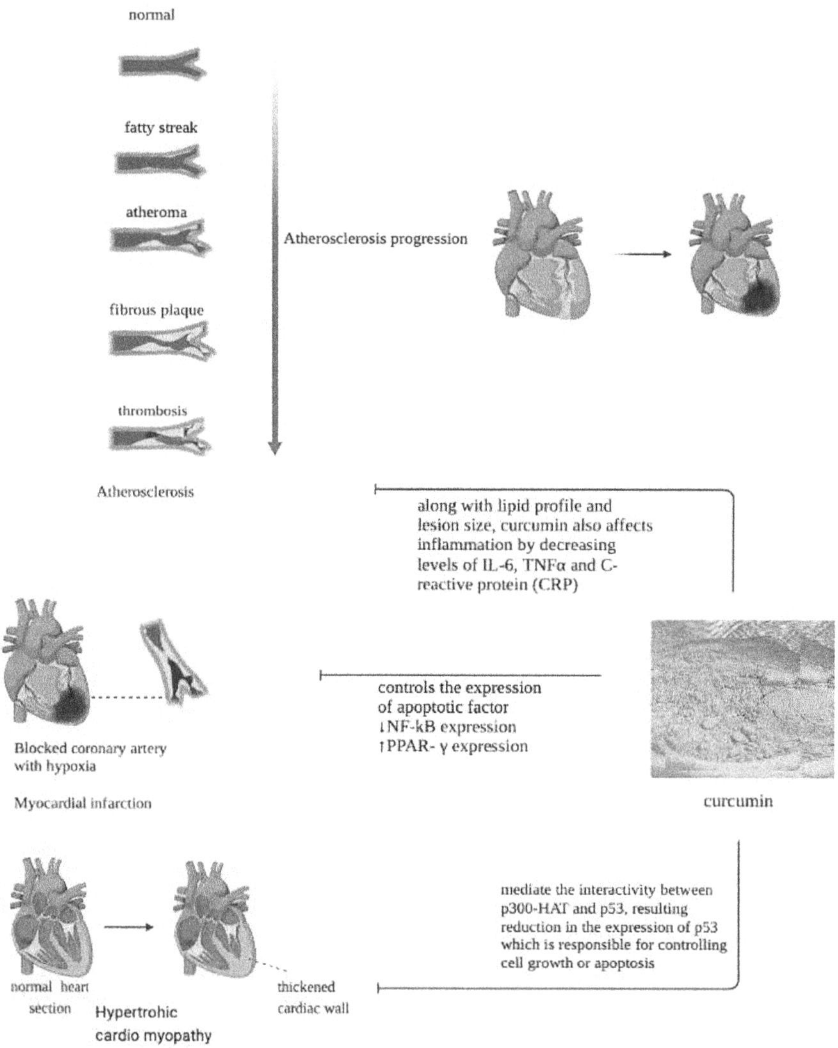

Figure 6.9 Cardiovascular disease progression and curcumin effects.

worsens, gradually reducing the vessel lumen. In the beginning of atherosclerosis, along with cholesterol deposition in the vessel walls and persistent inflammatory responses, endothelial dysfunction, which is generated in areas of arteries predisposed to plaque development and is marked by disturbed blood flow and a rise in reactive oxygen species (ROS), is one of the key events. In addition to damaging the vascular wall, oxidative stress in the vasculature also results in oxidation of lipid that leads to the formation and growth of atherosclerosis. In atherosclerosis, it appears that lipids and oxidative stress interact. Elevated formation of mitochondrial ROS in mice with low-density lipoprotein (LDL) receptors (Ldl-r), a model of atherosclerosis, demonstrates this. Moreover, the number of atherosclerotic lesions is considerably higher in apolipoprotein-e (apoe)-deficient mice because they have lower amounts or no amount of manganese superoxide dismutase (SOD2), the mitochondrial form of SOD, and more oxidative stress. In addition to affecting the peripheral vasculature, atherosclerosis is one of the principle cause of myocardial infarction (MI) and stroke. When a coronary artery is blocked, ischemia and reduced blood flow occur, and that causes MI.[47] Cardiac hypertrophy, also known as hypertrophic cardiomyopathy (HCM), is associated with abnormal heart muscle enlargement or thickening resulting from increased cardiomyocyte size.[48] Nuclear acetylation, which is regulated by p300, is a crucial process in the hypertrophy of cardiac cells. GATA-4 is one of the transcriptional factors which is responsible for hypertrophy and it has been noted that the cardiac myocytes begin to respond hypertrophically as the DNA binding activity of GATA-4 increases.[49]

Curcumin's anti-inflammatory and antioxidant capabilities, together with its ability to prevent vascular smooth muscle cell (VSMC) activation and proliferation, are believed to have an impact in its mechanistic function against atherosclerosis.[50] Curcumin's potential atheroprotective properties has been observed in a variety of atherosclerotic animal model systems, including Apo-e- and Ldl-r-deficient mice or a mixture of the two, along with using an atherogenic western diet. Turmeric extract was found to reduce the development of atherosclerosis in rabbits fed an atherogenic diet in early studies.[47] Curcumin has been shown to prevent increased matrix metalloproteinase activity, collagen deposition, increased collagen expression, and persistence of high volumes of myofibroblasts after MI. As a result, it has a positive effect on re-modelling and effectively reduces scar size. After MI, curcumin decreases the loss of cardiac function.[47] One of the most important mechanisms by which curcumin defends against cardiac hypertrophy is the inhibition of p300-specific histone acetyltransferase (HAT) activity (Table 6.8). The formation and differentiation of cardiomyocytes, which takes place before the advancement of cardiac hypertrophy and heart failure, are dependent on the transcriptional coactivator p300-HAT. This transcriptional coactivator involves GATA-binding factor 4 (GATA4).

6.4.3 Effects of Curcumin on Cancer

One of the most prevalent reasons for mortality in developed nations is cancer. The death rate has decreased recently due to early diagnosis and an expansion of therapy options.[52] Cancer illness and death are still rising globally despite ongoing advancements in the biomedical sciences. The occurrence of multiple malignancies is influenced by a variety of factors, such as diet, obesity, alcoholism, and overwork.[53] In this situation,

Table 6.8 Potential benefits of curcumin in various diseases.

Disease	Cause	Effect of curcumin	Model organism	Study/dose	Protective mechanism of curcumin	Ref.
Atherosclerosis	Chronic inflammatory disease that occurs due to injury of arterial wall sustained by risk factors such as diabetes, hypertension and other risk factors of cardiovascular disease, which results in the formation of macrophage and VSMC-derived foam cells, endothelial cell failure, immune cell activation, platelet activation and thrombus development	Regulates foam cell production, macrophage division, cholesterol outflow, and pro-inflammatory reactions	Apo-E and LDL-R-deficient mice	Vertebrate; 0.3 mg kg^{-1}, daily for four months	↓Cholesterol levels in the plasma ↓Triglycerides ↓LDL-cholesterol ↑HDL-cholesterol	47
			Atherosclerotic rabbits	Vertebrate	↓LDL peroxidation susceptibility ↓Plasma cholesterol ↓LDL-cholesterol ↓LDL-triglycerides ↓LDL- phospholipids	47 and 50
			Atherosclerotic rats	Vertebrate	↓Serum triglyceride levels ↓Total cholesterol ↓LDL-cholesterol ↑HDL-cholesterol	51
Myocardial infarction	Sudden decrease in the circulation of oxygenated blood induced by blood vessel constriction	Significant protection of cardiac function and reduction in the size of a cardiac infarction	Isoproterenol-treated rats	Vertebrate	Reduced heart morphological alterations	47
			MI rats	Vertebrate	JAK2/STAT3 signaling pathway activation ↓Oxidative stress ↓Cardiac muscle cell apoptosis	48

			Mice	Vertebrate	↓Expression of markers of M1 macrophages ↑Expression of markers in the myocardium indicating the M2 subtype	48
Cardiac hypertrophy	Alterations in the genes that thicken the cardiac muscle	Suppresses the activity of p-300 specific histone acetyl transferase (HAT)	H9c2 cardiac cells	*In vitro*	Reduces GATA-4's ability to bind DNA in the nuclear localization to avoid cardiac hypertrophy	48 and 49
			Rat cardio-myocyte	*In vivo*	Impeded GATA4's ability to acetylate and bind to DNA in response to hypertrophy	48 and 49

curcumin stands out as a prospective candidate for usage independently or in tandem with other drugs as an efficient anticancer medicine. According to a number of studies, curcumin has potent anticancer properties that inhibit cancer cell growth, migration, and metastasis, induce cancer cell apoptosis, obstruct tumour angiogenesis, and reverse multidrug resistance through the control of different signaling pathways. Many cancer growths have been shown to be significantly suppressed by curcumin *in vitro* and *in vivo* at all phases, including initiation, progression, and propagation.[54] *Via* PI3K signalling pathways, curcumin markedly reduces the growth and induction of apoptosis, as well as the invasion and migration of malignant cells.[55] The activator protein-1, which is linked to anti-apoptotic, mitogenic, and pro-angiogenic genes, is downregulated by curcumin. Curcumin also affects cyclin D1, a transcriptional co-regulator that controls cell cycle progression and has anticancer properties. Elevated concentrations of cyclin D1 have been found to be linked to the emergence and spread of cancer. Curcumin does, in fact, block cyclin D1 through suppressing NF-κB.[54]

Moreover, curcumin has the capacity to disrupt the cell signaling pathway of the EGFR family of receptor tyrosine kinases, which has been linked to cancer cell growth, adhesion, migration, and differentiation (Figure 6.10).[52] Additionally, curcumin inhibits NF-κB activity by blocking the activation of NF-κB kinase and preventing the phosphorylation of the p65 subunit.[54] Curcumin causes autophagy, suppresses the ubiquitin-proteasome pathway, and reduces Akt protein levels in breast cancer cells in time and dose dependent manners.[52] By suppressing DNMT (DNA methyltransferase) and HDAC (histone deacetylase) activity, curcumin inhibits the growth of cancer cells and triggers apoptosis, altering DNA methylation and histone

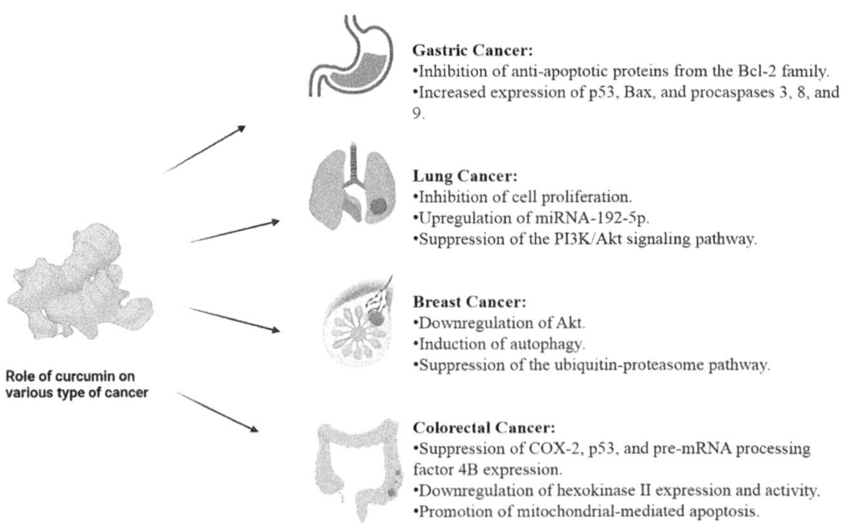

Figure 6.10 Role of curcumin on various types of cancer.

modification to have an anticancer impact.[53] According to *in vitro* studies curcumin significantly enhances the transcription of microRNA-34a in gastric cancer SGC-7901 cells, and inhibits the growth, migration, and invasion of SGC-7901 cells.[53]

6.4.4 Other Diseases

Curcumin is beneficial to human wellness since it has anti-inflammatory and antioxidant properties as well as immune-regulating properties.[56] Because of its multitarget plus multichannel qualities, it shows a variety of pharmacological activities. Along with cancer, neurodegenerative, and cardiovascular diseases, it is documented that curcumin affects blood sugar as well as platelet aggregation. Curcumin's ability to protect against oxidative stress has previously been documented in both *in vitro* and *in vivo*. For instance, its ability to scavenge ROS was primarily responsible for its biomembrane-protective function against peroxidative damage.[57] Numerous preclinical (both *in vitro* and *in vivo*) and clinical studies have demonstrated that curcumin (CUR) is an effective agent for the treatment and prevention of inflammatory diseases. By removing ROS and RNS as well as controlling the function of vital antioxidant enzyme systems like glutathione reductase, catalase, and superoxide dismutase, CUR can indeed increase its antioxidant activity.[58] The imbalance between formation and destruction of ROS can lead to oxidative stress, and ultimately the synthesis of free radicals and cell death or cellular damage.[57] Curcumin (200 mg per kg per day over 3 weeks) also had a protective impact on CCl_4-induced NASH (non-alcoholic steatohepatitis) in this situation.[57] According to an animal study, curcumin and its synthetic derivatives work by activating the peroxisome proliferator-activated receptor (PPAR-γ), which has a similar mechanism to the anti-diabetic medication thiazolidinedione.[59] As a result, curcumin might be useful in controlling cholesterol and glucose levels, which are important elements in the management of type 2 diabetes mellitus (T2DM).[59] Curcumin showed beneficial effects in hyperlipidemic type 2 diabetes mellitus patients who received turmeric daily for 8 weeks. Another clinical investigation revealed that T2DM patients with coronary heart disease who consumed curcumin had higher levels of PPAR-γ in their peripheral blood mononuclear cells.[60] Research on curcumin has revealed that it not only reduces the complications directly brought on by diabetes but also minimizes the indirect consequences brought on by atherosclerosis, stroke, and heart disease.[58] Several studies have suggested that curcumin possesses anti-viral properties. For example, research on all four strains of the dengue virus indicated that curcumin reduced plaque formation in

LLC-MK2 cells without causing cell destruction.[61] Influenza A virus (IAV) is also inhibited by curcumin, which probably has an effect at several stages of the virus lifecycle.[61] Curcumin's ability to inhibit viral haemagglutinin activity may be the cause of the lower infectivity observed when IAV is incubated with it. These result suggests that curcumin can enhance patients' quality of life, metabolic problems, and psychological state.[60] In animal tests, daily administrations of 1% curcumin gel lowered the cutaneous psoriasis-like inflammation by blocking of potassium channels produced in T cells and the suppression of interleukin-17A, -17F, -22, and some other pro-inflammatory cytokines in mouse ear samples.[62] In psoriatic mice, oral administration of curcumin (40 $mg\,kg^{-1}$ for 20 days) significantly decreased the concentration of IL-2, IL-12, IL-22, IL-23, IFN-γ, and TNF-α in serum, eventually lowering the inflammation associated with psoriasis and the hyperproliferation of keratinocytes.[62]

6.5 Food Formulations with Curcumin

Food supplements are products that contain one or more dietary ingredients such as vitamins, minerals, herbs, amino acids, or other substances. They are meant to supplement the daily diet and provide additional nutrients or health benefits. Food supplements can come in various forms such as capsules, tablets, powders, liquids, or gummies. One such food supplement being utilized traditionally in our routine life cycle is turmeric (Figure 6.11). Curcumin is the main and active constituent of turmeric, a popular and widely used food ingredient in Indian and Middle Eastern cuisine due to its bright yellow color, mild flavor, and potential health benefits,. Curcumin can be incorporated into various food formulations.

6.5.1 Curcumin in Spice Blends

Curcumin is a common ingredient in spice blends such as curry powder.[63] It can also be used to add color and flavor to other spice blends, such as garam masala or ras el hanout. Here are some popular spice blends that contain curcumin:

Curry powder: Curry powder is a blend of spices that typically includes turmeric, cumin, coriander, ginger, and other spices. The combination of spices in curry powder can vary depending on the region and the dish being prepared. Curry powder is often used in Indian cuisine and can be used to season meats, vegetables, and soups.

Garam masala: Garam masala is a spice blend commonly used in Indian cuisine that includes a variety of spices, including turmeric, cumin, coriander, cinnamon, cloves, and cardamom. Garam masala is typically used to flavor meat and vegetable dishes.

Ras el hanout: Ras el hanout is a Moroccan spice blend that includes a variety of spices, including turmeric, cumin, coriander, cinnamon, and

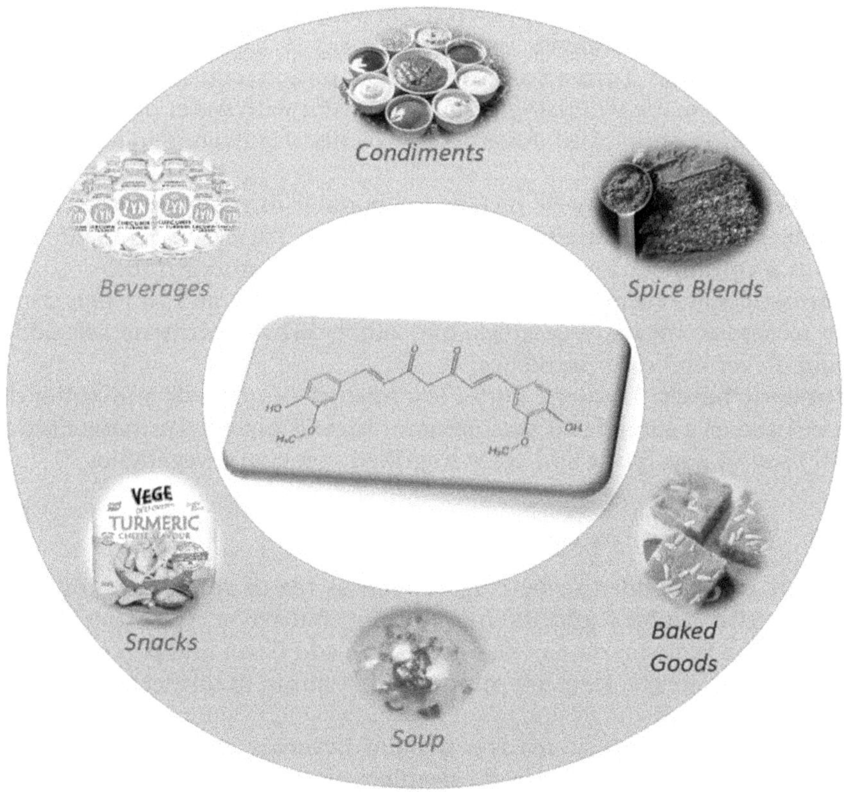

Figure 6.11 A generalized overview of different food formulations made from turmeric.

ginger. Ras el hanout is often used to season meat, fish, and vegetable dishes.

Berbere: Berbere is an Ethiopian spice blend that typically includes turmeric, cumin, coriander, cinnamon, and paprika, as well as other spices such as fenugreek, cloves, and cardamom. Berbere is often used to flavor stews and soups.

In addition to adding flavor and color to dishes, curcumin in these spice blends may provide some health benefits due to its antioxidant and anti-inflammatory properties. However, the amount of curcumin in spice blends is relatively small compared to the amount found in concentrated curcumin supplements, so it may not provide the same health benefits as taking curcumin supplements directly.

6.5.2 Curcumin in Condiments

Curcumin can be added to various condiments, such as mustard, mayonnaise, and ketchup, to give them a bright yellow color and a hint of flavor. Turmeric is often used as a condiment or seasoning to add flavor, color, and health benefits to various dishes. Here are some ways curcumin in turmeric can be used as a condiment:

Turmeric paste: Turmeric paste is a common condiment in Indian cuisine that is made by mixing turmeric powder with water or oil. It can be added to soups, stews, and other dishes to enhance flavor and add a bright yellow color.

Turmeric pickle: Turmeric pickles are popular in Indian cuisine and are made by soaking turmeric in vinegar and spices. Turmeric pickles are often used as a condiment for sandwiches, burgers, and other dishes.

Turmeric salt: Turmeric salt is a blend of turmeric and salt that can be used to season meats, vegetables, and other dishes. Turmeric salt adds a unique flavor and color to dishes.

Turmeric butter: Turmeric butter is a mixture of turmeric and butter that can be used as a spread or a seasoning for various dishes. Turmeric butter is often used to add flavor and color to grilled meats and vegetables.

6.5.3 Curcumin in Beverages

Turmeric can be added to beverages, such as tea or smoothies, for its potential health benefits and its unique color and flavor. Curcumin, a compound found in turmeric, can also be used to add flavor and potential health benefits to beverages. Here are some ways curcumin in turmeric can be used in beverages:

Turmeric tea: Turmeric tea is a popular beverage in Indian and Middle Eastern cultures that is made by steeping turmeric powder or grated turmeric root in hot water. Turmeric tea can be enjoyed plain or with added honey, lemon, or ginger for flavor.

Golden milk: Golden milk is a popular beverage in Ayurvedic medicine that is made by mixing turmeric powder with milk, honey, and other spices like ginger, cinnamon, and cardamom. Golden milk can be enjoyed hot or cold and is often used as a natural remedy for a variety of health issues.

Smoothies: Turmeric powder can be added to smoothies along with other ingredients like fruits, yogurt, and honey for a tasty and healthy beverage. Smoothies are a great way to incorporate turmeric into your diet, and the addition of other healthy ingredients can enhance the potential health benefits.

Juices: Turmeric powder or grated turmeric root can be added to fresh juices for a boost of flavor and potential health benefits. Turmeric pairs well with citrus fruits like oranges and lemons, as well as with apples, carrots, and ginger.

Curcumin in turmeric has antioxidant and anti-inflammatory properties that may provide health benefits when consumed in beverages.

6.5.4 Curcumin in Snacks

Turmeric can be added to snack foods, such as popcorn or roasted nuts, to give them a unique flavor and color. Curcumin, a compound found in turmeric, can also be used to add flavor and potential health benefits to snacks. Here are some ways curcumin in turmeric can be used in snacks:

Turmeric popcorn: Turmeric powder can be added to popcorn along with other seasonings like salt, pepper, and nutritional yeast for a healthy and flavorful snack. Turmeric popcorn is a great alternative to buttered or heavily salted popcorn.

Turmeric roasted nuts: Nuts like almonds, cashews, and pecans can be roasted with turmeric powder and other spices like cumin and paprika for a savory and healthy snack. Turmeric roasted nuts are a great source of protein and healthy fats.

Turmeric hummus: Hummus is a popular dip made from chickpeas, tahini, and olive oil, and it can be flavored with turmeric powder and other spices for a tasty and healthy snack. Turmeric hummus pairs well with raw vegetables like carrots, celery, and cucumber.

Turmeric roasted vegetables: Vegetables like sweet potatoes, carrots, and cauliflower can be roasted with turmeric powder, garlic, and olive oil for a healthy and flavorful snack. Turmeric roasted vegetables are a great source of vitamins, minerals, and fiber.

Curcumin in turmeric has antioxidant and anti-inflammatory properties that may provide health benefits when consumed in snacks.

6.5.5 Curcumin in Baked Goods

Turmeric can be added to baked goods such as bread or muffins to give them a golden color and a mild flavor. Curcumin, a compound found in turmeric, can also be used to add flavor and potential health benefits to baked goods. Here are some ways curcumin in turmeric can be used in baked goods:

Turmeric bread: Turmeric powder can be added to bread dough along with other ingredients like honey, olive oil, and yeast for a flavorful and healthy loaf of bread. Turmeric bread is a great alternative to plain white bread.

Turmeric muffins: Turmeric powder can be added to muffin batter along with other ingredients like almond flour, honey, and blueberries for a healthy and tasty snack. Turmeric muffins are a great way to incorporate turmeric into your diet.

Turmeric cookies: Turmeric powder can be added to cookie dough along with other ingredients like coconut oil, maple syrup, and oats for a healthy and delicious treat. Turmeric cookies are a great alternative to traditional sugar cookies.

Turmeric cake: Turmeric powder can be added to cake batter along with other ingredients like coconut milk, almond flour, and honey for a healthy and flavorful cake. Turmeric cake is a great option for special occasions or as a healthy dessert.

6.6 Conclusion

Recent studies have underscored curcumin's potential as an effective therapeutic agent for numerous diseases. Studies have shown that curcumin's anti-inflammatory and antioxidant properties make it a promising candidate for the treatment of Alzheimer's disease, Parkinson's disease, diabetes, and cardiovascular diseases. Furthermore, curcumin has been shown to have anticancer properties and may be useful in cancer prevention and treatment. Despite these promising findings, there are still several challenges in using curcumin as a therapeutic agent. One of the main challenges is its poor bioavailability, which limits its effectiveness in clinical applications. Researchers are exploring various strategies to enhance curcumin's bioavailability, such as using nanoparticle formulations or combining it with other compounds. In conclusion, recent advancements in curcumin research have shown its potential as a therapeutic agent in various diseases. Further studies are needed to fully understand its mechanisms of action and to develop effective strategies for its clinical application. With continued research and development, curcumin may offer new opportunities for the prevention and treatment of various diseases.

References

1. B. B. Aggarwal, A. Kumar and A. C. Bharti, Anticancer potential of curcumin: preclinical and clinical studies, *Anticancer Res.*, 2003, **23**(1/A), 363–398.
2. C. A. Damalas, Potential uses of turmeric ('Curcuma longa') products as alternative means of pest management in crop production, *Plant Omics*, 2011, **4**(3), 136–141.
3. P. Ravindran, K. N. Babu and K. Sivaraman, *Turmeric: the genus Curcuma*, CRC Press, 2007.
4. R. M. Borik, N. M. Fawzy, S. M. Abu-Bakr and M. S. Aly, Design, synthesis, anticancer evaluation and docking studies of novel heterocyclic derivatives obtained *via* reactions involving curcumin, *Molecules*, 2018, **23**(6), 1398.
5. T. S. A. T. Kamazeri, O. Abd Samah, M. Taher, D. Susanti and H. Qaralleh, Antimicrobial activity and essential oils of Curcuma aeruginosa, Curcuma mangga, and Zingiber cassumunar from Malaysia, *Asian Pac. J. Trop. Med.*, 2012, **5**(3), 202–209.
6. R. C. Padalia, R. S. Verma, V. Sundaresan, A. Chauhan, C. S. Chanotiya and A. Yadav, Volatile terpenoid compositions of leaf and rhizome of Curcuma amada Roxb. from Northern India, *J. Essent. Oil Res.*, 2013, **25**(1), 17–22.
7. S. Srivastava, C. Nitin, S. Srivastava, M. Dan, A. Rawat and P. Pushpangadan, Pharmacognostic Evaluation of Curcuma aeurigenosa Roxb, *Nat. Prod. Sci.*, 2006, **12**(3), 162–165.
8. P. Thongkhwan, T. Chaibunga, H. Kwanboonjan and O. Theanphong, Essential oil constituents of the fresh root and rhizome of Curcuma

angustifolia Roxb. from Thailand, *Bull. Heal. Sci. Technol.*, 2017, **15**, 52–53.
9. H. Xiang, L. Zhang, Z. Yang, F. Chen, X. Zheng and X. Liu, Chemical compositions, antioxidative, antimicrobial, anti-inflammatory and antitumor activities of Curcuma aromatica Salisb. essential oils, *Ind. Crops Prod.*, 2017, **108**, 6–16.
10. S. Y. Tsai, S.-J. Huang, C.-C. Chyau, C.-H. Tsai, C. C. Weng and J.-L. Mau, Composition and antioxidant properties of essential oils from Curcuma rhizome, *Asian J. Arts Sci.*, 2011, **2**(1), 57–66.
11. I. C. Herath, T. Wijayasiriwardene and G. Premakumara, Comparative GC–MS analysis of all Curcuma species grown in Sri Lanka by multivariate test, *Ruhuna J. Sci.*, 2017, **8**(2), 103–111.
12. S. M. Al-Reza, A. Rahman, M. Sattar, M. O. Rahman and H. M. Fida, Essential oil composition and antioxidant activities of Curcuma aromatica Salisb, *Food Chem. Toxicol.*, 2010, **48**(6), 1757–1760.
13. N. X. Dũng, P. X. Truong, P. T. Ky and P. A. Leclercq, Volatile constituents of the leaf, stem, rhizome, roof and flower oils of Curcuma harmandii Gagnep. from Vietnam, *J. Essent. Oil Res.*, 1997, **9**(6), 677–681.
14. S. N. Malek, C. K. Seng, Z. Zakaria, N. A. Ali, H. Ibrahim and M. N. Jalil, The essential oil of Curcuma inodora aff. Blatter from Malaysia, *J. Essent. Oil Res.*, 2006, **18**(3), 281–283.
15. S. Naz, S. Ilyas, Z. Parveen and S. Javed, Chemical analysis of essential oils from turmeric (Curcuma longa) rhizome through GC–MS, *Asian J. Chem.*, 2010, **22**(4), 3153.
16. F. D. Ferreira, C. Kemmelmeier, C. C. Arrotéia, C. L. da Costa, C. A. Mallmann and V. Janeiro, *et al.*, Inhibitory effect of the essential oil of Curcuma longa L. and curcumin on aflatoxin production by Aspergillus flavus Link, *Food Chem.*, 2013, **136**(2), 789–793.
17. I. A. Oyemitan, C. A. Elusiyan, A. O. Onifade, M. A. Akanmu, A. O. Oyedeji and A. G. McDonald, Neuropharmacological profile and chemical analysis of fresh rhizome essential oil of Curcuma longa (turmeric) cultivated in Southwest Nigeria, *Toxicol. Rep.*, 2017, **4**, 391–398.
18. S. Yasni, K. Imaizumi, K. Sin, M. Sugano and G. Nonaka, Identification of an active principle in essential oils and hexane-soluble fractions of Curcuma xanthorrhiza Roxb. showing triglyceride-lowering action in rats, *Food Chem. Toxicol.*, 1994, **32**(3), 273–278.
19. J.-L. Mau, E. Y. Lai, N.-P. Wang, C.-C. Chen, C.-H. Chang and C.-C. Chyau, Composition and antioxidant activity of the essential oil from Curcuma zedoaria, *Food Chem.*, 2003, **82**(4), 583–591.
20. W.-H. Lee, C.-Y. Loo, M. Bebawy, F. Luk, R. S. Mason and R. Rohanizadeh, Curcumin and its derivatives: their application in neuropharmacology and neuroscience in the 21st century, *Curr. Neuropharmacol.*, 2013, **11**(4), 338–378.
21. M. K. Oglah, Y. F. Mustafa, M. K. Bashir, M. H. Jasim and Y. F. Mustafa, Curcumin and its derivatives: A review of their biological activities, *Syst. Rev. Pharm.*, 2020, **11**(3), 472–481.

22. R. K. Verma, P. Kumari, R. K. Maurya, V. Kumar, R. Verma and R. K. Singh, Medicinal Properties of Turmeric (Curcuma Longa L.): A Review, *Int. J. Chem. Stud.*, 2018, **6**(4), 1354–1357.
23. S. Prasad and B. B. Aggarwal, Turmeric, the golden spice, *Herbal Medicine: Biomolecular and Clinical Aspects*, 2nd edn, 2011.
24. M. A. Tomeh, R. Hadianamrei and X. Zhao, A review of curcumin and its derivatives as anticancer agents, *Int. J. Mol. Sci.*, 2019, **20**(5), 1033.
25. M. T. El-Saadony, T. Yang, S. A. Korma, M. Sitohy, T. A. Abd El-Mageed and S. Selim, *et al.*, Impacts of turmeric and its principal bioactive curcumin on human health: Pharmaceutical, medicinal, and food applications: A comprehensive review, *Front. Nutr.*, 2023, **9**, 1040259.
26. F. Delgado-Vargas and O. Paredes-Lopez, *Natural colorants for food and nutraceutical uses*, CRC Press, 2002.
27. S. Fuloria, J. Mehta, A. Chandel, M. Sekar, N. N. I. M. Rani and M. Y. Begum, *et al.*, A comprehensive review on the therapeutic potential of Curcuma longa Linn. in relation to its major active constituent curcumin, *Front. Pharmacol.*, 2022, **13**, 820806.
28. E. J. Williams, S. Haque, C. Banks, P. Johnson, P. Sarsfield and N. Sheron, Distribution of the interleukin-8 receptors, CXCR1 and CXCR2, in inflamed gut tissue, *J. Pathol.*, 2000, **192**(4), 533–539.
29. J.-K. Son, Isolation and structure determination of a new tetralone glucoside from the roots of Juglans mandshurica, *Arch. Pharmacal Res.*, 1995, **18**, 203–205.
30. M. Kuroyanagi and S. Natori, Some observation on curcuminoids from Zingiberaceae plants, *Yakugaku Zasshi*, 1970, **90**(11), 1467–1470.
31. E. E. Nebrisi, Neuroprotective activities of curcumin in Parkinson's disease: a review of the literature, *Int. J. Mol. Sci.*, 2021, **22**(20), 11248.
32. L. Z. Racz, C. P. Racz, L.-C. Pop, G. Tomoaia, A. Mocanu and I. Barbu, *et al.*, Strategies for improving bioavailability, bioactivity, and physical-chemical behavior of curcumin, *Molecules*, 2022, **27**(20), 6854.
33. B. C. Walker and S. Mittal, Antitumor activity of curcumin in glioblastoma, *Int. J. Mol. Sci.*, 2020, **21**(24), 9435.
34. A. Zielińska, H. Alves, V. Marques, A. Durazzo, M. Lucarini and T. F. Alves, *et al.*, Properties, extraction methods, and delivery systems for curcumin as a natural source of beneficial health effects, *Medicina*, 2020, **56**(7), 336.
35. A. Mohammadi, A. Hosseinzadeh Colagar, A. Khorshidian and S. M. Amini, The functional roles of curcumin on astrocytes in neurodegenerative diseases, *Neuroimmunomodulation*, 2022, **29**(1), 4–14.
36. P. H. Reddy, M. Manczak, X. Yin, M. C. Grady, A. Mitchell and S. Tonk, *et al.*, Protective effects of Indian spice curcumin against amyloid-β in Alzheimer's disease, *J. Alzheimer's Dis.*, 2018, **61**(3), 843–866.
37. T. Benameur, G. Giacomucci, M. A. Panaro, M. Ruggiero, T. Trotta and V. Monda, *et al.*, New promising therapeutic avenues of curcumin in brain diseases, *Molecules*, 2021, **27**(1), 236.

38. F. Khanmohammadi, B. M. Razavizadeh and E. Fooladi, Application of electrochemical sensor modified by SBA-15/Fe_3O_4/polyaniline nanocomposite for determination of tyrosine in milk samples, *Sens. Bio-Sens. Res.*, 2023, **42**, 100602.
39. S. D. Voulgaropoulou, T. Van Amelsvoort, J. Prickaerts and C. Vingerhoets, The effect of curcumin on cognition in Alzheimer's disease and healthy aging: A systematic review of pre-clinical and clinical studies, *Brain Res.*, 2019, **1725**, 146476.
40. A. Margiotta, Role of SNAREs in neurodegenerative diseases, *Cells*, 2021, **10**(5), 991.
41. A. D. Gitler, P. Dhillon and J. Shorter, Neurodegenerative disease: models, mechanisms, and a new hope, *Dis. Models & Mech.*, 2017, **10**(5), 499–502.
42. A. R. Bland, J. C. Ashton, M. A. Kamal and A. Sahebkar, The current evidence for the therapeutic role of curcumin in Alzheimer's disease, *CNS Neurol. Disord.: Drug Targets*, 2023, **22**(3), 318–320.
43. Z. Chen, C. Wu, Z. Zhang, W. Wu, X. Wang and Z. Yu, Synthesis, functionalization, and nanomedical applications of functional magnetic nanoparticles, *Chin. Chem. Lett.*, 2018, **29**(11), 1601–1608.
44. M. Chen, Z.-Y. Du, X. Zheng, D.-L. Li, R.-P. Zhou and K. Zhang, Use of curcumin in diagnosis, prevention, and treatment of Alzheimer's disease, *Neural Regener. Res.*, 2018, **13**(4), 742–752.
45. T. T. Nguyen, M. D. Vuu, M. A. Huynh, M. Yamaguchi, L. T. Tran and T. P. T. Dang, Curcumin Effectively Rescued Parkinson's Disease-Like Phenotypes in a Novel Drosophila melanogaster Model with dUCH Knockdown, *Oxid. Med. Cell. Longevity*, 2018, **2018**(1), 2038267.
46. D. Ashtary-Larky, M. Rezaei Kelishadi, R. Bagheri, S. P. Moosavian, A. Wong and S. H. Davoodi, *et al.*, The Effects of Nano-Curcumin Supplementation on Risk Factors for Cardiovascular Disease: A GRADE-Assessed Systematic Review and Meta-Analysis of Clinical Trials, *Antioxidants (Basel)*, 2021, **10**(7), 1015.
47. F. F. Cox, A. Misiou, A. Vierkant, N. Ale-Agha, M. Grandoch and J. Haendeler, *et al.*, Protective effects of curcumin in cardiovascular diseases—Impact on oxidative stress and mitochondria, *Cells*, 2022, **11**(3), 342.
48. H. A. Zhang and D. D. Kitts, Turmeric and its bioactive constituents trigger cell signaling mechanisms that protect against diabetes and cardiovascular diseases, *Mol. Cell. Biochem.*, 2021, **476**(10), 3785–3814.
49. A. M. Pourbagher-Shahri, T. Farkhondeh, M. Ashrafizadeh, M. Talebi and S. Samargahndian, Curcumin and cardiovascular diseases: Focus on cellular targets and cascades, *Biomed. Pharmacother.*, 2021, **136**, 111214.
50. L. Singh, S. Sharma, S. Xu, D. Tewari and J. Fang, Curcumin as a natural remedy for atherosclerosis: a pharmacological review, *Molecules*, 2021, **26**(13), 4036.

51. B. Salehi, D. Calina, A. O. Docea, N. Koirala, S. Aryal and D. Lombardo, *et al.*, Curcumin's Nanomedicine Formulations for Therapeutic Application in Neurological Diseases, *J. Clin. Med.*, 2020, **9**, 430.
52. A. Giordano and G. Tommonaro, Curcumin and cancer, *Nutrients*, 2019, **11**(10), 2376.
53. T. Ming, Q. Tao, S. Tang, H. Zhao, H. Yang and M. Liu, *et al.*, Curcumin: An epigenetic regulator and its application in cancer, *Biomed. Pharmacother.*, 2022, **156**, 113956.
54. S. Hafez Ghoran, A. Calcaterra, M. Abbasi, F. Taktaz, K. Nieselt and E. Babaei, Curcumin-based nanoformulations: A promising adjuvant towards cancer treatment, *Molecules*, 2022, **27**(16), 5236.
55. H. Mirzaei, H. Bagheri, F. Ghasemi, J. M. Khoi, M. H. Pourhanifeh and Y. V. Heyden, *et al.*, Anti-cancer activity of curcumin on multiple myeloma, *Anti-Cancer Agents Med. Chem.*, 2021, **21**(5), 575–586.
56. S. Liu, J. Liu, L. He, L. Liu, B. Cheng and F. Zhou, *et al.*, A comprehensive review on the benefits and problems of curcumin with respect to human health, *Molecules*, 2022, **27**(14), 4400.
57. M. H. Farzaei, M. Zobeiri, F. Parvizi, F. F. El-Senduny, I. Marmouzi and E. Coy-Barrera, *et al.*, Curcumin in liver diseases: a systematic review of the cellular mechanisms of oxidative stress and clinical perspective, *Nutrients*, 2018, **10**(7), 855.
58. C. Quispe, N. Cruz-Martins, M. L. Manca, M. Manconi, O. Sytar and N. Hudz, *et al.*, Nano-Derived Therapeutic Formulations with Curcumin in Inflammation-Related Diseases, *Oxid. Med. Cell. Longevity*, 2021, **2021**(1), 3149223.
59. F. Pivari, A. Mingione, C. Brasacchio and L. Soldati, Curcumin and type 2 diabetes mellitus: prevention and treatment, *Nutrients*, 2019, **11**(8), 1837.
60. L. T. Marton, L. M. Pescinini-e-Salzedas, M. E. C. Camargo, S. M. Barbalho, J. F. D. S. Haber and R. V. Sinatora, *et al.*, The effects of curcumin on diabetes mellitus: a systematic review, *Front. Endocrinol.*, 2021, **12**, 669448.
61. M. R. Jennings and R. J. Parks, Curcumin as an antiviral agent, *Viruses*, 2020, **12**(11), 1242.
62. L. Vollono, M. Falconi, R. Gaziano, F. Iacovelli, E. Dika and C. Terracciano, *et al.*, Potential of curcumin in skin disorders, *Nutrients*, 2019, **11**(9), 2169.
63. R. F. Tayyem, D. D. Heath, W. K. Al-Delaimy and C. L. Rock, Curcumin content of turmeric and curry powders, *Nutr. Cancer*, 2006, **55**(2), 126–131.

CHAPTER 7

Resveratrol: Recent Advances and Applications in Functional and Nutraceutical Foods

F. S. SU AND Z. TACER-CABA*

Faculty of Engineering and Natural Sciences, Department of Molecular Biology and Genetics, Bahcesehir University, Besiktas, Istanbul, Turkey
*Email: zeynep.tacercaba@bau.edu.tr

7.1 Introduction

The scientific evidence elucidating the importance of diet in maintaining human health has led to an increasing demand on food products with a higher value than only satisfying hunger and providing nutrients. Functional foods are those foods that are created and consumed to provide benefits beyond the basic nutritional needs. These benefits include either the prevention of illnesses or boosting the wellbeing of the individuals. Although the functional foods can be categorized from traditional products to newer foods, the research on novel functional foods is quite dynamic and novel functional foods are currently expanding utilization of novel bioactive constituents like probiotics, polyunsaturated fatty acids, various phytoconstituents *etc.*, to make synergistic contributions to the health of individuals.[1]

The pleiotropic effects of bioactive ingredients in functional foods are significant, including antioxidant, anti-inflammatory, hypolipidemic, glycemic-regulating, cytoprotective, and neuroprotective functions.[2] A key worry about the fortification of functional substances in the free form is their low absorption, poor solubility, low temperature resistance, and lower gastrointestinal resistance during oral ingestion. Functional dietary elements

Food Chemistry, Function and Analysis No. 44
Functional Foods of the Future
Edited by Vijai Kumar Gupta, Minaxi Sharma, Smriti Gaur and Ramesh Chander Kuhad

Published by the Royal Society of Chemistry, www.rsc.org

such as polyphenols, vitamins, and probiotics must be absorbed by enteric epithelial cells, but they are poorly absorbed due to their limited water solubility.[3]

According to data from Fortune Business Insights, Statista (2022), the global worldwide functional food and beverage market size is forecast to grow from its current value of 281.14 billion U.S. dollars to over half a trillion U.S. dollars in 2028, with an annual growth rate of about 9.5 percent.

Nutraceuticals, on the other hand, can contain a blend of bioactive substances isolated or extracted from whole foods. As a result, they are considered to be on the cusp of nutrition and pharma. According to the European Nutraceutical Association (ENA), a nutraceutical is "a food or parts of food that provide medical or health benefits, including disease prevention and/or treatment." However, because there is no specific regulation for nutraceuticals in Europe, the United States, or Japan, the term 'nutraceutical' is incorrectly used to refer to a wide range of products, including isolated ingredients, nutrients, and food components, as well as food, processed food, functional foods, herbal products, and, most importantly, food supplements.[4]

Among different ingredients for functional foods, plant phytochemicals play a vital role in the formulations.

Plants synthesize a variety of compounds with different properties. Secondary metabolites of plants are the phytochemicals that are not directly necessary for the development of the organism but rather related with their interaction with the outer environment. Three main groups based on their structure and the metabolic pathways responsible for their biosynthesis can be used for the classification of secondary metabolites: phenolic compounds, nitrogen and/or sulfur-containing compounds, and terpenoids.[5] The synthesis of phenolic compounds is managed by pentose phosphate, shikimate, and phenylpropanoid pathways. Different phenolics can be classified into the subgroups of phenolic acids, flavonoids, stilbenes, tannins, lignans, coumarins, curcuminoids, and quinones based on their carbon skeleton.[6]

The secondary metabolites of plants have been used as drug components and for therapeutic purposes for years and recently they have also been utilized commonly as functional food ingredients. As recent research has revealed that diets rich in phenolics had preventive effects on a diverse range of diseases, such as cancers, diabetes, heart disease, neurodegenerative and cardiovascular diseases, and aging, these components have drawn attention also as significant components for the functional foods.

7.2 Resveratrol

Resveratrol (3,4′,5-trihydroxystilbene) is a stilbenoid that can be found as *cis*- and *trans*-resveratrol stereoisomers of the central ethylene moiety in its structure, and converts between them under specific conditions due to the presence of a C–C double bond. Exposure of the *trans* isomer to various environmental conditions, such as sunlight, artificial or natural UV radiation, and so on, are the primary sources of *cis* isomerization.[7,8]

The *trans* isomer is transformed into the *cis* isomer, which simulates the synthetic estrogen diethylstilbestrol in structure. As a result, resveratrol is considered a phytoestrogen. Its manufacturing process begins with a reaction catalyzed by the enzyme stilbene synthase between malonyl CoA and a coumaryl derivative. Stilbenes are secondary metabolites classified as phenolic compounds because they have two aromatic rings linked by an ethene bridge (diarylethene skeleton, C6–C2–C6 unit), and their structure is broadly characterized by substitution with hydroxyls (*e.g.* 3,5,4′-trihydroxystilbene, for resveratrol), methyl or methoxy groups, or bound with carbohydrates to provide glycosides.[9]

trans-Resveratrol, also known as 5-[(*E*)-2-(4-hydroxyphenyl)ethenyl]benzene-1,3-diol by the IUPAC, is a non-flavonoid phenolic molecule that belongs to the stilbenoid group and is more stable in nature than the *cis* isomer. Consequently, due to its hydroxyl groups, benzene ring, and C–C double bond, resveratrol can undergo a wide range of chemical reactions. *trans*-Resveratrol is generated and employed in plants to protect against pathogens, environmental effects, UV irradiation, fungal infections, and other threats.[8] It can exist in four different forms, *trans*- or *cis*-aglycone and *trans*- or *cis*-glycoside, as a result of enzymatic isomerisation and hydrolytic processes.[10]

Michio Takaoka, a Japanese researcher, discovered resveratrol in 1940. He termed a substance "resveratrol" after his investigations on plants because it can be isolated from the roots of *Veratrum grandiflorum*, the white hellebore plant, and contains the skeleton of a smaller molecule, resorcinol.

Resveratrol acts as a phytoalexin in plants, thus it is produced in response to injury or when the plant is under attack by pathogens; for example, it is primarily used as an antibiotic against the growth of parasites, the grape fungus *Botrytis cinerea,* and other infections.[11] It is important in the plant's defense against stress-inducing situations. When stressed, the plant normally initiates resveratrol production, increasing its content.[12]

Furthermore, it is efficient against environmental effects such as UV irradiation and fungal infections.[13] Exogenous stressors such as UV light and specific chemical signals from pathogenic fungi, for example, are among the main activators of stilbene synthase in red wine grapes, which is the terminal enzyme involved in the manufacture of resveratrol.[14] Stilbenes are responsible for the vivid blue fluorescence noticed on grape leaf surfaces or grape berries when exposed to long wavelength UV-light.[15]

Resveratrol exists in more than 70 different plant species such as grapes, peanuts, pistachios, bilberries, soy, blueberries, and Japanese knotweed (*Fallopia japonica*).[16] Grapes and giant knot weed (*Polygonnum cuspidatum*, a perennial shrub) have been reported as the richest sources. Giant knot weed can be utilized to treat inflammation, headaches, cancers, and amenorrhea.[9]

Particularly the skins of grapes are very rich sources of *cis*- and *trans*-resveratrol, in addition to resveratrol-3-*O*-β-D-glucopyranoside (resveratrol glucoside, piceid), piceatannol (3,4,3′,5′-tetrahydroxy-*trans*-stilbene) and resveratrol dimers (viniferins) and pterostilbene (a dimethylated derivative of resveratrol). Viniferins, most abundantly ε-viniferin, have also been discovered in grapevines due to infection or stress.[17] Moreover, foods such as wine and itadori tea also contain resveratrol. During the winemaking and particularly the fining practices, such as with bentonite, egg white, gelatin kieselsol or polyvinylpolypyrrolidone, *trans*-resveratrol levels have been reported to decrease significantly depending on the type and concentration of fining agent and grape variety. It is also found in Chinese and Japanese traditional medicine, in the form of extracts produced from *Polygonum cuspidatum*, which can be utilized to treat inflammation, headaches, malignancies, and amenorrhea.[9]

The high content of resveratrol in wine (concentrations of 0.1–14.3 $mg\,L^{-1}$) has been linked with the concept of the "French paradox". According to this phenomenon, although the diet of the southern French community contains high amounts of fat-rich foods, the incidence of heart diseases among these people is quite low. Since then, wine has been linked to the prevention of cardiovascular disease and healthy aging, as well as antioxidant, anti-inflammatory, antibacterial, anticancer, and antiangiogenic effects, as well as those of oxidative stress. The research on the activation potential of resveratrol dates back to 2003. It was demonstrated that resveratrol is one of the most powerful STACs (sirtuin-activating compounds) and for SIRT1 is an activator of sirtuin deacetylases and could lengthen the lifespan of *Saccharomyces cerevisiae*.[18] The influence on healthy aging is highly interrelated with the activation of SIRT1 proteins by resveratrol. SIRT1 proteins are one of the mammalian versions of the sirtuin family of proteins. SIRT1 is an NAD^+-dependent class III histone and protein deacetylase.[19]

The regulation of sirtuin expression and activity is thought to play a substantial part in the therapeutic activities of resveratrol.[20] The SIRT1-regulated pathway influences metabolism, stress resistance, cell survival, cellular senescence, inflammation-immune function, endothelial functions, and circadian rhythms.[19] With its high potential to activate SIRT1, resveratrol is used as a natural molecule in the form of nutraceuticals and functional food formulations to prevent a variety of disorders such as improper metabolic regulation, inflammation, and cell cycle abnormalities . Furthermore, clinical trials are being done around the world to determine its therapeutic efficacy for the treatment of various disorders.

Furthermore, supplements like NAD boosters can operate in tandem with resveratrol to improve mitochondrial function and protect against age-related disorders.[21] NAD^+ levels naturally diminish with age, resulting in a loss in cellular and organ function. NAD^+ boosters, on the other hand, have the potential to greatly improve mammalian health and longevity by raising NAD^+ levels. Elevated NAD^+ levels have been shown to improve cognitive and sensory function, liver gluconeogenesis, adipose tissue lipogenesis, pancreatic insulin production, and muscle insulin sensitivity. NAD is also important in immune function regulation, inflammation reduction, and protection against acute kidney injury. Furthermore, NAD activation of sirtuins has been hypothesized as a factor in both male and female fertility.[22] Resveratrol has also been shown to extend longevity and improve glucose homeostasis in mice by activating the NAD^+-dependent histone deacetylase SIRT1, which is part of the transcriptional coactivator PGC-1α.[23]

A search of "resveratrol" from Web of Science (http://apps.webofknowledge.com) conducted on 26th October 2023, returned a total 1517 results from 2023 and 2174 results from 2022.

7.3 Resveratrol and Anticancer

Clastogenic agents, for example radiation, cause in normal cells a variety of mutations, including deletions, point mutations, and conversions. Current cancer therapies such as chemotherapy, radiotherapy targeted agents, radiation, and surgery are linked to immunosuppression, an increased risk of carcinogenesis, the development of drug resistance, and cancer recurrence years after treatment.[9,24]

Resveratrol may be used against these effects, with numerous biological and medicinal activities such as antioxidant, anti-inflammatory, neuroprotective, immunomodulatory and anti-aggregatory activities[25] Resveratrol's cellular defensive qualities can be explained by its ability to directly neutralize reactive oxygen/reactive nitrogen species (ROS/RNS) or indirectly up-regulate the expression of cellular defense genes.[23] Exposure to resveratrol has been shown to be related to *in vitro* cytotoxic effects on a wide variety of human tumor cells, including myeloid and lymphoid cancer cells in addition to breast, skin, cervix, ovary, stomach, prostate, colon, liver, pancreatic, and thyroid carcinoma cells. Furthermore, resveratrol has been demonstrated to bind α- and β-estrogen receptors, and new research has highlighted resveratrol's ability to increase ROS production in cancer cells while serving as a cytotoxic agent.[26]

Resveratrol influences the various signal-transduction pathways that regulate cell growth and division, inflammation, death, metastasis, and angiogenesis. It was discovered that the OH radical induced the synthesis of cyclic guanosine monophosphate (cGMP). Resveratrol has some beneficial characteristics as well. The antioxidant activity of resveratrol can prevent apoptosis in endothelial cells.[27] It can improve endothelial function by modulating reduction/oxidation responses and attenuating inflammatory responses.[28] Furthermore, various mechanisms such as cellular communication and connections, cell cycle and apoptosis, and genetic activation of phase II enzymes are implicated.[13]

Antioxidant activity is the ability of antioxidants to eliminate free radicals in cells and food. We require modest amounts of reactive oxygen species (ROS) for a variety of biological processes, including gene expression regulation.[29] ROS comprise species such as the hydroxyl radical ($^{\bullet}OH$), whose reactivity is so great that it interacts violently at its site of creation, as well as the less reactive superoxide ($O_2^{\bullet -}$) and hydrogen peroxide (H_2O_2). Reactive nitrogen species (RNS) comprise nitric oxide ($NO^{\bullet}$), which is relatively unreactive, and its derivative peroxynitrite ($ONOO^-$), a potent oxidant capable of causing damage to many biological components.[30]

Uncontrolled overproduction of these ROS causes oxidative damage to cells. Antioxidants are able to scavenge free radicals *via* their hydroxyl substituents and their H-donating ability, and therefore they manage to protect DNA or cell walls from oxidative stress.[31,32] The presence of phenolic rings with three hydroxyl groups in positions 3, 4′, and 5, as well as a conjugated double bond and the likelihood of electron delocalization in the molecule, are thought to contribute to resveratrol's antioxidant properties.[23] The oxidative damage and ROS produced during aerobic cell respiration are related to many degenerative diseases, such as cancer, diabetes, autoimmune conditions, various respiratory diseases, eye diseases, schizophrenia, osteoporosis and aging.[31] Moreover, this pivotal function keeps the body's oxidative stress level below the critical point. ROS/RNS generation can be attributed to a variety of endogenous and exogenous sources, including mitochondria, peroxisomes, NADPH oxidases, xanthine oxidases, lipoxygenase, cyclooxygenase, and cytochrome P450, as well as UV radiation, pollutants, poisons, and medications.[30,33] Furthermore, generation of ROS by pro-oxidant enzymes may contribute to endothelial damage and the advancement of cardiovascular illnesses. A high glucose level, such as that found in diabetics, can induce the formation of ROS by NADPH oxidase, leading to the development of pathological diseases such as fibrosis.

RNS, like ROS, have a dual purpose in that they can be both damaging and beneficial to living systems. Nitric oxide, which was first identified as a signaling molecule in blood vessel modulation and is known as a regulator of essential physiological processes, and is biosynthesized by the nitric oxide synthase class of enzymes that metabolize arginine to citrulline in the presence of oxygen and NADPH or by reduction of inorganic nitrate, can mediate cellular toxicity by disrupting metabolic enzymes and producing peroxynitrite through an interaction with superoxide.[30]

Furthermore, resveratrol increases the activity of various antioxidant enzymes, including glutathione peroxidase, glutathione *S*-transferase, and glutathione reductase. Resveratrol also possesses antioxidant properties that help to preserve polyunsaturated fatty acids (PUFA).[11]

Wu *et al.* (2022) published a recent review that emphasized the benefits of resveratrol on a variety of cancer types, including breast, colorectal, pancreatic, prostate, lung, gastric, oral, thyroid, bladder, and ovarian cancers. The anticancer effects of resveratrol have been revealed to come from modulation of many different mechanisms at all three stages: initiation, promotion, and progression. For the use of resveratrol in functional foods, three main anticancer mechanisms might be highlighted: the effect of resveratrol on inflammation, effect on cancer cell proliferation, and effects on the gut microbiota.[34] The main effect on inflammation is through the effect on the nuclear factor κB (NFB), and leads to modification of the expression of numerous genes involved in apoptosis, DNA repair, oxidative stress resistance, antiaging processes, and lipid metabolism.[23,35] It has also been shown to inhibit the IGF-1R/Akt/Wnt pathways and activate p53, impacting cell growth and carcinogenesis.[36] Furthermore, it can inhibit the PI3K/Akt pathway, which controls cell differentiation, growth, and proliferation, as well as other variables.[37] Several studies have been conducted to investigate the various ways in which resveratrol stimulates the PI3K/Akt pathway.

To summarize, resveratrol participates in both direct and indirect antioxidant defense mechanisms, including resveratrol scavenging mechanisms, reduction of ROS/RNS production, lipid peroxidation, protein oxidation, and DNA oxidation, as well as intracellular antioxidant defense systems and signaling pathway regulation.[23]

7.4 Physicochemical Properties of Resveratrol

Existing research on resveratrol reveals its high biological value, however its usage in functional foods is also closely linked to its physicochemical properties. *trans*-Resveratrol has poor water solubility (21 $\mu g\,mL^{-1}$ at pH 7.4 and 37 °C), while it is mostly soluble in ethanol (50 $mg\,mL^{-1}$) and dimethyl sulfoxide (16 $mg\,mL^{-1}$). Resveratrol has a very low water solubility (<0.03 $mg\,mL^{-1}$). This might be related to its structure, containing two benzene rings with a total of three hydroxyl groups attached.[38] Therefore, including relatively large quantities of resveratrol in aqueous-based food items is problematic. Furthermore, resveratrol is susceptible to chemical degradation in dietary products.[39]

It is an off-white powder at room temperature and contributes to the flavor of the foods.[40] The stability and solubility of *trans*-resveratrol are mainly affected by external factors such as light, air exposure and oxidative enzymes, pH, and temperature.[41] Under acidic conditions, *trans*-resveratrol, for example, is stable at room or body temperature. However, as the pH rises, resveratrol degrades rapidly. Reduced pH and temperature, as well as reduced exposure to oxygen and light, may improve *trans*-resveratrol's stability in liquid dosing.[42]

Resveratrol concentration in fresh grape skins ranges from 50–100 mg per gram and is 0.2–5.8 $mg\,L^{-1}$ in wine depending on the kind of wine.[43] Resveratrol is rapidly metabolized after oral treatment, albeit the rate of resveratrol metabolization has been reported variously in different investigations. According to Intagliata *et al.* (2019), the first peak was observed 1 h after an oral dose of 25 mg was been taken and a second peak after 6 h.[26] According other studies, the total duration for the absorption of the orally taken resveratrol (25 mg) was found to be shorter (<30 min) with a peak plasma level of 2 M,[44] and the peak plasma resveratrol concentration was 30 minutes after the oral consumption of wine samples.[45] However, against to this rapid absorption, systemic bioavailability of resveratrol is less than 1%. This low bioavailability has been mostly related to the hydroxyl groups of resveratrol being substrates for conjugation reactions such as glucuronidation and sulfation in both intestine and liver.[23] Another concern is related to the limited stability of *trans*-resveratrol during storage: during wine aging, phenolics undergo considerable modifications, including self-association, polymerization, co-pigmentation, and precipitation reactions.[46]

Although oral administration of resveratrol as a preventative therapy is recommended, particularly for inclusion in functional foods, its poor water solubility and sensitivity to degradation need delivery strategies that boost its stability and bioavailability.[47]

Despite the positive effects of resveratrol, it does not have favorable pharmacokinetics due to low bioavailability and being extensively and rapidly degraded and metabolized. To address this issue, drug delivery systems that aim to protect and stabilize resveratrol while also increasing its bioavailability have been developed. Moreover, because of resveratrol's rapid metabolism and its low water solubility, together with its sensitivity to external factors such as light, air exposure and oxidative enzymes,[41] many different delivery systems have been studied to improve its stability and permeability through different biological barriers.

Davidov-Pardo and McClements (2014) provided a list of the possible reasons its low availability: (1) because resveratrol is locked within certain cellular structures, its bioavailability from natural sources is frequently poor; (2) the amount of resveratrol in a natural source might vary greatly based on species, climate, location, time of year, and other factors; (3) consuming the amount of a natural source required to have a noticeable biological effect is typically impractical; (4) resveratrol may be concentrated largely in sections of the fruit that are not generally ingested, such as the skin or seeds. For these reasons, it is frequently preferable to isolate resveratrol from a natural source before incorporating it into an appropriate delivery system. Thus, encapsulation strategies have been used to overcome the problems associated with adding resveratrol into functional food and beverage products.[39] Since then, expertise about encapsulating resveratrol for food applications has significantly evolved. Novel carrier systems, such as nanofibers and micro-sponges, have been created and tested for their efficacy in protecting and delivering resveratrol. These were attractive possibilities due to their ability to inhibit resveratrol (drug) degradation, improve solubility, and modulate resveratrol release kinetics (Xie and Chen, 2023). Edible delivery methods based on generally recognized as safe (GRAS) food-grade materials with biodegradable and biocompatible properties offer a wide range of uses in encapsulating functional chemicals like resveratrol.[49] Edible delivery systems of resveratrol are possible through molecular complexes (protein-based complexes and cyclodextrin-based complexes), particle based systems (natural polymer-based particles and lipid-based particles), nanofibers, micro-sponges and microgel systems.[48]

7.5 Resveratrol and Its Effects on Inflammation and Diabetes

Inflammation is a series of cellular and molecular events that help defend the body against infection. Resveratrol has anti-inflammatory properties, since it may decrease cyclooxygenase (COX)-2 production, which promotes inflammation, and it is active against viruses such as the *Herpesviridae* family. This family contains at least five species that are exceedingly common in humans, including HSV-1 and HSV-2 (herpes simplex virus types 1 and 2).

According to Lin *et al.* (2017) resveratrol exhibited significant antiviral activity against Middle East Respiratory Syndrome coronavirus (MERS-CoV) by suppressing the expression of nucleocapsid protein essential for MERS-CoV replication. Most of the studies involving the anti-SARS-CoV-2 activity of a molecule revolve around the binding of the main protease (Mpro) with the molecule.[50] Resveratrol also increases the expression of several genes, including *SIRT1*, *NfKb*, *PTEN*, and *TFEB*, as well as modifying the function of many molecules. Some pro-inflammatory cytokines (TNFα, IL-6, IL-17, and IL-1β) are raised in the liver of mice as they age; resveratrol reduces that increase, especially for IL-1β. As a result of the resveratrol treatment, transcriptional activation of the liver can be reduced, and as a result polyphenolic compounds such as resveratrol can be used in the diet of older liver mice with chronic inflammation.[51] Moreover, ageing and age-related diseases share some basic mechanistic pillars that largely converge on inflammation. During ageing, chronic, sterile, low-grade inflammation – called inflammaging – develops, which contributes to the pathogenesis of age-related diseases. In a recent study, the hypothesis that resveratrol can act as a neuroprotective molecule, mainly at a preventive level against oxidative stress and inflammation, was investigated. The effects are due to the action of resveratrol on the SIRT1 cell signaling pathway.[52]

Another important effect is against the diabetes. There are two main types of diabetes. Type 1 diabetes is distinguished by the autoimmune destruction of insulin-producing pancreatic beta cells, which results in impaired insulin output and loss of glycemic control. The death of pancreatic beta cells in type 1 diabetes is caused by inheritance, environmental factors, and autoimmunity. Type 2 diabetes, on the other hand, is linked to both genetic make-up and environmental factors, such as a high-calorie diet and a lack of physical activity. Obesity is the leading cause of type 2 diabetes insulin resistance and beta-cell failure.[53] In diabetics, glucosidase breaks down disaccharides to monosaccharides, and inhibition of these enzymes helps control blood glucose levels and thereby minimize postprandial plasma glucose levels. Diabetes is defined by hyperglycemia caused by insulin insufficiency, resistance, or both.[2] Resveratrol has been shown to reduce the risk of diabetes by decreasing oxidative or nitrosative stress, pro-inflammatory mediators/cytokines, and apoptosis.[55]

There are some functional food alternatives involving resveratrol in their formulation. Among them, the most commonly found one is functional wine. Although it is not possible according to current global food legislation, there have been some experimental studies of functional wines involving resveratrol.

Increasing the *trans*-resveratrol content in grapes with pre-harvest techniques has been studied in different trials, including use of stress factors during maturation of grapes (infections, ultrasonication, visible or UV irradiation) as well as changes in the nutrition of plants. Furthermore, use of alternative extraction methods, modification of vinification parameters, thermovinification, extended durations of maceration, must-freezing, and the addition of pectolytic enzymes have been applied to increase the resveratrol content from grape skins and seeds.[55,56] Beyond this, resveratrol presents some limitations, such as the fact that it is located in the seeds and skin of the fruit, whereby its consumption becomes more limited.[39]

Although research focusing on the different mechanisms of the health effects of resveratrol is quite common, the depth of consumer acceptance of the resveratrol is quite limited. Aschemann-Witzel and Grunert (2015) studied the consumer attitude for resveratrol as a new functional ingredient among 500 different consumers from each of Denmark and USA. According to their findings, previous familiarity was a significant factor in consumer perception of resveratrol as a novel functional ingredient. According to the results, 17.9% of participants in the United States and only 3.4% of respondents in Denmark said they had already experienced resveratrol or were currently taking resveratrol. In parallel to these results, 28.9% of U.S. participants mentioned that they were likely to take resveratrol supplements in the coming year. Since this supplement is not marketed in Denmark, only 14.5% said they were likely to take it in the future.[57] As a general output of the research, testing acceptance of resveratrol in different consumption contexts of various carrier products such as supplements *versus* different types of foods was presented as a suggestion for the future.

Encapsulation is a process that involves trapping one or more substances within a matrix. This matrix might be either solid or liquid, homogeneous or heterogeneous, microscopic or macroscopic. The element to be entrapped is known as the "active" or "core" material, whilst the matrix material is known as the "wall", "encapsulant", "shell", or "carrier" material.[58]

By using a resveratrol emulsion created with Arabic gum (an encapsulating agent) and spray drying technology, resveratrol became one of the most important ingredients in food and medicinal applications. The surface of all resveratrol microparticles had a rough, uneven shape, and spherical form, with an average size of 8.2 μm. The Weibull model suited the release profiles well, and the greatest encapsulation efficiency found was 87%.[59]

Recently, it was demonstrated that nanoparticles may accomplish regulated release of resveratrol and improve cellular uptake, which was attributable to its viscous wall materials and better resveratrol encapsulation efficiency (EE). Nanoencapsulated resveratrol was prepared by sonicating a resveratrol solution in ethanol and starch slurry (horse-chestnut, water-chestnut and lotus-stem starch particles) and incorporated into wheat flour to produce functional snacks. The snacks with nanoencapsulated resveratrol showed 43–53% resveratrol content after processing as compared to 5.42% in snacks containing free resveratrol. Similarly, snacks containing nanoencapsulated resveratrol had not only higher phenolic compounds but also higher antidiabetic, anti-obesity and antioxidant activities after extrusion than those fortified with free resveratrol. The findings of this research revealed the positive effects of nanoencapsulation on the thermal stability of resveratrol during processing and manufacture of nutraceuticals that are capable of being snacks.[60]

In another recent encapsulation study, resveratrol-filled freeze-dried potato microparticles were used as a novel delivery system to prevent degradation during storage and increase bioaccessibility during digestion. Results showed that the binary solvent of ethanol and polyethylene glycol 400 (40 : 60 v/v) was used to dissolve 30% w/v resveratrol for diffusion into potatoes, and resulted in high stability during ambient storage for 110 days, with a decrease in loading capacity of only 7%. Loading also protected the resveratrol from photodegradation.[47] In another recent study, a change in conformation was tried, novel conjugates of zein and resveratrol were produced by alkaline and free-radical grafting reactions. Compared with native zein (68.69 $\mu mol\,g^{-1}$), the sulfhydryl content of A-ZEIN-RES (alkaline) and R-ZEIN-RES (free radical grafting) conjugates reduced to 13.43 $\mu mol\,g^{-1}$ and 29.54 $\mu mol\,g^{-1}$, respectively, confirming that the amino acid side chain of zein was covalently bound to resveratrol to form stable C–S covalent bonds. The thermal stability, antioxidant activity, and emulsification activity was higher for conjugates than for native zein.[61] These new conjugates have the potential for food industry use. From a different point of view, liposomes, one of the first lipid-based drug nano-carriers for clinical use, are primarily composed of phospholipids and cholesterol, and may be manufactured in a variety of stimuli-responsive nanoplatforms to encapsulate both hydrophobic and hydrophilic drugs. Liposomes range in size from tens of nanometers to several microns. There are several distinct advantages, including excellent biocompatibility, quicker absorption, and trapping by enterocytes.[48] Liposomal nanoencapsulation was developed to overcome the limitations of the poor physicochemical stability and improve the antioxidant effects of curcumin and resveratrol. Liposomes co-encapsulating curcumin and resveratrol presented a lower particle size and greater encapsulation efficiency; in addition, they a higher stability during preparation, storage, heating and surfactant shock than those loaded with individual polyphenols.[62]

Particulate carriers (size >1000 nm), particularly polysaccharide-based particles, have been widely used in the oral delivery of resveratrol due to their relative nontoxicity, biocompatibility, and adhesivity. They are typically prepared through the granulation process of extrusion and desiccation following material compounding or cross-linking.[48]

Although these technologies have demonstrated successful results, instability of encapsulated resveratrol, high production costs, and inability to control the targeted release are some of the drawbacks limiting their practical applicability.[63] Furthermore, most delivery systems have concentrated on building nanoscale carriers, but the toxicity of these systems has not been explored.

As a functional food alternative, *trans*-resveratrol fortified Riesling and Cabernet Sauvignon (with 20 $mg\,L^{-1}$ and 200 $mg\,L^{-1}$ *trans*-resveratrol) were tested for 6 and 32 weeks of storage.[10] According to the results, the level of fortification and type of grape were the significant factors for the consumer acceptance as at the end of 32 weeks, significant differences were found for bitterness and color in Riesling (200 $mg\,L^{-1}$ *vs.* control), while no significant differences were detected at both levels for Cabernet Sauvignon. In contrast, the color was defined as statistically different at both levels of resveratrol in the same type of wine.[10] However, adding *trans*-resveratrol as a functional ingredient is not permitted by current food law legislations globally. In other studies, on the other hand, resveratrol detection threshold has been studied and, according to the results, the encapsulation of resveratrol significantly increased the taste detection threshold of the compound compared to unencapsulated resveratrol.[64] As another possible option for *trans*-resveratrol usage in wine, a recent review article focused on the usage of *trans*-resveratrol instead of sulfites as a healthy ingredient in wines.[56]

7.6 Conclusion

Plants produce a number of phytochemicals and resveratrol is one of the most important of these. Research on resveratrol will focus more on trials with different encapsulating materials in terms of encapsulation efficiency, stability, and bioavailability.

References

1. S. Basak and J. Gokhale, Immunity boosting nutraceuticals: Current trends and challenges, *J. Food Biochem.*, 2022, **46**(3), e13902.
2. J. Xie, J. Liang and N. Chen, Autophagy-associated signal pathways of functional foods for chronic diseases, *Food Sci. Hum. Wellness*, 2019, **8**(1), 25–33.
3. C. Bao, P. Jiang, J. Chai, Y. Jiang, D. Li and W. Bao, *et al.*, The delivery of sensitive food bioactive ingredients: Absorption mechanisms, influencing factors, encapsulation techniques and evaluation models, *Food Res. Int.*, 2019, **120**, 130–140.
4. L. Domínguez Díaz, V. Fernández-Ruiz and M. Cámara, The frontier between nutrition and pharma: The international regulatory framework of functional foods, food supplements and nutraceuticals, *Crit. Rev. Food Sci. Nutr.*, 2020, **60**(10), 1738–1746.
5. B. Kusznierewicz, Non-Nutritive Bioactive Compounds in Food of Plant Origin, in *Chemical and Functional Properties of Food Components*, CRC Press, 2023, pp. 497–534.
6. R. Mark, X. Lyu, J. J. Lee, R. Parra-Saldívar and W. N. Chen, Sustainable production of natural phenolics for functional food applications, *J. Funct. Food*, 2019, **57**, 233–254.
7. L. Camont, C.-H. Cottart, Y. Rhayem, V. Nivet-Antoine, R. Djelidi and F. Collin, *et al.*, Simple spectrophotometric assessment of the trans-/cis-resveratrol ratio in aqueous solutions, *Anal. Chim. Acta*, 2009, **634**(1), 121–128.

8. B. Tian and J. Liu, Resveratrol: a review of plant sources, synthesis, stability, modification and food application, *J. Sci. Food Agric.*, 2020, **100**(4), 1392–1404.
9. J.-H. Ko, G. Sethi, J.-Y. Um, M. K. Shanmugam, F. Arfuso and A. P. Kumar, *et al.*, The Role of Resveratrol in Cancer Therapy, *Int. J. Mol. Sci.*, 2017, **18**(12), 2589.
10. N. J. Gaudette and G. J. Pickering, Sensory and chemical characteristics of trans-resveratrol-fortified wine, *Aust. J. Grape Wine Res.*, 2011, **17**(2), 249–257.
11. S. Khattar, S. A. Khan, S. A. A. Zaidi, M. Darvishikolour, U. Farooq and P. P. Naseef, *et al.*, Resveratrol from dietary supplement to a drug candidate: An assessment of potential, *Pharmaceuticals*, 2022, **15**(8), 957.
12. M. Vestergaard and H. Ingmer, Antibacterial and antifungal properties of resveratrol, *Int. J. Antimicrob. Agents*, 2019, **53**(6), 716–723.
13. J. Yang and Y.-Y. Xiao, Grape phytochemicals and associated health benefits, *Crit. Rev. Food Sci. Nutr.*, 2013, **53**(11), 1202–1225.
14. U. Stervbo, O. Vang and C. Bonnesen, A review of the content of the putative chemopreventive phytoalexin resveratrol in red wine, *Food Chem.*, 2007, **101**(2), 449–457.
15. P. Goufo, R. K. Singh and I. Cortez, A reference list of phenolic compounds (including stilbenes) in grapevine (*Vitis vinifera* L.) roots, woods, canes, stems, and leaves, *Antioxidants*, 2020, **9**(5), 398.
16. S. Mukherjee, J. I. Dudley and D. K. Das, Dose-dependency of resveratrol in providing health benefits, *Dose-Response*, 2010, **8**(4), 478–500.
17. M. Careri, C. Corradini, L. Elviri, I. Nicoletti and I. Zagnoni, Direct HPLC analysis of quercetin and trans-resveratrol in red wine, grape, and winemaking byproducts, *J. Agric. Food Chem.*, 2003, **51**(18), 5226–5231.
18. K. T. Howitz, K. J. Bitterman, H. Y. Cohen, D. W. Lamming, S. Lavu and J. G. Wood, *et al.*, Small molecule activators of sirtuins extend Saccharomyces cerevisiae lifespan, *Nature*, 2003, **425**(6954), 191–196.
19. A. Kuno, M. Tanno and Y. Horio, The effects of resveratrol and SIRT1 activation on dystrophic cardiomyopathy, *Ann. N. Y. Acad. Sci.*, 2015, **1348**(1), 46–54.
20. A. Ungurianu, A. Zanfirescu and D. Margină, Sirtuins, resveratrol and the intertwining cellular pathways connecting them, *Ageing Res. Rev.*, 2023, **88**, 101936.
21. C. A. Caldeira, M. A. Santos, G. R. Araújo, R. C. Lara, F. N. Franco and M. M. Chaves, Resveratrol: Change of SIRT 1 and AMPK signaling pattern during the aging process, *Exp. Gerontol.*, 2021, **146**, 111226.
22. L. Rajman, K. Chwalek and D. A. Sinclair, Therapeutic potential of NAD-boosting molecules: the *in vivo* evidence, *Cell Metab.*, 2018, **27**(3), 529–547.
23. V. Truong, M. Jun and W. Jeong, Role of resveratrol in regulation of cellular defense systems against oxidative stress, *BioFactors*, 2018, **44**(1), 36–49.

24. S. Gorman, M. Tosetto, F. Lyng, O. Howe, K. Sheahan and D. O'Donoghue, *et al.*, Radiation and chemotherapy bystander effects induce early genomic instability events: telomere shortening and bridge formation coupled with mitochondrial dysfunction, *Mutat. Res., Fundam. Mol. Mech. Mutagen.*, 2009, **669**(1–2), 131–138.
25. G. Giovinazzo, I. Ingrosso, A. Paradiso, L. De Gara and A. Santino, Resveratrol biosynthesis: plant metabolic engineering for nutritional improvement of food, *Plant Foods Hum. Nutr.*, 2012, **67**, 191–199.
26. S. Intagliata, M. N. Modica, L. M. Santagati and L. Montenegro, Strategies to improve resveratrol systemic and topical bioavailability: An update, *Antioxidants*, 2019, **8**(8), 244.
27. A. M. Posadino, R. Giordo, A. Cossu, G. K. Nasrallah, A. Shaito and H. Abou-Saleh, *et al.*, Flavin oxidase-induced ROS generation modulates PKC biphasic effect of resveratrol on endothelial cell survival, *Biomolecules*, 2019, **9**(6), 209.
28. N. Parsamanesh, A. Asghari, S. Sardari, A. Tasbandi, T. Jamialahmadi and S. Xu, *et al.*, Resveratrol and endothelial function: A literature review, *Pharmacol. Res.*, 2021, **170**, 105725. Available from: https://www.sciencedirect.com/science/article/pii/S1043661821003091.
29. J. Checa and J. M. Aran, Reactive oxygen species: drivers of physiological and pathological processes, *J. Inflammation Res.*, 2020, 1057–1073.
30. S. Di Meo, T. T. Reed, P. Venditti and V. M. Victor, Role of ROS and RNS sources in physiological and pathological conditions, *Oxid. Med. Cell. Longevity*, 2016, **2016**, 1245049.
31. S. Bakir, G. Toydemir, D. Boyacioglu, J. Beekwilder and E. Capanoglu, apple, cherry, grape, oak, chestnut or strawberryg: Changes in content and *in vitro* bio-accessibility, *Int. J. Mol. Sci.*, 2016, **17**(10), 1658.
32. S. Rahaiee, M. Hashemi, S. Moini, S. Abbas Shojaosadati and S. H. Razavi, Comparison of phytochemical constituents and antioxidant activities of aqueous and alcoholic extracts of saffron, *Qual. Assur. Saf. Crops Foods*, 2015, **7**(4), 521–529.
33. J. Limón-Pacheco and M. E. Gonsebatt, The role of antioxidants and antioxidant-related enzymes in protective responses to environmentally induced oxidative stress, *Mutat. Res., Genet. Toxicol. Environ. Mutagen.*, 2009, **674**(1–2), 137–147.
34. S. X. Wu, R. G. Xiong, S. Y. Huang, D. D. Zhou, A. Saimaiti and C. N. Zhao, *et al.*, Effects and mechanisms of resveratrol for prevention and management of cancers: An updated review, *Crit. Rev. Food Sci. Nutr.*, 2022, 1–19.
35. J. A. Baur, Z. Ungvari, R. K. Minor, D. G. Le Couteur and R. De Cabo, Are sirtuins viable targets for improving healthspan and lifespan?, *Nat. Rev. Drug Discovery*, 2012, **11**(6), 443–461.
36. A. Y. Berman, R. A. Motechin, M. Y. Wiesenfeld and M. K. Holz, The therapeutic potential of resveratrol: a review of clinical trials, *npj Precis. Oncol.*, 2017, **1**(1), 35.
37. Y. He, M. M. Sun, G. G. Zhang, J. Yang, K. S. Chen and W. W. Xu, *et al.*, Targeting PI3K/Akt signal transduction for cancer therapy, *Signal Transduction Targeted Ther.*, 2021, **6**(1), 425.

38. H. Soto-Valdez, R. Auras and E. Peralta, Fabrication of poly(lactic acid) films with resveratrol and the diffusion of resveratrol into ethanol, *J. Appl. Polym. Sci.*, 2011, **121**(2), 970–978.
39. G. Davidov-Pardo and D. J. McClements, Resveratrol encapsulation: Designing delivery systems to overcome solubility, stability and bioavailability issues, *Trends Food Sci. Technol.*, 2014, **38**(2), 88–103.
40. J. Bié, B. Sepodes, P. C. B. Fernandes and M. H. L. Ribeiro, Polyphenols in Health and Disease: Gut Microbiota, Bioaccessibility, and Bioavailability, *Compounds*, 2023, **3**, 40–72.
41. A. Francioso, P. Mastromarino, R. Restignoli, A. Boffi, M. d'Erme and L. Mosca, Improved stability of trans-resveratrol in aqueous solutions by carboxymethylated (1,3/1,6)-β-D-glucan, *J. Agric. Food Chem.*, 2014, **62**(7), 1520–1525.
42. Š. Zupančič, Z. Lavrič and J. Kristl, Stability and solubility of trans-resveratrol are strongly influenced by pH and temperature, *Eur. J. Pharm. Biopharm.*, 2015, **93**, 196–204.
43. A. Carrizzo, M. Forte, A. Damato, V. Trimarco, F. Salzano and M. Bartolo, *et al.*, Antioxidant effects of resveratrol in cardiovascular, cerebral and metabolic diseases, *Food Chem. Toxicol.*, 2013, **61**, 215–226.
44. T. Walle, F. Hsieh, M. H. DeLegge, J. E. Oatis and U. K. Walle, High absorption but very low bioavailability of oral resveratrol in humans, *Drug Metab. Dispos.*, 2004, **32**(12), 1377–1382.
45. D. M. Goldberg, J. Yan and G. J. Soleas, Absorption of three wine-related polyphenols in three different matrices by healthy subjects, *Clin. Biochem.*, 2003, **36**(1), 79–87.
46. A. L. Waterhouse, Wine phenolics, *Ann. N. Y. Acad. Sci.*, 2002, **957**(1), 21–36.
47. I. Choi, N. Li and Q. Zhong, Enhancing bioaccessibility of resveratrol by loading in natural porous starch microparticles, *Int. J. Biol. Macromol.*, 2022, **194**, 982–992.
48. Z. Xie and X. Chen, Healthy benefits and edible delivery systems of resveratrol: a review, *Food Rev. Int.*, 2023, **39**(7), 3879–3905.
49. A. F. Esfanjani and S. M. Jafari, Biopolymer nano-particles and natural nano-carriers for nano-encapsulation of phenolic compounds, *Colloids Surf., B*, 2016, **146**, 532–543.
50. S.-C. Lin, C.-T. Ho, W.-H. Chuo, S. Li, T. T. Wang and C.-C. Lin, Effective inhibition of MERS-CoV infection by resveratrol, *BMC Infect. Dis.*, 2017, **17**(1), 1–10.
51. B. T. Tung, E. Rodríguez-Bies, E. Talero, E. Gamero-Estévez, V. Motilva and P. Navas, *et al.*, Anti-inflammatory effect of resveratrol in old mice liver, *Exp. Gerontol.*, 2015, **64**, 1–7.
52. R. C. Lara, F. N. Franco, G. R. de Araújo, B. N. M. Silva and M. M. Chaves, The preventive use of resveratrol increases its antioxidant effect by SIRT1 and subclinical anti-inflammatory action in Neuro-2A cells, *In Vitro Cell. Dev. Biol.: Anim.*, 2022, **58**(10), 979–986.

53. M. Hoca, E. Becer and H. S. Vatansever, The role of resveratrol in diabetes and obesity associated with insulin resistance, *Arch. Physiol. Biochem.*, 2023, **129**(2), 555–561.
54. A. C. Gonçalves, A. R. Nunes, J. D. Flores-Félix, G. Alves and L. R. Silva, Cherries and Blueberries-Based Beverages: Functional Foods with Antidiabetic and Immune Booster Properties, *Molecules*, 2022, **27**(10), 3294.
55. J. Cvejić and M. Atanacković, Effect of wine production techniques on wine resveratrol and total phenolics, in *Processing and impact on active components in food*, Elsevier, 2015, pp. 501–508.
56. E. Kontaxakis, E. Trantas and F. Ververidis, Resveratrol: A fair race towards replacing sulfites in wines, *Molecules*, 2020, **25**(10), 2378.
57. J. Aschemann-Witzel and K. G. Grunert, Resveratrol and health from a consumer perspective: perception, attitude, and adoption of a new functional ingredient, *Ann. N. Y. Acad. Sci.*, 2015, **1348**(1), 171–179.
58. A. Madene, M. Jacquot, J. Scher and S. Desobry, Flavour encapsulation and controlled release–a review, *Int. J. Food Sci. Technol.*, 2006, **41**(1), 1–21.
59. T. Cardoso, A. Gonçalves, B. N. Estevinho and F. Rocha, Potential food application of resveratrol microparticles: Characterization and controlled release studies, *Powder Technol.*, 2019, **355**, 593–601.
60. M. Ahmad and A. Gani, Development of novel functional snacks containing nano-encapsulated resveratrol with anti-diabetic, anti-obesity and antioxidant properties, *Food Chem.*, 2021, **352**, 129323.
61. G. Ren, J. Shi, S. Huang, C. Liu, F. Ni and Y. He, *et al.*, The fabrication of novel zein and resveratrol covalent conjugates: Enhanced thermal stability, emulsifying and antioxidant properties, *Food Chem.*, 2022, **374**, 131612.
62. M. Huang, C. Liang, C. Tan, S. Huang, R. Ying and Y. Wang, *et al.*, Liposome co-encapsulation as a strategy for the delivery of curcumin and resveratrol, *Food Funct.*, 2019, **10**(10), 6447–6458.
63. G. B. Celli, A. Ghanem and M. S.-L. Brooks, Bioactive encapsulated powders for functional foods—a review of methods and current limitations, *Food Bioprocess Technol.*, 2015, **8**, 1825–1837.
64. C. C. Koga, A. R. Becraft, Y. Lee and S. Lee, Taste detection thresholds of resveratrol, *J. Food Sci.*, 2015, **80**(9), S2064–S2070.

CHAPTER 8

Fortification of Phenols in Dairy Products

NIKITA SINGH, RISHIBHA GUPTA AND SMRITI GAUR*

Department of Biotechnology, Jaypee Institute of Information Technology, Noida, Uttar Pradesh, India
*Email: smritigaurjiit@gmail.com

8.1 Introduction

A nutritious diet and a balanced lifestyle are becoming more and more important to consumers. In the United Kingdom, 63% of adults claimed that they are committed to eating nutritious foods in order to stay healthy and enhance their overall wellness. This percentage rose to 75% in senior citizens (over 60) because of age-related medical conditions.[1] Polyphenols (PPs) are bioactive compounds that play an important role in the prevention of several diseases, including cancer, heart disease, and chronic illnesses. Therefore, dietary supplements that enhance people's health can be developed using PP extracts sourced from a variety of vegetables and fruits.[2] Due to their exceptional antimicrobial and antioxidant properties, phenolics are regarded as highly significant natural components in the human diet. Fats, lipids, proteins, lactose, nutrients, and vitamins are all found in milk in complex proportions.[3] There are many people who consume it worldwide. Due to their nutritional and physicochemical properties, milk proteins are frequently employed as elements in food preparations. They are a source of bioactive peptides, and the literature is replete with studies demonstrating their positive effects in areas like immunomodulation and improving heart health and metabolism. Milk and other dairy products have the potential to be excellent vehicles for delivering bioactive substances.[4] Due to their rich

Food Chemistry, Function and Analysis No. 44
Functional Foods of the Future
Edited by Vijai Kumar Gupta, Minaxi Sharma, Smriti Gaur and Ramesh Chander Kuhad

Published by the Royal Society of Chemistry, www.rsc.org

nutritional content and therapeutic benefits, yogurt, and other milk-based fermented products are among some of the most favored milk products worldwide.[5] PPs can be used in the production of dairy products with increased antioxidant properties.[6] Numerous investigations have attempted to determine the impact of the interaction between dairy proteins and PPs on their antioxidant effects.[7] PPs are known for their ability to interact with proteins through different types of bonds, which could influence the physical, functional, and structural characteristics of dairy products. There are a number of factors that influence the stability of PPs such as pH, temperature, and light. In dairy products, phenolic compounds serve vital functions for flavor, color, antioxidant activity, and nutritional value. We must carefully examine their distinctive chemical and physical characteristics in order to understand how to enhance their efficiency in dairy production. PPs have been added to a variety of milk and dairy products so far in an effort to boost their functional value.[8] Our understanding of how milk product processing affects PP mixtures in milk and the properties of the final products is still limited.[1] Until now, a range of dairy products has been enhanced with PPs to boost the overall nutritional content of the end product. Yogurt and other fermented milk products are globally popular for their rich nutrition and health advantages. Incorporating PPs as a food ingredient in dairy products requires understanding how they impact the end product's quality during standard processing steps.[5]

8.2 Polyphenols

PPs are among the most diverse and common metabolites in the plant kingdom. Plants yield PPs as secondary metabolites that raise their bitterness, giving them more protection against pathogens. Over the past few years, there has been increasing fascination with PPs due to their significant role in human health and their potential therapeutic properties.[9] Vegetables, fruits, beverages such as tea and coffee, liquor, cereal grains, and cherries are a few examples of plant-based foods that contain PPs.[10]

PPs have an aromatic ring linked to hydroxyl groups as their primary chemical structure. According to their structure, PPs can be divided into various groups depending on the number of phenolic rings and the structural components bound to these rings. Flavonoids, lignans, phenolic acids, stilbenes, and tannins are the primary classes of PPs.[11] Phenolic acids are phenols that have an attached carboxyl group. They primarily exist as hydroxycinnamic acids (C_6–C_3 skeleton) and hydroxybenzoic acids (C_6–C_1 skeleton) which are based on cinnamic acid and benzoic acid, respectively. While hydroxybenzoic acids are primarily found as glucosides, the hydroxycinnamic acids tend to occur as simple esters with hydroxycarboxylic acids. Ferulic acid is a common example of hydroxycinnamic acid, while gallic acid is the most prevalent hydroxybenzoic acid.[12] Tannins are primarily divided into two distinct categories: condensed tannins and hydrolyzable tannins.[13] Tannins have a substantially more complicated structure compared to phenolic acids. Tannins are large, water-soluble molecules with a molecular weight range of 600–20 000 daltons.[14] Concentrated tannins are polymer molecules of flavonols, whereas hydrolyzable tannins are made of glucose and gallic acid molecules. Gallotannins and ellagitannins are two groups of hydrolyzable tannins that can be distinguished based on their structural differences.[15]

8.3 Interactions of Polyphenols with Milk Proteins

Milk proteins are widely available, inexpensive, organic substances with excellent sensory qualities and high nutrient value. The hydrophobic regions of milk proteins allow them to link to multiple types of molecules with different levels of affinity. Milk protein chains connect to different kinds of PPs through a variety of interaction forces, including van der Waals forces, hydrogen bonds, and hydrophobic and ionic bonding,[13] as shown in Figure 8.1. Additionally, depending on how many phenolic molecules are attached to one protein or how many protein molecules bind to one phenolic group, the interactions are often referred to as multidentate or multisite.[16]

8.3.1 Covalent Linkage Between Polyphenols and Milk Proteins

Proteins and PPs typically interact covalently through C–S or C–N bonding. The alkaline reaction is one of the most popular non-enzymatic approaches for protein–PP combination.[13] PPs are susceptible to oxidation and form corresponding quinones in an alkaline environment at pH 9 when O_2 is

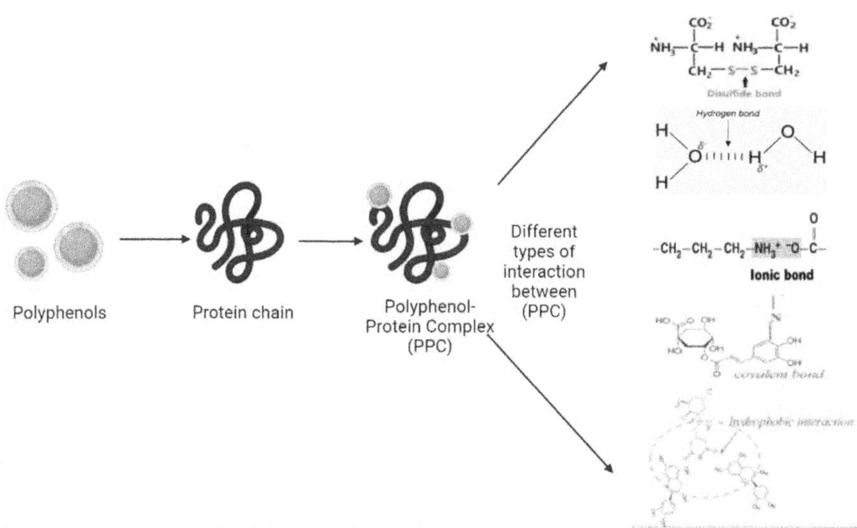

Figure 8.1 Different types of interactions between milk protein and polyphenols.[13]

present. During the formation of PPCs (polyphenol–protein complexes), the intermediate products of the enzymatic oxidation of PPs, like semi-quinone radicals, connect with certain amino acids of the protein to produce PPCs. The side chain residues of amino acids in proteins can easily interact with these intermediate products resulting in formation of covalent bonds.[17]

8.3.2 Non-covalent Linkage Between Polyphenols and Milk Proteins

Proteins and PPs can also interact non-covalently, which could result in reactions that are largely reversible and less strong than covalent interactions. The interaction that occurs between milk proteins and PPs is typically mediated by hydrogen bonding and hydrophobic interactions. In order to form hydrogen bonds, phenolic groups act as effective hydrogen donors that can link with the carbonyl groups of proteins. When the –OH groups of PPs come into contact with nitrogen or oxygen, specifically the $-NH_2$ and –OH groups of protein molecules, it results in the formation of hydrogen bonds.[17] The catechins (a type of phenolic compound) and amino acid residues of casein proteins interact to form hydrogen bonds. Protein molecules containing a number of amino acids, including leucine, tryptophan, glycine, cysteine, tyrosine, methionine, valine, alanine, and isoleucine phenylalanine that can serve as potential hydrophobic sites. These amino acid sites link up with inert or nonpolar aromatic rings of PPs. Amino acids with a positive charge, such as lysine, react with negatively charged –OH groups of PPs through a weak ionic bond.[13]

8.4 Polyphenols in the GI Tract: Bioavailability and Bioaccessibility

Digestion is a sophisticated process where the foods we consume are broken down through enzymes and physical actions. This transforms them into nutrients that can be absorbed into the bloodstream. We know that the interaction between helpful PP byproducts and gut bacteria can be good for digestion and the immune system. However, we are still figuring out exactly how it all works.[18] Studies revealed PPs and their derivatives play a positive role in gut health. They support good bacteria and limit the growth of harmful ones.[19] However, it is necessary to consider the chemical composition PPs in order to fully comprehend how phenols pass through and are digested inside the GI tract, because different phenol chemical compositions alter the digestion of food matrices.[20] Comparatively, PPs with two hydroxyl groups perform better and are more effective at eliminating free radicals than those with just one.[21] PP consumption boosts the beneficial effects of gut microbiota by stimulating the gastrointestinal immune system, increasing the release of short-chain fatty acids (SCFAs), and influencing various biological and physical processes. New findings show that the growth rate of gut bacteria is influenced by the PP dose, the structure of the diet, and the specific bacterial species or strain.[18]

When dietary polyphenols enter the GI tract, some of them are absorbed through the wall of the intestinal tract and get metabolised in tissues, while others are retained in the colon from the small intestine. Most dietary PPs remain in the colon as foreign compounds and interact with gut microbes during digestion. Gut bacteria usually break down the PPs that remain in the GI tract. The types of gut bacteria can affect how PPs are processed and used in the body. GI microbiota uses enzymatic activities like demethylation, glucosidase, decarboxylation, and others to break down PPs into different forms. These processes enhance the availability of PPs and contribute to better GI health benefits.[22]

The intestinal microflora and cells both can digest PPs. Colonic bacteria play a major role in the distribution and absorption of PPs throughout the gastrointestinal tract.[23] Understanding metabolic products as well as their biological and physiological activities is therefore crucial. The biological accessibility of food to the circulatory system following effective intake is one of the factors that impact PP bioaccessibility.[24] The percentage of bioactive compounds that emerge from food matrices following digestion, and are therefore available for absorption by the gastrointestinal tract, is another definition for "bioaccessibility". Bioavailability refers to the proportion of biologically active substances that are successfully absorbed into the circulatory system for use in metabolic processes. Hence, effective metabolism is required before the PPs are easily accessible for digestion. A small fraction of PPs will be immediately absorbed in the upper section of the gastrointestinal tract, while more than 90 percent are digested by the intestinal microbiota before absorption.[25] Hence, effective absorption relies on microbial metabolism. Difficulties in nutrient absorption in the intestine arise from inhibitors in the food, changes in compound combinations, nutritional components in the food, and all forms of metabolism.[26] Research has focused on understanding the effectiveness of PPs, considering their amount in food, absorption, and bioaccessibility. Many studies have aimed to determine how quickly PPs become accessible after digestion.[24] A schematic representation of the process is shown in Figure 8.2.

8.5 Application of PPs in Dairy

Fruits, vegetables, and cereals are some important sources of PPs in the diet. However, in recent years many food consumption surveys indicate that the entire repertoire of PPs in the human diet is usually obtained from limited food items such as tea and coffee, while the contribution from other sources is relatively low.[29,30] As a result, the vast spectrum of biological activities of PPs is often not fully exploited in our food consumption patterns. Therefore, many dairy-based products are being explored as potential carriers of PPs in our food system. A few such products are discussed below.

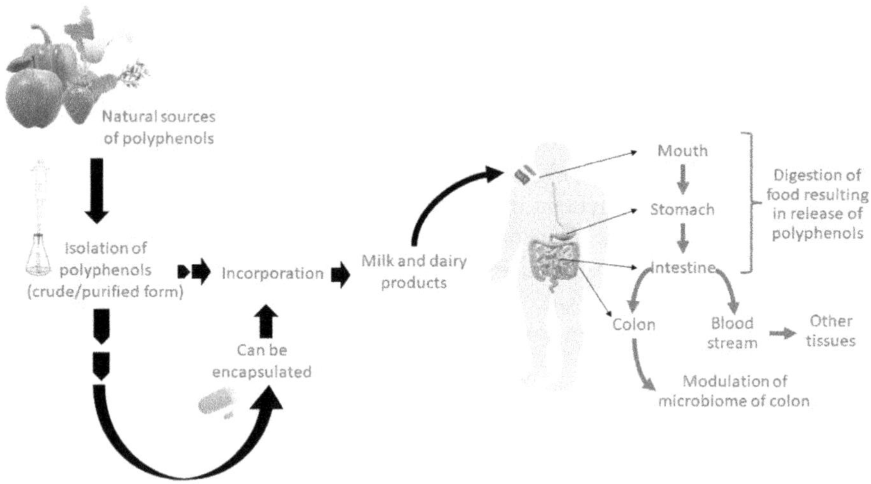

Figure 8.2 Extraction of polyphenols and their fate in GI tract.[27,28]

8.5.1 Yogurt

Yogurt is one of the most popular and widely consumed fermented food products throughout the world. Though it is a rich source of probiotics, vitamins, proteins, and minerals (particularly calcium) PPs are generally absent in plain yogurt.[31] Therefore, their fortification in yogurt is a cheap alternative strategy to improve the functional quality of our diet. PP-fortified yogurt is reported to have many health-beneficial properties such as enhanced antioxidant activity,[32,33] anti-hypertensive activity, anti-cancerous activity,[34] and many others. A few such studies are discussed in Table 8.1.

PPs can be added to yogurt either during the pre-fermentation stage as part of the yogurt ingredient mix or after fermentation similar to the addition of other colors and flavors.[35] They can be added in pure form or as crude mixtures depending on the desired end-use characteristics. It must be noted that, depending on the chemical nature of PPs, these compounds can have a significant effect on important characteristics of yogurt, such as pH, lactic acid bacterial growth, syneresis, shelf-life, flavor, aroma, *etc.*[36,37] Usually when PPs are added to yogurt, they bind to the proteins present in the yogurt matrix *via* covalent or non-covalent interactions. Casein protein, due to the presence of proline-rich sites, can form stable complexes with the hydroxyl groups of PPs.[37] These bonded PPs are then released during gastrointestinal digestion resulting in an increased bioaccessibility in the gut.[38] A study by Helal and Tagliazucchi[39] showed that the bioaccessibility of cinnamon PPs in the gastrointestinal tract was significantly enhanced by incorporation in yogurt compared to a water-based extract. However, as shown by Simonetti *et al.*,[38] not all PPs have a similar bioavailability index in the gut and many of these compounds can change while traversing the different microenvironments of the gut. Therefore, a careful evaluation of

Table 8.1 An overview of dairy products with added polyphenols or crude extract: composition, stage of incorporation and bioactivity.

	Polyphenol extract	Characteristic polyphenols	Method of incorporation	Bioactivity	Ref.
Yogurt	*Myrciaria jaboticaba* seed extract (JSE)	Castalagin, vescalagin, procyanidin A2, ellagic acid	Post-fermentation	• Inhibitory activity: ACE-1, α-amylase, α-glucosidase enzymes • Anti-hemolytic activity • Increased antioxidant activity • Regulation of gut microbiota dysbiosis in colon cancer-induced rats	48
Yogurt	Goldenberry juice	Gallic acid, ellagic acid, mangiferin, quercitin, caffeic acid, rutin[49]	Post-fermentation	• Hepatoprotective effect in hepatitis-induced rats	50
Yogurt	Noni juice	Quercetin, isoquercitrin, rutin, kaempferol glycoside, tricetin	Pre-fermentation	• Showed protective effects against ulcerative colitis in rats • Regulation of systemic inflammation induced due to colitis	51
Yogurt	*M. parviflora* extract nanoemulsion	Naringenin, apigenin-7-glucoside, *p*-coumaric acid, luteolin, cinnamic acid[52]	Pre-fermentation	• Reduction in oxidative stress and serum inflammatory markers in ulcerative colitis-induced rats	53
Yogurt	Moringa leaf powder water extract	Quercetin, kaempferol	Pre-fermentation	• A concentration-dependent increase in antioxidant activity • Reduced oxidative stress in human colorectal epithelial cell line HT-29 by increased expression of factors responsible for antioxidant response in the nucleus	54

Table 8.1 (*Continued*)

	Polyphenol extract	Characteristic polyphenols	Method of incorporation	Bioactivity	Ref.
Yogurt	Encapsulated olive fruit polyphenols	Verbascoside, hydroxytyrosol[55]	Pre-fermentation	• Human trials showed decreased levels of LDL cholesterol and malondialdehyde (an indicator of lipid peroxidation) after two weeks of consumption	56
Cheese from dromedary milk	*Allium roseum* leaf powder		Pre-fermentation	• PP addition improved the textural properties of full fat cheese	57
Labneh cheese	Pepper extract microencapsulated in sodium alginate beads		Post fermentation	• Increased radical scavenging activity	58
Low fat cheese	Catechin homogenized with PEG		Pre-fermentation	• Improved digestive recovery of PP in simulated GI conditions • Increased antioxidant activity	59
Manchego-type cheese	Nanoemulsified curcumin		Pre-fermentation	• Enhanced antioxidant activity • Overall acceptable sensorial qualities	60

the interaction of PPs in a food matrix must be done before claiming any health-beneficial properties. One common strategy to enhance the biological stability of PPs in yogurt is encapsulation.[40] A study by Šeregelj *et al.*[38] showed that raw pepper waste extract containing many bioactive PPs had higher retention in yogurt during storage when freeze-dry-encapsulated as compared to without encapsulation. Similarly, El-Messery *et al.*[41] showed that polycaprolactone-encapsulated olive leaf PPs are more stable in a gastrointestinal environment compared to uncoated samples. Encapsulation also masked the green color of the extract while decreasing the rate of syneresis in yogurt. Therefore, encapsulation as a technique can be explored further for enriching PPs in yogurt.

8.5.2 Cheese

Cheese is a dairy product formed by the coagulation of milk. Coagulation in milk is generally induced by means of renin or acid. This results in the formation of a *para*-β-casein network in milk which is separated from its whey (the aqueous fraction). Cheese mainly constitutes the protein network along with associated fats.[42] Therefore, similar to yogurt, proteins in cheese can interact with PPs and act suitable carriers for PPs in the diet. However, depending on the amount and composition of PPs, their incorporation can affect the internal physicochemical structure and quality attributes of cheese. The bitter and astringent taste of PPs often decreases the sensorial quality of cheese.[42] Himed-Idir *et al.*[43] showed that incorporation of rosemary powder (a natural source of PP) in cheese increased its antioxidant activity with no significant impact on proteins, fats, and pH. However, when used in high concentration it decreased its sensorial acceptability. Masmoudi *et al.*[44] showed that when a higher dose of PPs from *Arbutus unedo* L. is incorporated in milk during cheese-making it negatively affected its yield and gave low sensorial acceptability. Therefore, PP incorporation in cheese requires an understanding of the optimal dose of PP where cheese is not significantly different in its physical and sensorial properties but has improved functional properties.

8.5.3 Buttermilk

When butter is prepared from milk, the left-over liquid is often referred to as buttermilk. It mainly contains proteins, fats and lactose, and can be consumed as a beverage.[45] Buttermilk offers an attractive medium for carrying PPs that have low solubility in aqueous phases, such as resveratrol and curcuminoids. These PPs bind to the proteins and lipids present in buttermilk, thus increasing their stability and solubility.[46,47] Buttermilk can also increase their bioaccessibility in the gut.[45]

8.6 Health Applications

It is well known that phenolic compounds (PCs) have many positive health effects. Accessibility, biological activity, medical benefits, and efficiency of phenolic compounds have been reported in many studies.[18] They inhibit oxidative processes in milk that can be sparked by metallic chelators and free radicals. Their dissolution, composition, association with other PCs or other compounds, digestion, and therefore the way they are metabolized, influence PCs' ability to foster health. PPs have antioxidant and antibacterial activity, which influence numerous physiological functions. They can have an inhibitory impact on important cellular processes such as expansion, production, and death of cells. Due to their widely recognized health advantages, phenolic compounds like flavonoids, catechins, and their associated compounds have been used as medicines.[18]

The destruction of DNA from the effects of oxidative stress can result in mutations that could help cancer cells grow more rapidly in the future. Numerous studies have established that PPs are able to migrate to the less prominent sections of the digestive system, and as a result metabolism and bioaccessibility considerably rise. The anti-proliferative properties of PPs can be improved by milk proteins and (poly)phenolic interactions According to some research studies, PP reaction products with protein complexes in dairy products have anti-carcinogenic effects. Compared to free PPs, certain (poly)phenolic molecules and dairy protein complexes have been shown to lessen the harmful effects of cancer cells. A novel method to deliver drugs to the desired cancer cells may be made possible by the covalent formation of combinations with specific (poly)phenolic and dairy protein molecules.[61] The majority of phenolic substances that have antioxidative properties generally work by neutralizing the damaging effects of reactive oxygen species (ROS) produced during the metabolism of cells.[62]

8.7 Conclusion

In conclusion, dairy and dairy-based products can be suitable carriers for the incorporation of PPs. PPs are important functional compounds that have numerous health-beneficial roles. PPs obtained from natural sources are often added to milk or milk-based products such as yogurt, cheese, buttermilk, *etc.*, to improve their functional characteristics. However, depending on the chemical nature of PPs, their incorporation can significantly affect the overall physical, chemical, and sensory quality attributes of the final product. Besides, PPs often breakdown in the upper digestive tract or are poorly absorbed in the gut. Therefore, to increase their bioaccessibility and/or bioavailability they are generally encapsulated before their addition to a dairy product. Encapsulation can also mask the changes associated with certain PPs such as changes in taste, color, and/or sensorial characteristics of the final product.

References

1. M. Kilic Bayraktar, Incorporation of polyphenols into dairy matrices, Doctoral dissertation, University of Reading, 2018

2. O. L. Pop, R. Suharoschi, S. A. Socaci, E. Berger Ceresino, A. Weber, C. Gruber-Traub, D. C. Vodnar, A. C. Fărcaș and E. Johansson, Polyphenols–Ensured Accessibility from Food to the Human Metabolism by Chemical and Biotechnological Treatments, *Antioxidants*, 2023, **12**(4), 865.
3. S. Chauhan, P. Powar and R. Mehra, A review on nutritional advantages and nutraceutical properties of cow and goat milk, *Int. J. Appl. Res.*, 2021, 7, 101–105.
4. H. Punia, J. Tokas, A. Malik, S. Sangwan, S. Baloda, N. Singh, S. Singh, A. Bhuker, P. Singh, S. Yashveer and S. Agarwal, Identification and detection of bioactive peptides in milk and dairy products: Remarks about agro-foods, *Molecules*, 2020, **25**(15), 3328.
5. E. Hadjimbei, G. Botsaris and S. Chrysostomou, Beneficial effects of yoghurts and probiotic fermented milks and their functional food potential, *Foods*, 2022, **11**(17), 2691.
6. P. Pandey, K. Grover, T. S. Dhillon, A. Kaur and M. Javed, Evaluation of polyphenols enriched dairy products developed by incorporating black carrot (Daucus carota L.) concentrate, *Heliyon*, 2021, **7**(5).
7. A. Zeb, Concept, mechanism, and applications of phenolic antioxidants in foods, *J. Food Biochem.*, 2020, **44**(9), e13394.
8. H. Cao, O. Saroglu, A. Karadag, Z. Diaconeasa, G. Zoccatelli, C. A. Conte-Junior, G. A. Gonzalez-Aguilar, J. Ou, W. Bai, C. M. Zamarioli and L. A. de Freitas, Available technologies on improving the stability of polyphenols in food processing, *Food Front.*, 2021, **2**(2), 109–139.
9. A. Rashidinejad, E. J. Birch, D. Sun-Waterhouse and D. W. Everett, Total phenolic content and antioxidant properties of hard low-fat cheese fortified with catechin as affected by *in vitro* gastrointestinal digestion, *LWT-Food Sci. Technol.*, 2015, **62**(1), 393–399.
10. P. Cosme, A. B. Rodríguez, J. Espino and M. Garrido, Plant phenolics: Bioavailability as a key determinant of their potential health-promoting applications, *Antioxidants*, 2020, **9**(12), 1263.
11. A. M. Delgado, M. Issaoui and N. Chammem, Analysis of main and healthy phenolic compounds in foods, *J. AOAC Int.*, 2019, **102**(5), 1356–1364.
12. A. Bento-Silva, V. M. Koistinen, P. Mena, M. R. Bronze, K. Hanhineva, S. Sahlstrøm, V. Kitrytė, S. Moco and A. M. Aura, Factors affecting intake, metabolism and health benefits of phenolic acids: do we understand individual variability?, *Eur. J. Nutr.*, 2020, **59**, 1275–1293.
13. M. M. Tosif, A. Najda, A. Bains, T. C. Krishna, P. Chawla, M. Dyduch-Siemińska, J. Klepacka and R. Kaushik, A comprehensive review on the interaction of milk protein concentrates with plant-based polyphenolics, *Int. J. Mol. Sci.*, 2021, **22**(24), 13548.
14. M. Fraga-Corral, P. García-Oliveira, A. G. Pereira, C. Lourenço-Lopes, C. Jimenez-Lopez, M. A. Prieto and J. Simal-Gandara, Technological application of tannin-based extracts, *Molecules*, 2020, **25**(3), 614.

15. A. A. Watrelot and E. L. Norton, Chemistry and reactivity of tannins in Vitis spp.: A review, *Molecules*, 2020, **25**(9), 2110.
16. F. Shahidi and C. S. Dissanayaka, Phenolic-protein interactions: insight from in-silico analyses–a review, *Food Prod., Process. Nutr.*, 2023, **5**(1), 1–21.
17. M. Li, C. Ritzoulis, Q. Du, Y. Liu, Y. Ding, W. Liu and J. Liu, Recent progress on protein-polyphenol complexes: Effect on stability and nutrients delivery of oil-in-water emulsion system, *Front. Nutr.*, 2021, **8**, 765589.
18. M. S. Swallah, H. Fu, H. Sun, R. Affoh and H. Yu, The impact of polyphenol on general nutrient metabolism in the monogastric gastrointestinal tract, *J. Food Qual.*, 2020, **2020**, 1–2.
19. S. K. Ray and S. Mukherjee, Evolving interplay between dietary polyphenols and gut microbiota—An emerging importance in healthcare, *Front. Nutr.*, 2021, **8**, 634944.
20. M. Rudrapal, S. J. Khairnar, J. Khan, A. B. Dukhyil, M. A. Ansari, M. N. Alomary, F. M. Alshabrmi, S. Palai, P. K. Deb and R. Devi, Dietary polyphenols and their role in oxidative stress-induced human diseases: Insights into protective effects, antioxidant potentials and mechanism (s) of action, *Front. Pharmacol.*, 2022, **13**, 283.
21. M. A. Al-Mamary and Z. Moussa, Antioxidant activity: The presence and impact of hydroxyl groups in small molecules of natural and synthetic origin, *Antioxidants—Benefits, sources, mechanisms of action*, 2021, pp. 318–377.
22. W. Zhang, S. Qi, X. Xue, Y. Al Naggar, L. Wu and K. Wang, Understanding the gastrointestinal protective effects of polyphenols using foodomics-based approaches, *Front. Immunol.*, 2021, **12**, 671150.
23. T. Zhang, C. H. Jeong, W. N. Cheng, H. Bae, H. G. Seo, M. C. Petriello and S. G. Han, Moringa extract enhances the fermentative, textural, and bioactive properties of yogurt, *LWT*, 2019, **101**, 276–284.
24. T. Lippolis, M. Cofano, G. R. Caponio, V. De Nunzio and M. Notarnicola, Bioaccessibility and bioavailability of diet polyphenols and their modulation of gut microbiota, *Int. J. Mol. Sci.*, 2023, **24**(4), 3813.
25. K. Wojtunik-Kulesza, A. Oniszczuk, T. Oniszczuk, M. Combrzyński, D. Nowakowska and A. Matwijczuk, Influence of *in vitro* digestion on composition, bioaccessibility and antioxidant activity of food polyphenols—A non-systematic review, *Nutrients*, 2020, **12**(5), 1401.
26. L. Arfaoui, Dietary plant polyphenols: Effects of food processing on their content and bioavailability, *Molecules*, 2021, **26**(10), 2959.
27. S. M. Aravind, S. Wichienchot, R. Tsao, S. Ramakrishnan and S. J. Chakkaravarthi, Role of dietary polyphenols on gut microbiota, their metabolites and health benefits, *Food Res. Int.*, 2021, **142**, 110189.
28. J. Grgić, G. Šelo, M. Planinić, M. Tišma and A. Bucić-Kojić, Role of the encapsulation in bioavailability of phenolic compounds, *Antioxidants*, 2020, **9**(10), 923.
29. Q. Huang, B. H. Braffett, S. J. Simmens, H. A. Young and C. L. Ogden, Dietary polyphenol intake in US adults and 10-year trends: 2007–2016, *J. Acad. Nutr. Diet.*, 2020, **120**(11), 1821–1833.

30. C. Taguchi, Y. Fukushima, Y. Kishimoto, N. Suzuki-Sugihara, E. Saita, Y. Takahashi and K. Kondo, Estimated dietary polyphenol intake and major food and beverage sources among elderly Japanese, *Nutrients*, 2015, **7**(12), 10269–10281.
31. A. K. Rashwan, A. I. Osman and W. Chen, Natural nutraceuticals for enhancing yogurt properties: a review, *Environ. Chem. Lett.*, 2023, **21**(3), 1907–1931.
32. A. B. Shori, Storage quality and antioxidant properties of yogurt fortified with polyphenol extract from nutmeg, black pepper, and white pepper, *Electron. J. Biotechnol.*, 2022, **57**, 24–30.
33. D. Sahingil and A. A. Hayaloglu, Enrichment of antioxidant activity, phenolic compounds, volatile composition and sensory properties of yogurt with rosehip (*Rosa canina* L.) fortification, *Int. J. Gastron. Food Sci.*, 2022, **28**, 100514.
34. M. Abdel-Hamid, Z. Huang, T. Suzuki, T. Enomoto, A. M. Hamed, L. Li and E. Romeih, Development of a multifunction set yogurt using *Rubus suavissimus S. lee* (Chinese sweet tea) extract, *Foods*, 2020, **9**(9), 1163.
35. D. Sun-Waterhouse, J. Zhou and S. S. Wadhwa, Drinking yoghurts with berry polyphenols added before and after fermentation, *Food Control*, 2013, **32**(2), 450–460.
36. V. Šeregelj, V. Tumbas Šaponjac, S. Lević, A. Kalušević, G. Ćetković, J. Čanadanović-Brunet, V. Nedović, S. Stajčić, J. Vulić and A. Vidaković, Application of encapsulated natural bioactive compounds from red pepper waste in yogurt, *J. Microencapsulation*, 2019, **36**(8), 704–714.
37. A. C. Vital, P. A. Goto, L. N. Hanai, S. M. Gomes-da-Costa, B. A. de Abreu Filho, C. V. Nakamura and P. T. Matumoto-Pintro, Microbiological, functional and rheological properties of low fat yogurt supplemented with *Pleurotus ostreatus* aqueous extract, *LWT-Food Sci. Technol.*, 2015, **64**(2), 1028–1035.
38. A. Simonetti, A. Perna, G. Grassi and E. Gambacorta, In vitro phenols bioaccessibility and antioxidant activity of goat milk yogurt fortified with Rhus coriaria leaf powder, *J. Food Sci.*, 2021, **86**(4), 1400–1409.
39. A. Helal and D. Tagliazucchi, Impact of *in-vitro* gastro-pancreatic digestion on polyphenols and cinnamaldehyde bioaccessibility and antioxidant activity in stirred cinnamon-fortified yogurt, *Lwt*, 2018, **89**, 164–170.
40. K. B. Petrotos, F. K. Karkanta, P. E. Gkoutsidis, I. Giavasis, K. N. Papatheodorou and A. C. Ntontos, Production of novel bioactive yogurt enriched with olive fruit polyphenols, *World Acad. Sci. Eng. Technol.*, 2012, **64**, 867–872.
41. T. M. El-Messery, E. Aly, R. López-Nicolas, T. Sánchez-Moya and G. Ros, Bioaccessibility and antioxidant activity of PCL-microencapsulated olive leaves polyphenols and its application in yogurt, *J. Food Sci.*, 2021, **86**(10), 4303–4315.
42. U. Picciotti, A. Massaro, A. Galiano and F. Garganese, Cheese fortification: Review and possible improvements, *Food Rev. Int.*, 2022, **38**(sup1), 474–500.

43. H. Himed-Idir, K. Mouhoubi, E. H. Siar, H. Boudries, H. Mansouri, N. Adjeroud, K. Madani and L. Boulekbache-Makhlouf, Effect of rosemary (*Rosmarinus officinalis* L.) supplementation on fresh cheese: Physicochemical properties, antioxidant potential, and sensory attributes, *J. Food Process. Preserv.*, 2021, **45**(1), e15057.
44. M. Masmoudi, I. Ammar, H. Ghribi and H. Attia, Physicochemical, radical scavenging activity and sensory properties of a soft cheese fortified with *Arbutus unedo* L. extract, *Food Biosci.*, 2020, **35**, 100579.
45. S. Fu, M. A. Augustin, Z. Shen, K. Ng, L. Sanguansri and S. Ajlouni, Bioaccessibility of curcuminoids in buttermilk in simulated gastrointestinal digestion models, *Food Chem.*, 2015, **179**, 52–59.
46. S. Fu, Z. Shen, S. Ajlouni, K. Ng, L. Sanguansri and M. A. Augustin, Interactions of buttermilk with curcuminoids, *Food Chem.*, 2014, **149**, 47–53.
47. J. H. Ye, E. Thomas, L. Sanguansri, Y. R. Liang and M. A. Augustin, Interaction between whole buttermilk and resveratrol, *J. Agric. Food Chem.*, 2013, **61**(29), 7096–7101.
48. M. Fidelis, J. S. Santos, G. B. Escher, R. S. Rocha, A. G. Cruz, T. M. Cruz, M. B. Marques, J. B. Nunes, M. A. do Carmo, L. A. de Almeida and T. Kaneshima, Polyphenols of jabuticaba [*Myrciaria jaboticaba (Vell.) O. Berg*] seeds incorporated in a yogurt model exert antioxidant activity and modulate gut microbiota of 1, 2-dimethylhydrazine-induced colon cancer in rats, *Food Chem.*, 2021, **334**, 127565.
49. G. O. Corazza, D. Bilibio, O. Zanella, A. L. Nunes, J. P. Bender, N. Carniel, P. P. dos Santos and W. L. Priamo, Pressurized liquid extraction of polyphenols from Goldenberry: Influence on antioxidant activity and chemical composition, *Food Bioprod. Process.*, 2018, **112**, 63–68.
50. M. R. Shahein, E. S. Atwaa, H. A. Radwan, A. A. Elmeligy, A. A. Hafiz, A. Albrakati and E. K. Elmahallawy, Production of a yogurt drink enriched with Golden berry (*Physalis pubescens* L.) juice and its therapeutic effect on hepatitis in rats, *Fermentation*, 2022, **8**(3), 112.
51. S. H. Kwon, D. Kothari, H. I. Jung, J. M. Lim, W. L. Kim, H. C. Kwon, S. G. Han, S. M. Seo, Y. K. Choi and S. K. Kim, Noni juice-fortified yogurt mitigates dextran sodium sulfate-induced colitis in mice through the modulation of inflammatory cytokines, *J. Funct. Foods*, 2021, **86**, 104652.
52. E. A. Abd El-Salam and N. F. Morsy, Optimization of the extraction of polyphenols and antioxidant activity from Malva parviflora L. leaves using Box–Behnken design, *Prep. Biochem. Biotechnol.*, 2019, **49**(9), 876–883.
53. M. E. El-Naggar, J. Hussein, S. M. El-sayed, A. M. Youssef, M. El Bana, Y. A. Latif and D. Medhat, Protective effect of the functional yogurt based on *Malva parviflora* leaves extract nanoemulsion on acetic acid-induced ulcerative colitis in rats, *J. Mater. Res. Technol.*, 2020, **9**(6), 14500–14508.
54. T. Zhang, C. H. Jeong, W. N. Cheng, H. Bae, H. G. Seo, M. C. Petriello and S. G. Han, Moringa extract enhances the fermentative, textural, and bioactive properties of yogurt, *LWT*, 2019, **101**, 276–284.

55. R. Ciriminna, F. Meneguzzo, A. Fidalgo, L. M. Ilharco and M. Pagliaro, Extraction, benefits and valorization of olive polyphenols, *Eur. J. Lipid Sci. Technol.*, 2016, **118**(4), 503–511.
56. K. Georgakouli, A. Mpesios, D. Kouretas, K. Petrotos, C. Mitsagga, I. Giavasis and A. Z. Jamurtas, The effects of an olive fruit polyphenol-enriched yogurt on body composition, blood redox status, physiological and metabolic parameters and yogurt microflora, *Nutrients*, 2016, **8**(6), 344.
57. H. El Hatmi, Z. Jrad, W. Mkadem, A. Chahbani, O. Oussaief, M. B. Zid, S. Zaidi, S. Khorchani, K. Belguith and N. B. Mihoubi, Fortification of soft cheese made from ultrafiltered dromedary milk with Allium roseum powder: Effects on textural, radical scavenging, phenolic profile and sensory characteristics, *LWT*, 2020, **132**, 109885.
58. T. Balabanova, N. Petkova, M. Ivanova and N. Panayotov, Design of Labneh cheese fortified with alginate-encapsulated pepper (*Capsicum annuum*) extracts, *Emirates J. Food Agric.*, 2020, 559–566.
59. A. Rashidinejad, E. J. Birch, D. Sun-Waterhouse and D. W. Everett, Total phenolic content and antioxidant properties of hard low-fat cheese fortified with catechin as affected by *in vitro* gastrointestinal digestion, *LWT – Food Sci. Technol.*, 2015, **62**(1), 393–399.
60. M. Sardiñas-Valdés, J. A. Hernández-Becerra, H. S. García, A. J. Chay-Canul, J. R. Velázquez-Martínez and A. A. Ochoa-Flores, Physicochemical and sensory properties of Manchego-type cheese fortified with nano-emulsified curcumin, *Int. Food Res. J.*, 2021, **28**(2), 326–336.
61. E. Brglez Mojzer, M. Knez Hrnčič, M. Škerget, Ž. Knez and U. Bren, Polyphenols: Extraction methods, antioxidative action, bioavailability and anticarcinogenic effects, *Molecules*, 2016, **21**(7), 901.
62. J. Bié, B. Sepodes, P. C. Fernandes and M. H. Ribeiro, Polyphenols in health and disease: Gut microbiota, bioaccessibility, and bioavailability, *Compounds*, 2023, **3**(1), 40–72.

CHAPTER 9

Prebiotic Foods: An Emerging Therapeutic Approach

ADARSH KUMAR SHUKLA, NEERU, RITU SINGH, SUNNY YADAV, VAISHALI AND ASHWANI KUMAR*

Department of Nutrition Biology, Central University of Haryana, Mahendergarh, Jant-Pali- 123031, Haryana, India
*Email: ashwanindri@gmail.com

9.1 Introduction

Manning and Gibson (2004)[1] defined a prebiotic as a "non-digestible food ingredient that beneficially affects the host by selectively stimulating the growth and/or activity of one or a limited number of bacteria in the colon and thus improves the host health". Prebiotics include fibres that are non-digestible components present in various foods, which can beneficially influence the host health by promoting the growth and actions of some species of colonic microbiota, usually *Lactobacillus* spp. and *Bifidobacterium* spp.[2] Some examples of prebiotics include arabinose, fructooligosaccharides (FOS), galactose, inulin, xylooligosaccharides (XOS), palatinose, galacto-oligosaccharides (GOS), stachyose, *etc.*[3,4] The main prebiotics commercially available are galactooligosaccharides/transgalactosylated-oligosaccharides (GOS/TOS), inulin, and fructooligosaccharides (FOS).[5] Prebiotics have shown their involvement in improving the bioavailability and uptake of minerals, reducing the possibility of diseases associated with the cardiovascular system, and stimulating satiety and loss of body weight and therefore inhibiting the occurrence of obesity.[6] They are present in vegetables like onions, tomatoes, bananas, *etc.*

Food Chemistry, Function and Analysis No. 44
Functional Foods of the Future
Edited by Vijai Kumar Gupta, Minaxi Sharma, Smriti Gaur and Ramesh Chander Kuhad

Published by the Royal Society of Chemistry, www.rsc.org

Prebiotic properties include preservation of intestinal microflora and stimulation of intestinal transportation, removing excess substances like cholesterol and glucose, as well as supporting the absorption of only essential substances,[7] promoting the production and absorption of B vitamins (B1, B2, B3, B6, B9, and B12) and contributing to controlling and decreasing the risk of diseases like osteoporosis and obesity.

This chapter presents a detailed overview of prebiotics in terms of types, sources, foods, therapeutic potentials and mechanisms of action. Molecular informatics has revealed the significant therapeutic role of prebiotics in humans. The beneficial effect of prebiotics alone or in combination with probiotics is also discussed in this chapter.

9.2 Types of Prebiotics

9.2.1 Fructans

This type of prebiotic constitutes an oligosaccharide or oligofructose and inulin. They have fructose subunits linked by a β (2-1) linkage into a chain, which generally ends with glucose units in a β (2-1) linkage. The degree of polymerization (DP) of inulin is up to 60, while that of fructooligosaccharide (FOS) is less than 10.[8] Some studies showed that the criteria for bacterial fermentation depend on the chain length of fructans.[9] Earlier it was thought that the stimulation of lactic acid bacteria by fructans is done selectively.

9.2.2 Oligosaccharides Derived from Glucose and Starch

There is a type of starch called resistant starch (RS) because of its resistance to upper gut digestion. RS is known to produce a high level of butyrate; therefore, it is classified as a prebiotic.[9] An experimental study showed that *Ruminococcus bromii* and *Bifidobacterium adolescentis* can degrade RS. However, without *R. bromii*, RS degradation is not achievable in mixed faecal and bacterial incubation.[10]

9.2.3 Other Oligosaccharides

Pectin is a polysaccharide that also results in the formation of oligosaccharides. These oligosaccharides are called pectin oligosaccharides (POS), which are a group of acidic polysaccharides based on homogalacturonan chains. POS can be partly methyl-esterified to form xylogalacturonan.[11] Methyl esterification modified the carboxyl groups and the structures can be acetylated either at C_2 or C_3. Side chains are linked by different types of sugars like galactose, arabinose, and xylose.

9.2.4 Lactulose

Lactulose consists of disaccharide units of both galactose and fructose and is present in products that are formed by heat treating milk and dairy products. It helps to enhance the growth of *Lactobacillus* and *Bifidobacterium* in the large intestine and increases calcium and magnesium bioavailability.[12] Lactulose is generally used in the field of pharmaceutical sciences that help relieve constipation.

9.2.5 Fructooligosaccharides (FOS)

Natural undigestible oligosaccharides that fulfill the specifications of a colonic diet and prebiotics are used as food additives.[13] Fructooligosaccharides are short and intermediate-length β-D-fructan chains in which the fructosyl units are bound by a β 2-1 glycosidic bond. FOS falls into that group of oligosaccharides which are obtained from plants. These substances contain two to three monosaccharide units that are joined together by α-glycosidic bonds (1-2) between terminally attached glucose and fructose.[14]

9.2.6 Polydextrose (PDX)

PDX contains polysaccharide units which are formed by arbitrary polymerization of glucose ($C_6H_{12}O_6$) in the presence of a small amount of sorbitol along with an acid catalyst (respectively in proportion of 90 : 10 : 1) under high temperature and maintaining a partial vacuum.[15] The large intestine is the primary source of partial fermentation of PDX, where faecal matter is increased in high amounts, transit time is reduced, and pH level decreases. As a result, fermentation helps to increase the growth of intestinal microbiota, and the removal of metabolites that could be carcinogenic.

9.2.7 Xylooligosaccharides (XOS)

Several xylose units are combined to form XOS. They are indigestible for human beings and are non-caloric. Lignin, cellulose, and hemicellulose are the three polymers that contribute to form lignocellulosic materials (LCM). XOS has high stability and is used as a food ingredient in industries. It is stable under a wide range of pH (2.5 to 8.0) and temperature. These oligosaccharides help in enhancing the quality of food products, alter the flavor, and stimulate the intestinal activity of *Bifidobacterium.* Xylooligosaccharides function as prebiotics in the digestive tract[16] by selectively feeding healthy bacteria. Many clinical trials for XOS have been performed, showing a range of health benefits including a decrease in blood sugars and lipids, digestive health effects, laxation, and positive changes to immune markers. Xylooligosaccharides are produced in plant fibre from the xylan fraction. Their C_5 structure is fundamentally different from that of other prebiotics based on C_6 sugars.

9.2.8 Galactooligosaccharides (GOS)

Galactooligosaccharides are found mainly in both human and cow's milk as digestion-resilient oligosaccharides.[17] They can also be produced from unique microbial lactose fermentation or synthesized by the enzyme β-galactosidase and lactose syrup. *Bifidobacterium* and *Lactobacillus* are selectively increased by GOS. They occur naturally in human milk. Galactooligosaccharides (GOS), an integral component derived from lactose, are categorized into two subgroups: GOS with excess galactose linked at the C_3, C_4, or C_6 positions, and GOS produced from lactose through enzymatic transglycosylation. The end result of this reaction is mainly a mixture of tri- to pentasaccharides of galactose bound β (1→6) and β (1→4), which are also known as transgalactooligosaccharides (TOS). Bifidobacteria and lactobacilli are greatly stimulated by GOS. They also stimulate enterobacteria, Bacteroidetes, and Firmicutes to a limited extent. The structure of GOS consists of galactose molecules attached to lactose, and comprise two to five units of galactose joined by β-glycosidic bonds. GOS can be described as those substances that are not utilized by their host and that reach the intestinal tract and are metabolized by *Bifidobacterium*.[18]

9.2.9 Inulin

The structure of inulin comprises a linear polymer of D-fructose held by β-glycosidic bonds. It belongs to the class of dietary fibres known as fructans and is produced and found in a variety of plants.[19,20] The concentration of inulin varies in different varieties of plants and depends on the time of their harvesting, storage conditions, and uses. Inulin proved to be useful in enriching food products produced from plants with fibre, while maintaining the aroma and taste of the standard formulations. Inulin helps in lowering the levels of harmful bacteria present in the intestine of the GI tract, reducing the risk of diseases like osteoporosis, relieving constipation, enhancing bowel movement frequency, enhancing weight of stools, lowering faecal pH, *etc.* It also exhibits some anticarcinogenic properties. The evaluation of glomerular filtration rate through intravenous injection using inulin has been proven to be very helpful. It is a kind of practice that becomes a standard procedure with no toxicity or side effects. Between 2 g and 12 g of inulin are consumed on average per person in Europe, while in Spain it is between 7 g and 12 g per person.[21]

Inulin sources, edible parts, and concentration in foods have been shown in Table 9.1.

9.3 Natural Plant-based Sources of Prebiotics

Prebiotic content in natural foods changes during processing like cooking, boiling, germination, soaking, *etc.* Different food consumption by humans leads to development of different microbiota in the gut due to different prebiotics present in natural foods. There are several prebiotic sources that may be obtained naturally. An overview of the sources, types, structure, and health benefits of prebiotics is given in Figure 9.1.

Table 9.1 Source, edible parts, and concentration of inulin in foods.

Edible parts of plants	Inulin (%)
Bulb of onion	2–6
Bulb of garlic	9–16
Root of chicory	15–20
Fruit of banana	0.3–0.7
Cereal of barley	0.5–1.5
Cereal of wheat	1–4
Cereal of rye	0.5–0.1
Leaves of dandelion	12–15

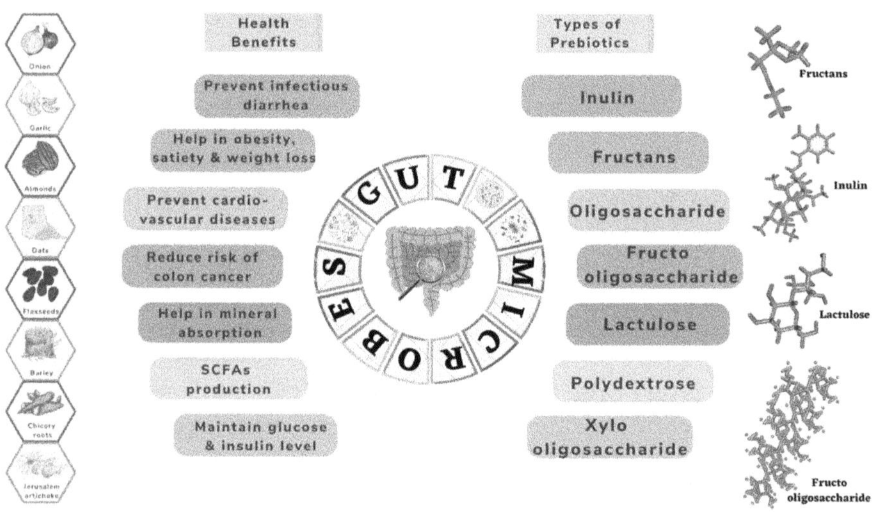

Figure 9.1 An overview of sources, types, structures, and health benefits of prebiotics.

9.3.1 Garlic

Allium sativum, commonly called garlic, is used mostly in making Indian curries and garlic breads. It is also helpful in the treatment of different diseases such as flu and gastrointestinal disorders.[22] Garlic has fibre which belongs to the class of fructans and helps in the growth of health-promoting microflora.[23] Other sources of FOS are fruits, vegetables and tubers like onion, *etc.* Garlic differentiated into cloves has higher fructans (76.24%) than white onions (25.86%) and non-differentiated garlic (71.13%). Chandrashekar *et al.* (2011)[24] reported a comparable yield of differentiated garlic (80%). Garlic contains a significant amount of FOS, which helps to protect the gastrointestinal tract (GIT) and prevent several ailments. 75% of the dry weight of garlic is due to fructans, which promote the growth of beneficial gut microbiota like bifidobacteria; however, it inhibits the growth of clostridia species, which cause harmful diseases in animals.[25]

9.3.2 Onion

Onion has the scientific name *Allium cepa* and belongs to the family Liliaceae[26] and has nutritional and therapeutic potential. Onion has carbohydrates, dietary fibre, vitamins, and minerals. The important carbohydrates in onion are monosaccharides and fructooligosaccharides.[27] Onion, like garlic, has a high amount of fructans which act as prebiotics; therefore, it can result in the growth of health-promoting microbiota.[23] It was reported that button forest onion *Eleutherine bulbosa* (BFOE) has a significant amount of inulins (2.1%), fructooligosaccharides (FOS) (10%), galactooligosaccharides (GOS) (1%), and raffinoses (7.5%). It was discovered that the mixed oligosaccharides might promote the development of *P. piscicida* 1Ub and *Bacillus* spp. NP5. In order to promote the growth of probiotics, the ideal BFOE concentration was 1.25 $mg\,mL^{-1}$.[23]

9.3.3 Chicory Root

Chicory root, also called *Cichorium intybus*, belongs to the Asteraceae family and has therapeutic properties.[28] A prebiotic fibre inulin that encourages the growth of good bacteria in the gut is abundant in chicory root. Jerusalem artichokes, chicory root, and garlic had the highest concentration (g per 100 g) of fructans, *i.e.* 8.9 ± 0.238, 8.40 ± 0.186, 7.51 ± 0.209, respectively. Scallion and white onion have low fructans; however, they have higher oligofructans, *i.e.* 2.24 ± 0.092 and 4.10 ± 0.147, respectively. Dried chicory has the highest amount of inulin (98%) and only 2% of other compounds; however, fresh chicory has inulin (68%), sucrose (14%), protein (6%), cellulose (5%), ash (4%), and other compounds (3%). Minerals (Ca, P, K), vitamins (A, C), and phenolic compounds are also present. The addition of inulin obtained from chicory root improves the nutritional properties of low-fat synbiotic yogurt.[28]

9.3.4 Jerusalem Artichoke

Jerusalem artichoke, scientifically known as *Cynara scolymus*, has high fibre content and low calories, carbohydrates (6.8%), and nitrogen compounds (2.9%), and also has therapeutic potential.[29] The inulin in Jerusalem artichokes makes them a superior prebiotic food that boosts the growth of beneficial gut flora.[30] The majority of oligomers found in artichokes cannot be digested by stomach enzymes; however, they are absorbed in the large intestine after reaching the colon and have a prebiotic impact by encouraging the proliferation of probiotic microorganisms. Additionally, it includes inulin and polyphenols, which are known to have antioxidant, hepatoprotective, and anticancer properties.[31] It has been suggested to combine probiotic microorganisms with inulin from artichokes as a prebiotic source to create synbiotic food items. In fact, throughout the process of manufacture and storage these prebiotics significantly boost probiotic viability.

9.3.5 Almonds

Almond is called *Prunus amygdalus* (formerly *Amygdalus communis*) and belongs to the genus *Prunus* and the family Rosaceae.[32] Almonds have 50% weight due to lipids.[33] Almonds also contain monounsaturated fatty acids (MUFA), polyunsaturated fatty acids (PUFA), vitamin E, arginine, and magnesium. Almonds also have indigestible carbohydrates, dietary fibre, polysterols, and polyphenols which affect gut health.[34] Almonds were reported to have a high prebiotic index (4.43), promote the growth of *Eubacterium rectale* and bifidobacteria, and increase the metabolism of healthy gut microbiota.[33] Another study on the effect of almonds on gut microbiota showed an increased number of *Lactobacillus* and *Bifidobacterium* and inhibition of harmful microbes like *Clostridium perfringens*. Decreases in nitroreductase, azoreductase, and glucuronidase activities as well as increases in galactosidase activity are caused by changes in the makeup of the gut microbiota. Raw almonds were reported to increase intestinal lipase as well as galactosidase activity and reduce azoreductase and glucuronidase activity.[35]

9.3.6 Flax Seeds

Flax seed, also called linseed, has beneficial properties and, therefore, is regarded as a functional food. It also has prebiotic properties due to the presence of soluble and insoluble fibre which helps in the growth of beneficial microbiota.[36] Flax seeds also have soluble fibre, linolenic acid, proteins, and phenolic compounds.[37] Flax seed consumption may change the microenvironment of the colon by enhancing *Prevotella* species up to 20 times and suppressing *Akkermansia muciniphila* by 30 times.[38] A different study showed that the consumption of flax seeds reduces the growth of Proteobacteria and Porphyromonadaceae in the stomach and may also improve the alcoholic liver.[39]

9.3.7 Barley

Barley, termed *Hordeum vulgare*, belongs to the Poaceae family and has high fibre, protein and vitamins, and low fat.[40] β-Glucan, a soluble fibre found in barley, functions as a prebiotic and promotes the development of beneficial gut flora. Barley has vitamins, minerals (Ca, Fe, Zn), polysaccharides, and oligosaccharides. The main saccharide component is β-glucan which influences the gut microbiota. Barley also enhances immunity, controls blood sugar, and lowers blood cholesterol.[41]

9.3.8 Oats

Oats, also called *Avena sativa*, has high polysaccharide, fibre, minerals, vitamins, and proteins.[42] The main polysaccharide component of oats is glucan. A viscous solution of glucans can promote the health of the human gut.[43] Oats are helpful in treating insulin resistance, obesity, hypertension, and dyslipidemia.[44]

9.4 Prebiotics in Food Products

Prebiotics are incorporated into foods in today's life which results in the formulation of functional foods that may promote the growth and activity of beneficial gut microbiota. Prebiotics have been incorporated in both dairy and non-dairy fermented foods. Dairy foods include yogurts, frozen desserts like ice cream, kefir, fermented cheese, flavored whey beverages, synbiotic shakes, kheer, *etc.* Prebiotics are also incorporated into non-dairy foods like fruit juices, fruit jelly, ready-to-eat fruit products, snack bars, granola bars, breads, cookies, vegetable milks like fermented soy milk, *etc.* These foods can be considered functional foods due to the incorporation of specific prebiotics which can confer specific health benefits to the host. The details of prebiotic food products along with beneficial effects are given in Table 9.2.

9.5 Therapeutic Approaches to Prebiotics

The primary purpose of prebiotics is to promote the development of specific probiotic bacteria in the GIT that are beneficial to the host's health. This is accomplished through competition for nutrients and epithelial adhesion sites, as well as antagonism (the creation of antimicrobial substances), which also serves as a barrier for the pathogens. Mostly short-chain fatty

Table 9.2 Details of prebiotic food products and their beneficial effects.

Sr. no.	Prebiotic food product	Beneficial effects	Ref.
1.	Fermented soy milk with fructooligosaccharides	Increase natural folate level	13
2.	Water kefir with coconut water	Low-calorie and high-lipid value product	45
3.	Sheep milk yogurts with inulin	Increased oleic acid	46
4.	Sheep milk ice creams with prebiotics	Low calorie, high fibre without loss in protein, improved sensory properties	47
5.	Yogurts with oligosaccharides from lactulose	Bifidogenic effect, enhanced iron absorption, anti-inflammatory effect	48
6.	Prebiotic whey beverage with inulin	Increased stability of milk proteins, rheological properties	49
7.	Vanilla ice cream with galactooligosaccharide	Improved sensory properties and positive impact on physiochemical parameters	50

acids (SCFAs) such as butyric acid, acetic acid, *etc.* are the final products of the digestion of carbohydrates, which are used as an energy source by the host.[51]

There are various health benefits of prebiotics. Prebiotics are helpful in decreasing the prevalence of antibiotic-associated diarrhea and decreasing the symptoms of inflammatory bowel disease (IBD).[52] Prebiotic food substances provide protective effects against the occurrence of colon cancers and increase the absorption of minerals like Ca, Mg, *etc.*, decrease the risk factor of cardiovascular diseases, and promote satiety and weight loss, hence preventing obesity.[53]

9.5.1 Infectious Diarrhea

The consumption of inulin and FOS mixture showed a substantial reduction in disease severity indices due to a decrease in calprotectin, an abundant neutrophil protein that is substantially increased in patients with inflammatory intestinal disease.[54]

9.5.2 Bioavailability and Calcium Uptake

Studies have shown that prebiotic intake, mainly fructans, enhances calcium absorption. This is due to lowered pH resulting from the production of SCFAs after fermentation of prebiotics. This decreases pH shifts and calcium solubility, and therefore increases bioavailability. An increase in calcium absorption and eventually bone density was seen in a study of 100 adults for 12 months with an intake of short- and long-chain inulin fructans of about 80 g per day.[55]

9.5.3 Obesity, Satiety, and Weight Loss

Obesity is correlated with an increased concentration of Firmicutes bacteria and a decreased concentration of bacteria from the Bacteroidetes phylum. Studies have shown that the energy balance in lean and obese mice is affected by the intestinal microbiota.[56] Monozygotic and dizygotic female twins were assessed based on their faecal microbial population. The effect of host environmental exposure, genotype, and host adiposity on gut microbiota was identified by Carmody *et al.* (2015).[57] Obesity was linked to reduced bacterial diversity, changes in the phylum type of the microbiota, and transforming metabolic pathways and gene representation.[58]

9.5.4 Cardiovascular Disease (CVD)

Although there is extensive evidence from potential epidemiological studies that dietary fibre has a beneficial effect on CVDs, the dietary fibre components that provide this benefit are unknown. The Dietary Reference Intake Committee concluded that cereal fibres are the most efficient fibres.[59]

Soluble and viscous fibre, C-reactive protein (CRP) and LDL cholesterol, appears to change the biomarkers of CVD.

9.5.5 Empowering Natural Colon Transit Period

Prebiotics increase the faecal mass, usually by raising the faecal microbiota mass and optimizing stool consistency. They boost faecal bulk, promote passage through the colon and shorten the transit time.[60] Colonic water retention decreases, stools become smoother and bulkier, and also the frequency of stool increases. These factors relieve constipation and improve the evacuation of the colon.

9.5.6 Production of Short-chain Fatty Acids (SCFAs)

Short-chain fatty acids perform an essential role in the development and metabolism of intestinal matter and even some systemic metabolic activity.[61] Prebiotics are an important source of energy for good and healthy intestinal bacteria, and are fermented into SCFAs. The increase in fatty acids is due to metabolic inter-feeding. Many of the prebiotics are fermented by other bacteria, such as *Bifidobacterium*, resulting in increased concentrations and abundance of SCFAs.[62] Acetate is the most common SCFA in the colon, accompanied by almost equal amounts of butyrate and propionate. Acetate is an important source of energy for the human body and is reabsorbed by the skeletal muscle, brain, and heart.[63] Fermentation of healthy microbiota is the primary source of acetate in the human body.

9.5.7 Enhancement of Mineral Absorption

Galactooligosaccharides, isomaltooligosaccharides, lactisol, lactulose, and inulin-type fructans significantly improve the absorption of minerals, particularly calcium and magnesium. Addition of inulin and oligofructose has been shown to increase Mg and Ca absorption more than oligosaccharide alone. In addition to increasing calcium and magnesium absorption, fructans in the form of inulin have been shown to prevent animals from developing magnesium deficiency-related symptoms and to correct osteopenia.[64]

9.5.8 Favorable Modulation of Lipid Levels

Inulin-type fructans regulate lipid digestion, secretion and metabolism resulting in reduced serum cholesterol level as well as favorable lipid redistribution among the different lipoproteins. Animal results showed that inulin-type fructans reduced both fasting and postprandial plasma triglyceride levels, mainly due to decreased plasma VLDL-triglyceride concentrations in the post-absorbent state.[65] An increase in the generation of SCFAs in the large intestine appears to be the primary mechanism of action for hypotriglyceridaemia, which results in a reduction in liver lipogenesis. SCFA production resulted in increased concentrations of propionate relative to acetate, inhibiting lipogenesis in hepatocytes.[66] In animals, the impact of prebiotics on cholesterol levels is less stable, and decreases are usually mild.[67] In humans, inulin is more effective in lowering triglyceride and cholesterol levels than oligofructose and the effects in diabetic and hyperlipidemic subjects are more pronounced.

9.5.9 Influence on Glucose and Insulin Levels

Prebiotics can have a variety of ways to favorably influence serum glucose and insulin levels. Prebiotics may decrease the amount of glucose to absorbed into the blood. The mechanism by which prebiotics can modulate glycemia and insulinemia is bacterial fermentation that produces SCFAs. Glycolysis is enhanced and gluconeogenesis is reduced by propionate or butyrate, which is known for glucagon-like peptide-1 (GLP-1) induction.[68]

9.5.10 Reduction in Colon Cancer Risk

Carcinogenic substances are incorporated from exogenous food components into the intestinal tract or formed endogenously by gut microflora. *In vitro* and *in vivo* studies have shown the ability of prebiotics to improve the mechanism of detoxification in colon cells and defend against the creation of colonic tumors. In animal studies, it has been shown that inulin-type fructans, xylooligosaccharides, and galactooligosaccharides inhibited chemically-induced colorectal cancer and pre-cancerous colon lesions.[69] At the same time, the growth of bacteria increases organic matter and therefore stools in bulk and accelerates colonic transit time, reducing the exposure of the colon to possible carcinogens. Prebiotics minimized both the occurrence and variety of abnormal crypt foci and colon tumors in animal studies and indicated that synbiotics have an even more pronounced preventive impact.[70]

Colon cancer incidences were reduced by supplementation of oligofructose. Inulin decreased the occurrence of medically induced breast cancer and tended to inhibit the growth of implanted tumors, reduce implanted cancer metastates and increase the efficacy of cancer chemotherapy.[71]

9.6 Mechanism of Action of Prebiotics

The primary purpose of prebiotics is to promote the growth and development of specific probiotic bacteria in the GIT that provide a beneficial impact on the host's health. The mechanism by which this is achieved involves antagonism (antimicrobial material production) and competition for nutrients and epithelial adhesion sites. They also act as barriers for pathogens.[72] SCFAs (butyric acid, acetic acid, *etc.*) are mostly the final products of the digestion of carbohydrates, which are used as an energy source by the host.[73] Since the carbohydrates are fermented, *Bifidobacterium* or *Lactobacillus* may generate certain substances that inhibit the production of gastrointestinal pathogens, and also reduce the pH of the intestine.[74]

The primary action of prebiotics is that they enhance the number of beneficial bacteria present in the intestinal gut and affect the health of the organism. They have a strong impact on intestinal microflora and induce several health benefits to their host. These are generally non-digestible substances; therefore, they are broken down through the process of fermentation by gut bacteria present in the intestine.[75] Products made by the process of fermentation enhance the availability of numerous genes and their transcriptional factors and therefore stimulate the intestinal microbiome to increase their population. They have the potential to raise lymphocyte and leukocyte count in gut-associated lymphoid tissue (GALT) and the peripheral system.[76] The mechanism of action of prebiotics is depicted in Figure 9.2.

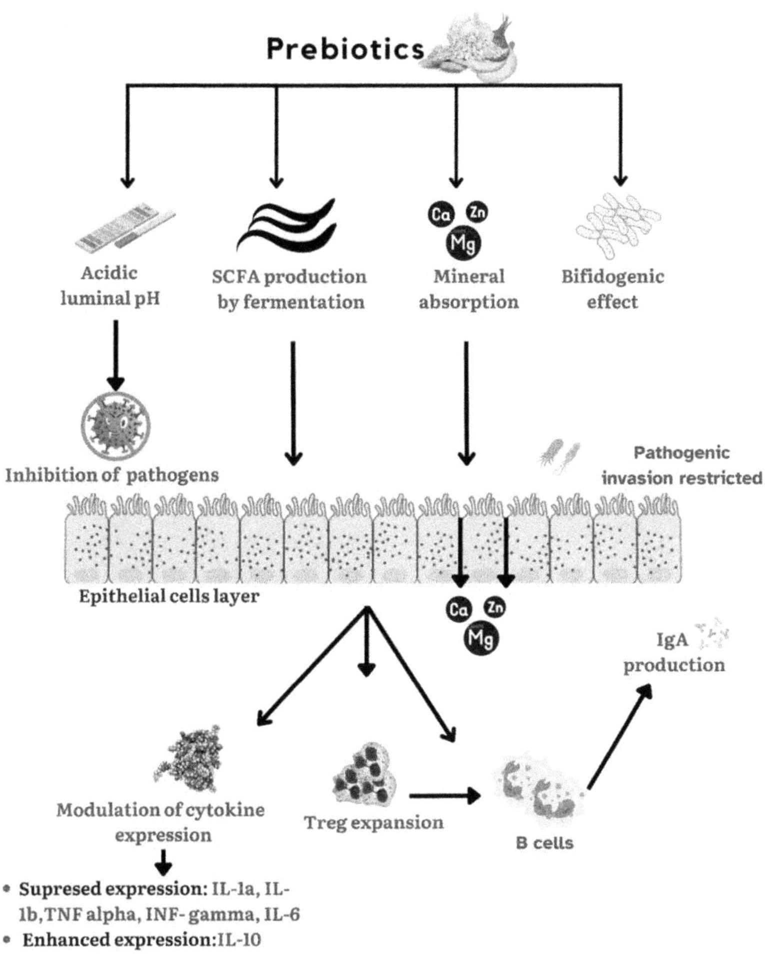

Figure 9.2 Mechanism of action of prebiotics.

9.7 Molecular Informatics of Prebiotic Substances with Human Genes

The STITCH database (Search Tool for Interactions of Chemicals) is a database that provides information on interactions between small molecules and proteins, including human genes.[77–79] It contains information on over 10 million interactions between more than 400 000 small molecules and over 9000 proteins.[80] Prebiotics are non-digestible carbohydrates that are digested by gut bacteria to produce short-chain fatty acids (SCFAs) and other metabolites with potential health benefits.[13] However, prebiotics themselves do not interact directly with human genes; rather, the beneficial effects of prebiotics are mediated by the gut microbiota, which can interact with human genes through various mechanisms.[81,82] The gut microbiota can produce metabolites that act as ligands for nuclear receptors, which can then modulate gene expression.[83] The STITCH database includes the interactions between small molecules and proteins, including those involved in the gut microbiota–host interactions that mediate the effects of prebiotics on human health.[84] Some metabolites that are present in prebiotics such as lactulose, fructans, fructose, and inulin interacted and were mapped with the human genes (Figure 9.3 and Table 9.3).

There are several human proteins that are known to interact with fructose, including enzymes involved in fructose metabolism and transporters that help to move fructose into cells. Fructokinase is an enzyme that converts fructose to fructose-1-phosphate, which can then be further metabolized to produce energy.[84] α-Glucosidase plays a key role in the final stage of carbohydrate digestion. Hexokinase 1 is expressed in many different tissues, including the liver, muscle, and brain.[85] Many human proteins are connected with fructans that

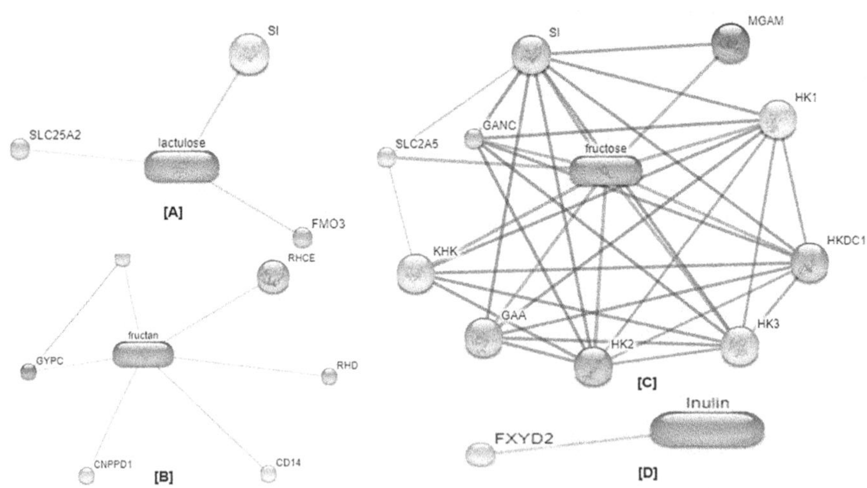

Figure 9.3 Network of prebiotic metabolites: [A] lactulose, [B] fructans, [C] fructose, and [D] inulin, with the human proteins.

Table 9.3 Description of prebiotics associated human genes.

Symbol	Description	Transcribed protein length	Bonding score
Fructans			
CD14	• Fixes to monomeric lipopolysaccharide and carries it to the MD-2/TLR4 complex • Facilitates the innate immunity to bacterial lipopolysaccharide (LPS) • Increases the expression of cell surface molecules, such as adhesion molecules	375	0.553
RHCE	• "Rh blood group, CcEe antigens"	417	0.515
CNPPD1	• "Cyclin Pas1/PHO80 domain containing"	410	0.493
RHD	• "Rh blood group, D antigen"	417	0.475
RHAG	• "Rh-associated glycoprotein" • Linked to the expression of the rhesus blood group antigen	409	0.435
GYPC	• "Glycophorin C (Gerbich blood group)" • Plays a significant part in controlling the integrity of red blood cells	128	0.431
Lactulose			
SI	• "Sucrase-isomaltase (alpha-glucosidase)" • Crucial for the breakdown of carbohydrates in the human body	1827	0.630
FMO3	• "Flavin-containing monooxygenase-3" • Takes part in the oxidative metabolism of numerous xenobiotics, including medications and pesticides	532	0.579
SLC25A2	• "Solute carrier family 25 (mitochondrial carrier; ornithine transporter) member-2" • Ornithine transfer from the cytoplasm to the matrix over the inner mitochondrial membrane	301	0.430
Inulin			
FXYD2	• "FXYD domain-containing ion transport regulator-2" • Contributes to the formation of the receptor site for cardiac glycoside binding or may modify sodium ATPase transport activity	66	0.700
Fructose			
KHK	• "Ketohexokinase (fructokinase)" • Helps in phosphorylation of fructose to form fructose-1-phosphate, which can then be further metabolized to create energy	298	0.996
SI	• "Sucrase-isomaltase (alpha-glucosidase)" • Significant impact on the last phase of digestion of carbohydrates	1827	0.985
HK1	• "Hexokinase-1" • An enzyme that is crucial for glucose metabolism	921	0.945

Table 9.3 (*Continued*)

Symbol	Description	Transcribed protein length	Bonding score
HK2	• "Hexokinase-2" • Insulin, the hormone that aids in controlling blood glucose levels, is regulated through *HK2*	917	0.945
HK3	• "Hexokinase-3 (white cell)" • In particular, hexokinase 3 in the pancreatic β-cells aids in triggering the release of insulin when blood glucose levels rise	923	0.942
HKDC1	• "Hexokinase domain containing-1" • Key role in the metabolism of glucose	917	0.932
SLC2A5	• "Solute carrier family 2 (facilitated glucose/fructose transporter)" • Carrier sensitive to cytochalasin B, appears to have a primary role in the transfer of fructose	501	0.920
MGAM	• "Maltase-glucoamylase (alpha-glucosidase)" • May have a special function in the digestion of malted dietary oligosaccharides used in the production of food	1857	0.918
GANC	• "Glucosidase, alpha; neutral C" • Has α-glucosidase activity	914	0.908
GAA	• "Glucosidase, alpha acid" • Important for the lysosomal process that converts glycogen to glucose	952	0.907

play an important role in mediating the innate immune response to bacterial lipopolysaccharide (LPS). Inulin is associated with the FXYD2 protein, which is a key player in the formation of receptor sites for cardiac glycoside binding or may modulate the transport function of sodium ATPase.[86]

9.8 Role of Prebiotics for the Development of Synbiotic Preparation

Synbiotics are defined as "a mixture of probiotics and prebiotics that beneficially affects the host by improving the survival and implantation of live microbial dietary supplements in the gastrointestinal tract, by selectively stimulating the growth and/or by activating the metabolism of one or a limited number of health-promoting bacteria, and thus improving host welfare".[87]

The use of prebiotics in the development of synbiotic foods, which are products that contain both probiotics and prebiotics, is becoming increasingly popular as a means of promoting gut health and preventing or treating various diseases.[88] By providing a source of nutrients for these beneficial bacteria, prebiotics can help to promote their growth and activity, leading to improved gut health and a range of other health benefits. The synbiotic effects are synergistic effects of prebiotics and probiotics, where the primary function of prebiotics is to promote probiotic implantation and survival. The metabolization of the prebiotic oligosaccharide is the main function of the probiotic and provides selective enrichment. Synbiotics are thought of as single products.

9.8.1 Procedure for Evaluating the Effectiveness of Probiotic and Prebiotic Components of a Synbiotic Formulation

Every ingredient and microorganism to be put in a synbiotic preparation has to be characterized fully with the help of the latest technologies. Firstly *in vitro* and *in vivo* efficacy of the microbes used will have to be examined.[89] Then possible growth indication of probiotics on specific media is obtained *via* a genomic scan of the microorganism's glycosidase spectra, which would allow the potential probiotic selection. Cell yield and growth rate information of the probiotic is achieved through growth curve experiments.[90] These types of experiments provide limited information, and not the interactions or affinity of probiotics with commensal microbiota. The prebiotic and probiotic components of the synbiotic formulation should be tested on their own and alongside the synbiotic components.[91] This process provides information on and compares the action of each component of faecal microbiota when each is used separately as well as in combination. Specific information on the synbiotic activity mechanism (synergistic or complementary) can be obtained using molecular methodologies such as probiotic survival in the mixed faecal environments during fermentation. The efficacy of the synbiotic produced must be determined in placebo-controlled human trials. Then a hybrid approach should be adopted, in which the efficacy of health biomarkers is monitored for each probiotic, prebiotic, and synbiotic procedure.[92]

A milk product containing a combination of *L. acidophilus* (7 $\log_{10}$ cfu per ml), *B. lactis* (7 $\log_{10}$ cfu per ml), and 2% Raftiline (DP10 inulin from Orafti) was investigated in 26 healthy adults in a randomized, placebo-controlled study.[93] Upon synbiotic administration, a noticeable increase in faecal bifidobacteria and lactobacilli was observed, which were evaluated by microbial culture techniques. In another study of 20 healthy humans, the impact of synbiotic on the metabolism of colonic nitrogen-protein was studied for 16 weeks.[94] Two groups of volunteers were assigned randomly, each testing a different combination of probiotics, either *Bifidobacterium breve* (Yakult) or *L. casei* Shirota. During every 4th week of treatment, volunteers ingested the synbiotic, prebiotic (2×10 g per day FOS), and the probiotic. Two weeks of washout were taken after each treatment. Urinary and faecal excretion of *p*-cresol and ammonia were analysed. This indicated fermentation in the colon. Prebiotics alone mediated a significant decrease in urinary *p*-cresol and ammonia and increases in faecal bifidobacteria concentration. No additive effect regarding proteolytic biomarkers was observed.

9.8.2 Health Benefits of Synbiotics

Synbiotics have several beneficial effects on human health such as the count of *Bifidobacterium* and *Lactobacillus* in the gut microflora being significantly

Table 9.4 Overview of efficacy of synbiotics in human studies.

Synbiotic	Study type	Duration	Results	Ref.
L. acidophilus, *B. longum* and FOS	16 healthy individuals who did not receive any antimicrobial therapy in the last 3 months	Two times a day for 30 days	Activity of intestinal microbiota was significantly influenced	96
L. rhamnosus GG", oligofructase-enriched inulin and *B. lactis* BB-12	A placebo-controlled, double-blind trial in 43 polypectomized individuals and 37 cancer patients	12 weeks	Improved intestinal barrier function and significant reduction in colorectal cell proliferation and genotoxicity	97
B. longum with a mixture of inulin and FOS	Placebo-controlled trial including 18 patients with ulcerative colitis (UC)	1 month	Increased gut bifidobacteria and hence improved ulcerative colitis symptoms	98
L. casei, *B. breve* and galactooligosaccharides	7 malnourished individuals with refractory enterocolitis	>1 year	Enhanced intestinal anaerobic bacteria, increased SCFAs in faeces, and suppressed pathogenic bacterial residence	99
Synbiotic yogurt having *L. acidophilus* 145, *B. longum* 913, and oligofructose	Placebo-controlled, randomized, and cross-over study consisting of 29 women	21 weeks	Improved ratio of LDL/HDL, increased serum HDL by 0.3 $mmol\ L^{-1}$	100
Combination of GOS/FOS (9 : 1) and *B. breve* M16-V	Placebo-controlled, double-blind multi-center trial (90 infants having atopic dermatitis)	12 weeks	Elevated levels of IgE, reduced the severity of atopic dermatitis	101
B. bifidum, *L. acidophilus*, oligofructose	Randomized, placebo-controlled, double-blind study on 20 individuals (50–60 years) having type 2 diabetes	2 weeks (daily two doses of 100 ml)	Enhanced HDL-cholesterol and reduced fasting glucose level	102

increased, and improvement in the functioning of the cirrhotic liver. Prebiotic foods help to improve immunoregulatory capabilities, prevent the translocation of bacteria, and reduce the occurrence of nosocomial infections in post-surgical procedures and related treatments.[95] Some of the studies of the efficacy of synbiotics in humans are summarized in Table 9.4.

The overall beneficial effects of synbiotics can be described as implanting the probiotics in the colon and improving their survival, improving the composition of microbes in the gastrointestinal tract, and stimulating the growth of selective health-promoting bacteria in the colon. Alander *et al.* (2001)[12] studied the effects of *Bifidobacterium* spp. and inulin in healthy volunteers. As a result, an enhanced count of *Bifidobacterium* spp. was observed in faecal samples. However, there was no significant increase in the bifidobacterial population with prebiotics alone. In another study, an increase in faecal SCFA levels, an improved rate of body weight gain, and an increase in concentrations of *Lactobacillus* and *Bifidobacterium* in faecal material were observed in patients with short bowel syndrome by the synbiotic intervention of *L. casei* Shirota, *Bifidobacterium breve* and GOS for over 1 year.[103] *In vitro*, synbiotic effects of probiotic *Lactobacillus* strains and cereal β-glucans were studied by Lam and Cheung (2013).[103] The effects on immune regulation were assessed for the host cells. The result of this study showed collaborative effects that modulated the transcriptional levels of various immune-related genes and hence led to the overall enhancement of anti-inflammatory effects.

9.9 Conclusion

Prebiotics are non-digestible food components that support the development and activity of healthy gut microbes. They can be found in various foods, such as chicory root, onions, garlic, and bananas, and can also be obtained through dietary supplements. Prebiotic foods have the potential to provide therapeutic benefits in a range of conditions, including gastrointestinal disorders, metabolic disorders, and immune system disorders. They may help to improve gut health, reduce inflammation, and modulate the immune system. Molecular informatics forecasts the more significant role of prebiotics in humans. Combining prebiotics with probiotics in synbiotic foods can provide even greater health benefits. However, further research is needed to fully understand the mechanisms underlying these effects and to optimize the use of prebiotics in therapeutic interventions.

References

1. T. S. Manning and G. R. Gibson, Prebiotics, *Best Pract. Res., Clin. Gastroenterol.*, 2004, **18**(2), 287–298.
2. J. M. Laparra and Y. Sanz, Interactions of gut microbiota with functional food components and nutraceuticals, *Pharmacol. Res.*, 2010, **61**(3), 219–225.

3. D. Bosscher, J. Van Loo and A. Franck, Inulin and oligofructose as prebiotics in the prevention of intestinal infections and diseases, *Nutr. Res. Rev.*, 2006, **19**(2), 216–226.
4. M. H. Saier Jr, R. N. Hvorup and R. D. Barabote, Evolution of the bacterial phosphotransferase system: from carriers and enzymes to group translocators, *Biochem. Soc. Trans.*, 2005, **33**(1), 220–224
5. J. K. Malik, A. H. Ahmad, S. Kalpana, A. Prakash and R. C. Gupta, Synbiotics: Safety and toxicity considerations, in *Nutraceuticals*, Academic Press, 2016, pp. 811–822.
6. K. Pokusaeva, G. F. Fitzgerald and D. van Sinderen, Carbohydrate metabolism in Bifidobacteria, *Genes Nutr.*, 2011, **6**, 285–306.
7. N. Kaur and A. K. Gupta, Applications of inulin and oligofructose in health and nutrition, *J. Biosci.*, 2002, **27**, 703–714.
8. B. R. Galdón, C. T. Rodríguez, E. R. Rodríguez and C. D. Romero, Fructans and major compounds in onion cultivars (Allium cepa), *J. Food Compos. Anal.*, 2009, **22**(1), 25–32.
9. K. P. Scott, J. C. Martin, S. H. Duncan and H. J. Flint, Prebiotic stimulation of human colonic butyrate-producing bacteria and bifidobacteria, *in vitro*, *FEMS Microbiol. Ecol.*, 2014, **87**(1), 30–40.
10. X. Ze, S. H. Duncan, P. Louis and H. J. Flint, Ruminococcus bromii is a keystone species for the degradation of resistant starch in the human colon, *ISME J.*, 2012, **6**(8), 1535–1543.
11. Z. Košťálová, Z. Hromádková and A. Ebringerová, Structural diversity of pectins isolated from the Styrian oil-pumpkin (Cucurbita pepo var. styriaca) fruit, *Carbohydr. Polym.*, 2013, **93**(1), 163–171.
12. M. Alander, J. Mättö, W. Kneifel, M. Johansson, B. Kögler, R. Crittenden, T. Mattila-Sandholm and M. Saarela, Effect of galacto-oligosaccharide supplementation on human faecal microflora and on survival and persistence of Bifidobacterium lactis Bb-12 in the gastrointestinal tract, *Int. Dairy J.*, 2001, **11**(10), 817–825.
13. M. A. Albuquerque, R. Bedani, J. G. LeBlanc and S. M. Saad, Passion fruit by-product and fructooligosaccharides stimulate the growth and folate production by starter and probiotic cultures in fermented soymilk, *Int. J. Food Microbiol.*, 2017, **261**, 35–41.
14. V. R. Konasani, Structure-function studies of processing alpha-glucosidase-i, Doctoral dissertation, University of British Columbia.
15. Z. Jie, L. Bang-yao, X. Ming-Jie, L. Hai-wei, Z. Zu-kang, W. Ting-song and S. A. Craig, Studies on the effects of polydextrose intake on physiologic functions in Chinese people, *Am. J. Clin. Nutr.*, 2000, **72**(6), 1503–1509.
16. J. Rui, S. Deng, A. Arazi, A. L. Perdigoto, Z. Liu and K. C. Herold, β cells that resist immunological attack develop during progression of autoimmune diabetes in NOD mice, *Cell Metab.*, 2017, **25**(3), 727–738.
17. K. M. Tuohy, G. C. Rouzaud, W. M. Bruck and G. R. Gibson, Modulation of the human gut microflora towards improved health using prebiotics-assessment of efficacy, *Curr. Pharm. Des.*, 2005, **11**(1), 75–90.

18. H. W. Modler, Bifidogenic factors—sources, metabolism and applications, *Int. Dairy J.*, 1994, **4**(5), 383–407.
19. N. C. Carpita, J. Kanabus and T. L. Housley, Linkage structure of fructans and fructan oligomers from Triticum aestivum and Festuca arundinacea leaves, *J. Plant Physiol.*, 1989, **134**(2), 162–168.
20. M. Roberfroid, Dietary fiber, inulin, and oligofructose: a review comparing their physiological effects, *Crit. Rev. Food Sci. Nutr.*, 1993, **33**(2), 103–148.
21. S. Valdemiro Carlos, The importance of prebiotics in functional foods and clinical practice, *Food Nutr. Sci.*, 2011, **2**(2), 133–144.
22. P. Mikaili, S. Maadirad, M. Moloudizargari, S. Aghajanshakeri and S. Sarahroodi, Therapeutic uses and pharmacological properties of garlic, shallot, and their biologically active compounds, *Iran. J. Basic Med. Sci.*, 2013, **16**(10), 1031.
23. P. Gullón, B. Gullón, A. Romaní, G. Rocchetti and J. M. Lorenzo, Smart advanced solvents for bioactive compounds recovery from agri-food by-products: A review, *Trends Food Sci. Technol.*, 2020, **101**, 182–197.
24. P. M. Chandrashekar, K. V. H. Prashanth and Y. P. Venkatesh, Isolation, structural elucidation and immunomodulatory activity of fructans from aged garlic extract, *Phytochemistry*, 2011, **72**(2–3), 255–264.
25. M. M. Tawfick, H. Xie, C. Zhao, P. Shao and M. A. Farag, Inulin fructans in diet: Role in gut homeostasis, immunity, health outcomes and potential therapeutics, *Int. J. Biol. Macromol.*, 2022, **208**, 948–961.
26. K. Singh Bora and A. Sharma, Phytoconstituents and therapeutic potential of Allium cepa Linn. A Review, *Pharmacogn. Rev.*, 2009, **3**, 170.
27. B. R. Galdón, C. T. Rodríguez, E. R. Rodríguez and C. D. Romero, Fructans and major compounds in onion cultivars (Allium cepa), *J. Food Compos. Anal.*, 2009, **22**(1), 25–32.
28. W. M. El-Kholy, R. A. Aamer and A. N. Ali, Utilization of inulin extracted from chicory (Cichorium intybus L.) roots to improve the properties of low-fat synbiotic yoghurt, *Ann. Agric. Sci.*, 2020, **65**(1), 59–67.
29. H. Wieser, P. Koehler and K. A. Scherf, *Wheat-An exceptional crop: Botanical features, chemistry, utilization, nutritional and health aspects*, Elsevier, 2020.
30. N. Tiengtam, S. Khempaka, P. Paengkoum and S. Boonanuntanasarn, Effects of inulin and Jerusalem artichoke (Helianthus tuberosus) as prebiotic ingredients in the diet of juvenile Nile tilapia (Oreochromis niloticus), *Anim. Feed Sci. Technol.*, 2015, **207**, 120–129.
31. M. Van Hul and P. D. Cani, Targeting carbohydrates and polyphenols for a healthy microbiome and healthy weight, *Curr. Nutr. Rep.*, 2019, **8**, 307–316.
32. B. Nawade, M. Yahyaa, H. Reuveny, L. Shaltiel-Harpaz, O. Eisenbach, A. Faigenboim, I. Bar-Yaakov, D. Holland and M. Ibdah, Profiling of volatile terpenes from almond (Prunus dulcis) young fruits and characterization of seven terpene synthase genes, *Plant Sci.*, 2019, **287**, 110187.

33. G. Mandalari, C. Nueno-Palop, G. Bisignano, M. S. Wickham and A. Narbad, Potential prebiotic properties of almond (Amygdalus communis L.) seeds, *Appl. Environ. Microbiol.*, 2008, **74**(14), 4264–4270.
34. S. C. Li, Y. H. Liu, J. F. Liu, W. H. Chang, C. M. Chen and C. Y. Chen, Almond consumption improved glycemic control and lipid profiles in patients with type 2 diabetes mellitus, *Metabolism*, 2011, **60**(4), 474–479.
35. M. K. Sharif, M. S. Butt and H. R. Sharif, Role of nanotechnology in enhancing bioavailability and delivery of dietary factors, in *Nutrient delivery*, Academic Press, 2017, pp. 587–618.
36. K. A. Baritugo, A. Bakhsh, B. Kim and S. Park, Perspectives on functional foods for improvement of canine health and treatment of diseases, *J. Funct. Foods*, 2023, **109**, 105744.
37. C. K. Dzuvor, J. T. Taylor, C. Acquah, S. Pan and D. Agyei, Bioprocessing of functional ingredients from flaxseed, *Molecules*, 2018, **23**(10), 2444.
38. M. J. Liu, J. Y. Yang, Z. H. Yan, S. Hu, J. Q. Li, Z. X. Xu and Y. P. Jian, Recent findings in Akkermansia muciniphila-regulated metabolism and its role in intestinal diseases, *Clin. Nutr.*, 2022, **41**(10), 2333–2344.
39. N. Zafari, M. Velayati, M. Fahim, M. Maftouh, G. Pourali, M. Khazaei, M. Nassiri, S. M. Hassanian, M. Ghayour-Mobarhan, G. A. Ferns and M. A. Kiani, Role of gut bacterial and non-bacterial microbiota in alcohol-associated liver disease: Molecular mechanisms, biomarkers, and therapeutic prospective, *Life Sci.*, 2022, **305**, 120760.
40. *Breakfast cereals and how they are made: raw materials, processing, and production*, ed. A. A. Perdon, S. L. Schonauer and K. Poutanen, Elsevier, 2020.
41. A. K. Sinha, V. Kumar, H. P. Makkar, G. De Boeck and K. Becker, Non-starch polysaccharides and their role in fish nutrition–A review, *Food Chem.*, 2011, **127**(4), 1409–1426.
42. P. Rasane, A. Jha, L. Sabikhi, A. Kumar and V. S. Unnikrishnan, Nutritional advantages of oats and opportunities for its processing as value added foods-a review, *J. Food Sci. Technol.*, 2015, **52**, 662–675.
43. M. Henrion, C. Francey, K. A. Lê and L. Lamothe, Cereal B-glucans: The impact of processing and how it affects physiological responses, *Nutrients*, 2019, **11**(8), 1729.
44. D. El Khoury, C. Cuda, B. L. Luhovyy and G. H. Anderson, Beta glucan: health benefits in obesity and metabolic syndrome, *J. Nutr. Metab.*, 2012, **2012**(1), 851362.
45. V. Alves, T. Scapini, A. F. Camargo, C. Bonatto, F. S. Stefanski, E. P. de Jesus, L. G. Diniz, L. C. Bertan, R. R. Maldonado and H. Treichel, Development of fermented beverage with water kefir in water-soluble coconut extract (Cocos nucifera L.) with inulin addition, *LWT*, 2021, **145**, 111364.
46. C. F. Balthazar, C. C. Júnior, J. Moraes, M. P. Costa, R. S. Raices, R. M. Franco, A. G. Cruz and A. C. Silva, Physicochemical evaluation of sheep milk yogurts containing different levels of inulin, *J. Dairy Sci.*, 2016, **99**(6), 4160–4168.

47. C. F. Balthazar, H. L. Silva, R. N. Cavalcanti, E. A. Esmerino, L. P. Cappato, Y. K. Abud, J. Moraes, M. M. Andrade, M. Q. Freitas, C. Sant'Anna and R. S. Raices, Prebiotics addition in sheep milk ice cream: A rheological, microstructural and sensory study, *J. Funct. Foods*, 2017, **35**, 564–573.
48. P. Delgado-Fernández, O. Hernández-Hernández, A. Olano, F. J. Moreno and N. Corzo, Probiotic viability in yoghurts containing oligosaccharides derived from lactulose (OsLu) during fermentation and cold storage, *Int. Dairy J.*, 2020, **102**, 104621.
49. J. T. Guimaraes, E. K. Silva, A. L. Costa, R. L. Cunha, M. Q. Freitas, M. A. Meireles and A. G. Cruz, Manufacturing a prebiotic whey beverage exploring the influence of degree of inulin polymerization, *Food Hydrocolloids*, 2018, **77**, 787–795.
50. C. F. Balthazar, H. L. Silva, R. M. Celeguini, R. Santos, G. M. Pastore, C. C. Junior, M. Q. Freitas, L. C. Nogueira, M. C. Silva and A. G. Cruz, Effect of galactooligosaccharide addition on the physical, optical, and sensory acceptance of vanilla ice cream, *J. Dairy Sci.*, 2015, **98**(7), 4266–4272.
51. L. Wang, S. Cen, G. Wang, Y. K. Lee, J. Zhao, H. Zhang and W. Chen, Acetic acid and butyric acid released in large intestine play different roles in the alleviation of constipation, *J. Funct. Foods*, 2020, **69**, 103953.
52. C. Mantegazza, P. Molinari, E. D'Auria, M. Sonnino, L. Morelli and G. V. Zuccotti, Probiotics and antibiotic-associated diarrhea in children: A review and new evidence on Lactobacillus rhamnosus GG during and after antibiotic treatment, *Pharmacol. Res.*, 2018, **128**, 63–72.
53. R. Catanzaro, M. Sciuto and F. Marotta, Lactose intolerance: An update on its pathogenesis, diagnosis, and treatment, *Nutr. Res.*, 2021, **89**, 23–34.
54. R. Gyawali, N. Nwamaioha, R. Fiagbor, T. Zimmerman, R. H. Newman and S. A. Ibrahim, The role of prebiotics in disease prevention and health promotion, in *Dietary interventions in gastrointestinal diseases*, Academic Press, 2019, pp. 151–167.
55. J. W. Finley, A. Soto-Vaca, J. Heimbach, T. P. Rao, L. R. Juneja, J. Slavin and G. C. Fahey, Safety assessment and caloric value of partially hydrolyzed guar gum, *J. Agric. Food Chem.*, 2013, **61**(8), 1756–1771.
56. R. Kübeck, C. Bonet-Ripoll, C. Hoffmann, A. Walker, V. M. Müller, V. L. Schüppel, I. Lagkouvardos, B. Scholz, K. H. Engel, H. Daniel and P. Schmitt-Kopplin, Dietary fat and gut microbiota interactions determine diet-induced obesity in mice, *Mol. Metab.*, 2016, **5**(12), 1162–1174.
57. R. N. Carmody, G. K. Gerber, J. M. Luevano, D. M. Gatti, L. Somes, K. L. Svenson and P. J. Turnbaugh, Diet dominates host genotype in shaping the murine gut microbiota, *Cell Host Microbe*, 2015, **17**(1), 72–84.
58. C. A. Lozupone, J. I. Stombaugh, J. I. Gordon, J. K. Jansson and R. Knight, Diversity, stability and resilience of the human gut microbiota, *Nature*, 2012, **489**(7415), 220–230.

59. E. Cabré, Clinical Nutrition University: Nutrition in the prevention and management of irritable bowel syndrome, constipation and diverticulosis, *e-SPEN, Eur. e-J. Clin. Nutr. Metab.*, 2011, **6**(2), e85–e95.
60. M. Wang, S. Wichienchot, X. He, X. Fu, Q. Huang and B. Zhang, In vitro colonic fermentation of dietary fibers: Fermentation rate, short-chain fatty acid production and changes in microbiota, *Trends Food Sci. Technol.*, 2019, **88**, 1–9.
61. L. X. Zheng, X. Q. Chen and K. L. Cheong, Current trends in marine algae polysaccharides: The digestive tract, microbial catabolism, and prebiotic potential, *Int. J. Biol. Macromol.*, 2020, **151**, 344–354.
62. A. Jiao, B. Yu, J. He, J. Yu, P. Zheng, Y. Luo, J. Luo, H. Yan, Q. Wang, H. Wang and X. Mao, Sodium acetate, propionate, and butyrate reduce fat accumulation in mice *via* modulating appetite and relevant genes, *Nutrition*, 2021, **87**, 111198.
63. H. Whelton, Functional foods and oral health, in *Food Constituents and Oral Health*, Woodhead Publishing, 2009, pp. 488–528.
64. N. C. Habib, S. M. Honoré, S. B. Genta and S. S. Sánchez, Hypolipidemic effect of Smallanthus sonchifolius (yacon) roots on diabetic rats: Biochemical approach, *Chem.-Biol. Interact.*, 2011, **194**(1), 31–39.
65. K. Weitkunat, S. Schumann, K. J. Petzke, M. Blaut, G. Loh and S. Klaus, Effects of dietary inulin on bacterial growth, short-chain fatty acid production and hepatic lipid metabolism in gnotobiotic mice, *J. Nutr. Biochem.*, 2015, **26**(9), 929–937.
66. K. M. Tuohy, H. M. Probert, C. W. Smejkal and G. R. Gibson, Using probiotics and prebiotics to improve gut health, *Drug Discovery Today*, 2003, **8**(15), 692–700.
67. Y. Ma, E. Lee, H. Yoshikawa, T. Noda, J. Miyamoto, I. Kimura, R. Hatano and T. Miki, Phloretin suppresses carbohydrate-induced GLP-1 secretion via inhibiting short chain fatty acid release from gut microbiome, *Biochem. Biophys. Res. Commun.*, 2022, **621**, 176–182.
68. A. Yasmin, M. S. Butt, M. Afzaal, M. van Baak, M. T. Nadeem and M. Z. Shahid, Prebiotics, gut microbiota and metabolic risks: Unveiling the relationship, *J. Funct. Foods*, 2015, **17**, 189–201.
69. I. Aggeletopoulou, C. Konstantakis, S. F. Assimakopoulos and C. Triantos, The role of the gut microbiota in the treatment of inflammatory bowel diseases, *Microb. Pathog.*, 2019, **137**, 103774.
70. O. Brunser and M. Gotteland, Probiotics and prebiotics in human health: an overview, *Bioact. Foods Promot. Health*, 2010, 73–93.
71. T. S. Manning and G. R. Gibson, Prebiotics, *Best Pract. Res., Clin. Gastroenterol.*, 2004, **18**(2), 287–298.
72. A. Monteagudo-Mera, R. A. Rastall, G. R. Gibson, D. Charalampopoulos and A. Chatzifragkou, Adhesion mechanisms mediated by probiotics and prebiotics and their potential impact on human health, *Appl. Microbiol. Biotechnol.*, 2019, **103**, 6463–6472.

73. S. I. Mussatto and I. M. Mancilha, Non-digestible oligosaccharides: A review, *Carbohydr. Polym.*, 2007, **68**(3), 587–597.
74. P. Brandtzaeg, H. Kiyono, R. Pabst and M. W. Russell, Terminology: nomenclature of mucosa-associated lymphoid tissue, *Mucosal Immunol.*, 2008, **1**(1), 31–37.
75. Deepika, A. K. Shukla, A. Kumari and A. Kumar, Gut brain regulation using psychobiotics for improved neuropsychological illness, *Dev. Psychobiol.*, 2023, **65**(5), e22404.
76. P. Yadav, A. K. Shukla, T. Dhewa and A. Kumar, In silico investigation of antioxidant interaction and effect of probiotic fermentation on antioxidant profiling of pearl millet-based rabadi beverage, *J. Appl. Nat. Sci.*, 2021, **13**(4), 1531–1544.
77. P. Yadav, A. K. Shukla, A. Kumari, T. Dhewa and A. Kumar, Nutritional evaluation of probiotics enriched rabadi beverage (PERB) and molecular mapping of digestive enzyme with dietary fibre for exploring the therapeutic potential, *Food Humanity*, 2024, **2**, 100221.
78. D. Szklarczyk, A. Santos, C. Von Mering, L. J. Jensen, P. Bork and M. Kuhn, STITCH 5: augmenting protein–chemical interaction networks with tissue and affinity data, *Nucleic Acids Res.*, 2016, **44**(D1), D380–D384.
79. A. S. Neish, Microbes in gastrointestinal health and disease, *Gastroenterology*, 2009, **136**(1), 65–80.
80. A. K. Shukla and A. Kumar, Virtual screening of orally active lead compounds of pearl millet and their structural activity against target protein of COVID-19, *Russ. J. Bioorg. Chem.*, 2023, **49**(Suppl 1), S53–S70.
81. O. Chávez-Talavera, A. Tailleux, P. Lefebvre and B. Staels, Bile acid control of metabolism and inflammation in obesity, type 2 diabetes, dyslipidemia, and nonalcoholic fatty liver disease, *Gastroenterology*, 2017, **152**(7), 1679–1694.
82. C. Zhang, L. Li, Y. Zhang and C. Zeng, Recent advances in fructose intake and risk of hyperuricemia, *Biomed. Pharmacother.*, 2020, **131**, 110795.
83. M. M. Adeva-Andany, M. González-Lucán, C. Donapetry-García, C. Fernández-Fernández and E. Ameneiros-Rodríguez, Glycogen metabolism in humans, *BBA Clin.*, 2016, **5**, 85–100.
84. K. L. Lansbery, *Characterization of the FXYD family: Trafficking, regulation and function*, Washington University in St. Louis, 2001.
85. E. Silva and P. Soares-da-Silva, New insights into the regulation of Na^+, K^+-ATPase by ouabain, *Int. Rev. Cell Mol. Biol.*, 2012, **294**, 99–132.
86. C. J. Ziemer and G. R. Gibson, An overview of probiotics, prebiotics and synbiotics in the functional food concept: perspectives and future strategies, *Int. Dairy J.*, 1998, **8**(5–6), 473–479.
87. R. G. Kerry, J. K. Patra, S. Gouda, Y. Park, H. S. Shin and G. Das, Benefaction of probiotics for human health: A review, *J. Food Drug Anal.*, 2018, **26**(3), 927–939.

88. S. Kolida and G. R. Gibson, Synbiotics in health and disease, *Annu. Rev. Food Sci. Technol.*, 2011, **2**(1), 373–393.
89. S. A. Ali, P. Singh, S. K. Tomar, A. K. Mohanty and P. Behare, Proteomics fingerprints of systemic mechanisms of adaptation to bile in Lactobacillus fermentum, *J. Proteomics*, 2020, **213**, 103600.
90. F. Gaggìa, P. Mattarelli and B. Biavati, Probiotics and prebiotics in animal feeding for safe food production, *Int. J. Food Microbiol.*, 2010, **141**, S15–S28.
91. R. Mishra, S. Mohanty, S. Mahapatra and P. Prasad, Overview of microbial therapeutics in immunological disorders, in *Microbiome Therapeutics*, Academic Press, 2023, pp. 289–353.
92. M. C. Casiraghi, E. Canzi, R. Zanchi, E. Donati and L. Villa, Effects of a synbiotic milk product on human intestinal ecosystem, *J. Appl. Microbiol.*, 2007, **103**(2), 499–506.
93. B. Guida, R. Germanò, R. Trio, D. Russo, B. Memoli, L. Grumetto, F. Barbato and M. Cataldi, Effect of short-term synbiotic treatment on plasma p-cresol levels in patients with chronic renal failure: a randomized clinical trial, *Nutr., Metab. Cardiovasc. Dis.*, 2014, **24**(9), 1043–1049.
94. R. Kumari, A. Singh, A. N. Yadav, S. Mishra, A. Sachan and S. G. Sachan, Probiotics, prebiotics, and synbiotics: Current status and future uses for human health, in *New and future developments in microbial biotechnology and bioengineering*, Elsevier, 2020, pp. 173–190.
95. M. Ndagijimana, L. Laghi, B. Vitali, G. Placucci, P. Brigidi and M. E. Guerzoni, Effect of a synbiotic food consumption on human gut metabolic profiles evaluated by ^{1}H Nuclear Magnetic Resonance spectroscopy, *Int. J. Food Microbiol.*, 2009, **134**(1–2), 147–153.
96. J. Rafter, M. Bennett, G. Caderni, Y. Clune, R. Hughes, P. C. Karlsson, A. Klinder, M. O'Riordan, G. C. O'Sullivan, B. Pool-Zobel and G. Rechkemmer, Dietary synbiotics reduce cancer risk factors in polypectomized and colon cancer patients, *Am. J. Clin. Nutr.*, 2007, **85**(2), 488–496.
97. E. Furrie, S. Macfarlane, A. Kennedy, J. H. Cummings, S. V. Walsh, D. A. O'neil and G. T. Macfarlane, Synbiotic therapy (Bifidobacterium longum/Synergy 1) initiates resolution of inflammation in patients with active ulcerative colitis: a randomised controlled pilot trial, *Gut*, 2005, **54**(2), 242–249.
98. Y. Kanamori, M. Sugiyama, K. Hashizume, N. Yuki, M. Morotomi and R. Tanaka, Experience of long-term synbiotic therapy in seven short bowel patients with refractory enterocolitis, *J. Pediatr. Surg.*, 2004, **39**(11), 1686–1692.
99. G. Kiessling, J. Schneider and G. Jahreis, Long-term consumption of fermented dairy products over 6 months increases HDL cholesterol, *Eur. J. Clin. Nutr.*, 2002, **56**(9), 843–849.
100. L. B. Van Der Aa, H. S. Heymans, W. M. Van Aalderen, J. H. Sillevis Smitt, J. Knol, K. Ben Amor, D. A. Goossens, A. B. Sprikkelman and

Synbad Study Group, Effect of a new synbiotic mixture on atopic dermatitis in infants: a randomized-controlled trial, *Clin. Exp. Allergy*, 2010, **40**(5), 795–804.
101. C. Moroti, L. F. Souza Magri, M. de Rezende Costa, D. C. Cavallini and K. Sivieri, Effect of the consumption of a new symbiotic shake on glycemia and cholesterol levels in elderly people with type 2 diabetes mellitus, *Lipids Health Dis.*, 2012, **11**, 1–8.
102. N. Zhang, T. Liang, Q. Jin, C. Shen, Y. Zhang and P. Jing, Chinese yam (Dioscorea opposita Thunb.) alleviates antibiotic-associated diarrhea, modifies intestinal microbiota, and increases the level of short-chain fatty acids in mice, *Food Res. Int.*, 2019, **122**, 191–198.
103. K. L. Lam and P. C. Cheung, Non-digestible long chain beta-glucans as novel prebiotics, *Bioact. Carbohydr. Diet. Fibre*, 2013, **2**(1), 45–64.

CHAPTER 10

Microbial Functional Foods and Nutraceuticals

ANANYA RANA,[a] VIKRAM KUMAR,[a] NEETU KUMRA TANEJA[a] AND TEJPAL DHEWA*[b]

[a] Department of Basic and Applied Sciences, National Institute of Food Technology and Entrepreneurship Management, Kundli, Sonepat, Haryana, 131028, India; [b] Department of Nutrition Biology, Central University Of Haryana, Jant-Pali, Mahendergarh, Haryana, 123031, India
*Email: tejpaldhewa@gmail.com

10.1 Introduction

Functional foods are defined as conventional foods or dietary ingredients that supply vital nutritional constituents, such as carbohydrates, proteins, fats, vitamins, and minerals.[1] They might provide other health advantages that can be used to prevent or reverse a diseased state. Bioactive substances, naturally occurring molecules produced from plants, animals, or microorganisms and advantageous for human health, may be present in functional foods. Since ancient times, cyanobacteria have been used as potential nutritional supplements.[2] *Spirulina maxima* was first used for food purposes in 1521 when it was taken from Lake Texcoco and sold dry in Mexico City's markets. The first spirulina products were produced commercially there in the 1970s as well. Since 1940, members of the Kanembu tribe in Africa who reside near Lake Chad have utilised spirulina as a protein supplement.[3] The use of bacteria, yeast and microalgae can accelerate the production of food components, enzymes, and nutraceuticals. Recent interest in microbial flavour, colours, and bioprocessing employing enzymes has increased due to the current vogue for natural ingredients.[4] The importance of microbial

Food Chemistry, Function and Analysis No. 44
Functional Foods of the Future
Edited by Vijai Kumar Gupta, Minaxi Sharma, Smriti Gaur and Ramesh Chander Kuhad

Published by the Royal Society of Chemistry, www.rsc.org

synthesis of chemicals, including organic acids, enzymes, proteins, vitamins, antibiotics, and hydrocolloids, has not changed. It has been shown that lactic acid bacteria (LAB), especially *Lactococcus lactis*, are the optimal cell factories for the manufacture of these vital nutraceuticals.[5] Genetic engineering and mutations may be used for some food-grade microorganisms for overexpression of nutraceuticals, improvement of nutraceuticals, or introducing nutraceutical production in microbes for the first time as newly engineered strains. Microbes are responsible for the generation of food flavours, flavonoids, carotenoids, terpenoids, enzymes, organic acids that are utilized in food, food preservatives such as bacteriocins, amino acids, live probiotic cells, nutraceuticals, and other derivatives used in foods and pharmaceuticals.[6] They are also used to produce microbial polysaccharides in food, *e.g.* xylitol and other polyols such as prebiotics and oligosaccharides. Scientists recently discovered that microorganisms could potentially be a significant source of terpenes, which are a key component of many pharmaceuticals and food additives.[7] A previous study identified 262 gene sequences in genomic databases of different bacteria that encode terpene synthases, enzymes that catalyse the synthesis of terpenes. The researcher subsequently isolated 13 previously unidentified bacterial terpenes using a number of those enzymes.[8] The terpene compounds were produced using a bio-refinery made of genetically modified *Streptomyces* bacteria.

Furthermore, for conditions like protein energy malnutrition (PEM), anaemia, diarrhoea, cancer, irritable bowel syndrome, ulcerative colitis and gluten-treatment-resistant celiac disease, microorganisms can be used as adjuvant therapy.[9] Numerous diseases, including physiologic and lifestyle disorders like diabetes, high blood pressure, cardiovascular disease, neurological disease, *etc.*, are largely caused by free radicals. Free radicals catalyse harmful oxidative processes that result in the generation of toxic lipid peroxides. Additionally, they prevent the mitochondrial respiratory chain enzymes from functioning, damage DNA and proteins, and render the cell fatal.[10] Therefore, looking for newer, less-harmful alternatives to allopathic medications is critical. With the aid of their racemase enzymes, bacteria synthesise amino acids and contribute to lowering D-amino acid toxicity. Despite the fact that D-amino acids are dangerous to life on Earth, they are constantly produced through bacterial growth and geological racemisation.[11]

10.2 Functional Foods and Nutraceuticals

Functional foods are believed to offer advantages above and beyond basic nutrition and may help to lower or minimise the risk of developing certain illnesses and other health issues. In addition to providing essential nutrients (such as vitamins and minerals), functional foods are believed to have additional health benefits.[12] When consumed regularly as a part of a varied diet, functional foods may help lower or prevent the risk of developing certain diseases and other health conditions (Table 10.1). The idea first

Table 10.1 Functional food components, natural sources, and beneficial effects on human health.[17]

Natural food sources	Functional food components	Health benefits
Spinach, tomatoes, mango, pumpkin, carrots	β-Carotene	Reduced the risk of developing cancer or heart-related disease
Papaya, watermelon, red cabbage, mango, tomato and its products	Lycopene	Reduced chronic eye disease (associated with age-related macular degeneration)
Citrus fruit, corn, spinach, papaya, broccoli, cabbage, collards, turnip greens	Zeaxanthin	Reduced the risk of breast and prostate cancer arising due to misbalancing cellular signaling pathways
Millet, sorghum, triticale, quinoa, cereal grains	Whole grains	Reduced the chances of the coronary heart disease (CHD), contributed to maintaining healthy weight
Bran, nuts, fruit skins, vegetables (potatoes, green beans)	Insoluble fibers	Contributed to maintaining healthy digestive system, reduced the risk of breast and colonic cancer
Cheese, beef, some mushroom species (*Agaricus bisporus*), meat products	Conjugated linoleic acid	Contributed to maintaining healthy digestive system and maintaining weight
Fish oil, algal oil, fatty fish (salmon)	Long-chain ω-3 fatty acids	Reduced the risk of autoimmune disease like rheumatoid arthritis and CHD, enhanced infant cognition
Kidney beans, beetroot, fruits, tomatoes	Anthocyanins (cyanidin, malvidin)	Helped improve cholesterol level as well as blood sugar
Citrus fruits	Flavanones	Protection against free radicals, reduced the risk of many chronic diseases as well as cancer, prevention against cardiovascular disease
Low fat dairy products, fish, leafy vegetables	Calcium	Provided bone strength, helped to maintain body weight
Onions, plums, pears, coffee, apples, whole grains	Caffeic acid	Protected against free radicals, improved heart health
Cereal grain, meat, pork	Thiamin	Improved mental health, metabolism regulation
Fish, dairy products, meat, nuts	Riboflavin	Increased cellular growth, precursor for coenzymes FMN and FAD
Broccoli, beans, green leafy vegetables, carrots	Folate or folic acids	Reduced the risk of neural tube defects, helps to reduce anemia and skin disorders
Eggs, almonds, nuts, dairy products, oysters	Biotin	Helped to treat skin disorders, metabolic regulation, enhanced hair growth
Meat, beef liver, fish, nuts, bananas	Pyridoxine	Regulated immune system, enhanced energy level

emerged in Japan in the 1980s, when the Ministry of Health and Welfare established a regulatory framework to certify specific meals with proven health advantages for the country's ageing population.[13] Now known as items for specified health use, these foods are qualified to display a distinctive seal. Functional foods are any modified food or food ingredients that may give health benefits in addition to the traditional nutrients it provides, according to the Food and Nutrition Board of the US National Academy of Sciences, which issued this definition in 1994.[12]

In 1989, the Foundation of Innovation Medicine (FIM), Crowford, New Jersey, combined the words "nutrition" and "pharmaceutical" to coin the phrase "nutraceutical". Health Canada has altered its meaning, defining "nutraceutical" as a substance that has been separated or purified from foods, typically offered in therapeutic forms unrelated to food, and proven to provide physiological benefits or offer protection from chronic disease, for example, β-carotene and lycopene.[14] Functional foods and nutraceuticals both contain active ingredients with physiological actions that promote a better and happier way of life. While the administration differs between nutraceuticals and functional foods, both share the same biological and chemical processes.[15] The body requires precise amounts of vitamins, lipids, proteins, carbohydrates and other nutrients from functional food to ensure a healthy existence. Functional food is referred to as nutraceutical if it helps with the prevention or disorders other than anaemia. Citrus fruits and fortified dairy products (like milk) are examples of nutraceuticals (*e.g.* orange juice).[16]

Over the millennia, our forefathers have advocated the idea that certain diets might fend off disease. In the Old Testament, Proverbs 24:13, the wise Solomon emphasised the primary significance of honey in nourishment and as a treatment for health issues, saying, "Eat honey, my son, since it is good".[17] Like cinnamon, valued more highly than gold in ancient Egypt, ginseng has been used as a traditional medicine for thousands of years in China. Due to the focus on economic development, the standard of living in terms of earnings, expenses, and lifestyle has significantly increased today.[18] The nutraceutical market has grown due to the notion that they are essential to wellness and a viable alternative to contemporary medications. Numerous studies in recent years have proposed the use of nutraceuticals as therapeutic tools due to their beneficial effects, with the ultimate goal of validating their efficacy in extending life expectancy, improving the quality of life for the elderly, and, last but not least, lowering the overall cost of health care.[19]

10.3 Microbes as a Potential Source for the Production of Nutraceutical Ingredients

Nutraceuticals, as opposed to functional foods and drugs, are now more focused on both structural and functional aspects of diverse bioactive

Table 10.2 Microbial production of nutraceuticals in different metabolically engineered strains.

Nutraceutical	Microbial strain	Production concentration (mg L^{-1})	Ref.
Apigenin (flavones)	*Escherichia coli*	13	77
Curcumin (curcuminoid)	*Escherichia coli*	0.6	78
Caffeic acid	*Escherichia coli*	766.68	79
Catechin (flavanonol)	*Escherichia coli*	910.9	80
Genistein glucosides	*Escherichia coli*	37.29	81
Genistin	*Saccharomyces cerevisiae*	Not defined	82
Diadzin	*Saccharomyces cerevisiae*	Not defined	82
Galactooligosaccharides	*Lactococcus lactis*	197 000	83
Heparosan	*Escherichia coli*	1880	84
Hyaluronan	*Streptomyces albulus*	6200	85
Scleroglucan	*Sclerotium rolfsii*	22 320	86
Cyanophycin	*Escherichia coli*	120 000	77
Zeaxanthin	*Escherichia coli*	43.46	87
Poly-γ-glutamic acid	*Bacillus subtilis* BL53	17 000	88
Poly-ε-L-lysine	*Streptomyces* sp. M-Z 18	35 140	89

compounds which have long-term therapeutic or physiological benefits, in addition to merely nutritional or direct pharmacological effects.[20] They can come from a variety of sources, including plants (such as phytochemicals and vitamins), animals (such as polysaccharides), microbes (such as poly-amino-acids) (Table 10.2), and marine sources (such as glucosamine and chitosan).[21] Numerous people use nutraceuticals to promote health or prevent disease, particularly in the prevention of ageing-related illnesses such as cancer, diabetes, inflammation, oxidative stress, depression, osteoporosis, gastrointestinal disorders, and cardiovascular diseases.[22] The interest in nutritional supplements to maintain human wellness has significantly fueled the market's expansion.[22] The effectiveness of inventory control, the low amount and purity of the nutraceuticals, the price and access to raw materials, and direct extraction procedures are all factors that restrict their use. Although a different method, chemical synthesis can only produce simple biochemicals and is impractical for synthesising complicated biochemicals, especially ones that are chemically unfavourable.[23] Microbes and microbial-based metabolic engineering provide a sustainable solution to resolve these problems.

10.4 Microbes as Producers of Amino Acids

According to studies, bacteria are capable of producing some important amino acids. The productivity of strains that produce amino acids is constantly increasing *via* metabolic engineering, particularly *Corynebacterium glutamicum* and *Escherichia coli* strains. Combining intracellular metabolite sensing with traditional mutagenesis and screening has sped up both processes.[24] In order to actualise a variable feedstock concept, synthetic biology technologies have made it possible to access additional carbon sources. New amino acid synthesis mechanisms and fermentative strategies are producing novel substances made from amino acids or their metabolic substrates.[25] These include amino acids, dipeptides, acetylated amino acids, diamines, diacids, and keto acids. Numerous bacteria have been shown to synthesise various types of amino acids. MAAs are amino acids that resemble mycosporines and include porphyra and shinorine.[24] Marine microbes, fungi, cyanobacteria, and microalgae produce mycosporines, glycine, and threonine, a substance that absorbs UV light at 334 nm.[26] These species can be protected from UV radiation in the environment using MAAs. In other studies, it was discovered that a cytosolic tyrosine participates in a mechanism resembling that of microbes that helps plants produce the amino acid phenylalanine: phosphoryl pyruvate amino-transferases or PPAT.[26] All living things need phenylalanine as a component of their proteins, and plants use it as a starting point for thousands of other metabolites. Phenylalanine cannot be synthesised by animals. Thus, they must either directly or indirectly acquire it from plants.[24]

10.5 Microbes as the Source of Vitamins

A high vitamin intake is related to better health, as humans and other organisms require various vitamins to sustain metabolic processes. Their primary function is to act as co-enzymes and enter biological reactions.[27] The B-group or B-complex vitamins are produced by microorganisms and include thiamin (B1), riboflavin (B2), niacin (B3), pyridoxine (B6), pantothenic acid (B5), cobalamin (B12), biotin (B7 or H), and folate (B9 or M).[28] These compounds are water-soluble and significantly impact metabolic processes, specifically metabolism of proteins and fats (riboflavin and pyridoxine). B-group vitamins are typically found in a variety of foods, but because cooking and food processing may easily remove or destroy them, it is usual for people to be deficient in them.[29] Due to this, a number of nations have passed legislation requiring fortification of various foods with particular vitamins and minerals. Microbes possess the simplicity and adaptability required for efficient vitamin synthesis. Enzymes involved in the manufacturing pathway of different vitamins can be over-expressed in a variety of microorganisms by genetic engineering.[29] In addition to having a higher yield than chemical synthesis and the capacity to grow on inexpensive feedstock, microbial production of vitamins also claims a significantly reduced energy input need. Lactic acid bacteria (LAB) used in foods to produce, release, or increase particular nutraceutical substances is made possible by some of its key characteristics,[30] including their biosynthetic ability and metabolic plasticity.

An intriguing method for creating innovative fermented foods with enhanced nutritional and/or health-enhancing properties is the careful selection and utilisation of LAB that produce nutraceuticals. LAB-promoted biosynthesis can produce fermented milk or bread with higher concentrations of B-group vitamins (such as folate and riboflavin).[31] In order to produce innovative fermented foods by nutraceutical-producing LAB, vitamin synthesis by lactobacilli is increasingly being focused on.[32,33] Numerous lactobacilli include the genes necessary for riboflavin (vitamin B2) production; however, the genetic ability to do so varies depending on the species and/or strain. It has been demonstrated that some bacteria, such as *L. plantarum* WCFS1, have interruptions in the *rib* operon, which prevent them from producing riboflavin.

On the other hand, due to their ability to synthesise riboflavin, *Lactobacillus plantarum* NCDO 1752, *Lactobacillus plantarum* JDMI, and other strains recently isolated from products produced from cereals can survive without this vitamin.[34] The generation of vitamins by microorganisms still faces several significant problems. Allosteric inhibition from the vitamin or intermediates in the synthesis pathway, for example, can lower yield in engineered microorganisms expressing every enzyme required to produce the needed vitamin.[35]

10.6 Microbial Production of Polysaccharides

Numerous bacteria, yeasts, and fungi may synthesise polysaccharides, which are sugar polymers with a variety of structures. They are also found in plant and animal tissues. Due to their favourable health effects, polysaccharides produced from microbes have been recognised as a source of nutraceuticals.[36] Xanthan, gellan, dextran, and alginate are a few examples of bacterial polysaccharides that can be produced and purified by microbes for commercial use. Exo-polysaccharides (EPS) are widely used in dairy products and are preferentially synthesised through metabolic engineering of *Lactococcus* and *Streptococcus.*[37] The wide range of immunostimulating, anticancer, antibacterial, antioxidant, hypercholesterolemic, and hypoglycemic effects of bioactive fungal polysaccharides, in particular, suggest tremendous promise in both nutraceuticals and pharmaceutical applications. Scleroglucan is a potential anticancer and anti-viral polysaccharide secreted by *Sclerotium rolfsii* mycelia, and its yields may be increased by addition of L-lysine with uridine monophosphate (UMP). A number of animal polysaccharides, such as hyaluronic acid (HA),[38] chondroitin and heparosan, are synthesised by microbial guests rather than being extracted from animal tissues. Bacteria, including *Escherichia coli*, *Lactococcus lactis*, and *Streptomyces albulus*, have produced hyaluronic acid in a variety of hosts. Engineered *Escherichia coli* may manufacture medicinally significant polysaccharides like heparosan and chondroitin, reaching comparatively high titers of 1.88 $g\,L^{-1}$ and 2.4 $g\,L^{-1}$, respectively.[21]

10.7 Alkaloids

Alkaloids are nitrogenous chemicals produced from amino acids that have significant medicinal benefits, such as anticancer and anti-malarial properties. Due to their complicated structures and protracted biosynthesis pathways, plants have been the only source of alkaloids for a very long period.[39] The top two recognised types of alkaloids are monoterpene indole alkaloids (MIAs), synthesised from glucosinolates and tryptophan, and benzylisoquinoline alkaloids (BIAs), synthesised from tyrosine. In recent years, explication of BIA pathways in both *Escherichia coli* and *Saccharomyces cerevisiae* was made possible by the reconstruction of numerous BIA biosynthetic pathways.[40] Bacterial consortia containing *Escherichia coli* produced (*S*)-reticuline, and 7.2 $mg\,L^{-1}$ of magnoflorine and 8.3 $mg\,L^{-1}$ of scoulerine were produced by *Saccharomyces cerevisiae* expressing either CYP80G2 monooxygenase or the BBE (berberine bridge enzyme). *Saccharomyces cerevisiae* also helped to create (*R*,*S*)-reticuline, and it was further developed to manufacture salutaridine from (*S*)-reticuline as well as scoulerine, tetrahydrocolumbamine and tetrahydroberberine from (*R*)-reticuline.[41] However, in microorganisms, there is little metabolic engineering of MIA alkaloids. By inserting 21 new genes and removing three others from the yeast genome, *de novo* manufacture of the MIA alkaloid strictosidine has just been accomplished in yeast. Glucosinolates are natural substances generated from amino acids that are sulfur-rich. Indolylglucosinolate (IG), derived from tryptophan, has recently been synthesised in *Saccharomyces cerevisiae*[42] by introducing eight plant genes into the genome. This is a rare instance of glucosinolate production in microorganisms. Although plant platforms are better suited for producing scalable amounts of alkaloids, metabolic engineering in microorganisms has demonstrated possibilities for producing these plant-derived compounds at a low cost.[43]

10.8 Probiotic-based Functional Foods

There are already more than 500 probiotic food products that have been available in the market on a global scale in the last several years, and this number is steadily rising. Probiotic foods made from the fermentation of fruits and vegetables, grains, and animal products are attracting interest from both consumers and scientists.[44] Probiotics are now commercially available in items including ice cream, milk, sour milk, and fruit juices. The most recent probiotic-containing foods are mayonnaise spreads, cheese and cheese-based dips, and meat-based goods. The probiotic bacteria are artificially infused into the food during the manufacture of probiotic foods.[45] Most culture preparations are produced for various uses as either highly concentrated freeze-dried cultures or freeze-dried powders, both of which are commercially available in extremely concentrated forms. Due to the production of many metabolic components, including the synthesis of acetic acid during fermentation by *Bifidobacterium* spp., the scent and flavour of food products may change when probiotic microorganisms are added.[46] The amount of active cells or total viable cells per ml per g of the product at the moment of consumption by the individual determines how successful probiotic food products are as medicines. To retain consumer confidence in probiotic food products, we must make sure that a high probiotic survival rate is maintained during production and over the product's shelf life.[47] The survival of probiotic bacteria in food products throughout production, processing and storage is influenced by a number of factors. Chemicals like bacteriocins, hydrogen peroxide, colouring agents, and artificial flavours are among the factors that have been identified. Titratable acidity, water activity, molecular oxygen, and the amount of sugar and salt are a few examples of food parameters.[48] Processing parameters include incubation temperature, the rate at which food products cool after heat treatment, storage procedures, and packing materials and methods. Microbiological parameters include the percentage of inoculums and the symbiotic count.

Infants that received probiotic supplements of the *Lactobacillus casei* bacterial strain had high levels of circulating immunoglobulin A, which is correlated with a shorter duration of rotavirus-induced diarrhoea.[49] Additionally, nonspecific immunological phagocytic activity of circulating blood granulocytes is noticeably enhanced by the addition of probiotic foods containing *Bifidobacterium bifidum* and *Lactobacillus acidophilus.* According to a study, probiotic yoghurt consumption can increase blood mononuclear cells' ability to produce cytokines.[50] Only when the food contains the necessary minimum viable microbe count at the moment of intake can probiotics' claimed health benefits be realised. The food sector frequently uses the lowest advised level, which is 10^6 CFU ml^{-1} at the time of ingestion.[51] The United States Food and Drug Administration (US FDA) recommends probiotic food items include a minimum of 10^6 CFU ml^{-1}. A daily dose of 10^8 to 10^9 CFU ml^{-1} probiotic bacteria is necessary to accomplish probiotic action in humans, depending on the amount consumed and considering storage conditions.[52]

10.9 Health-promoting Properties of Probiotic Functional Foods

Probiotics should be consumed first by healthy individuals because they improve general health and protect against various ailments. Improving health is the most sensible and preventive move we can take to safeguard ourselves against various illnesses.[53] However, how do probiotics enhance health? The topic of how probiotics directly or indirectly benefit our health will be discussed in the following paragraphs. The various microorganisms found in human bodies collaborate to carry out various tasks. Our digestive system's functions are their most important ones.[54] These microbes improve the consumption and digestion of particular foods and significantly lighten the load on our digestive system. These bacteria simplify the processes that the human body must go through in order to change complicated dietary structures. The following facts and points can be used to summarise how probiotics relate to improving human health:

1. Probiotic microbes are functional and user-friendly.
2. Probiotic microorganisms aid in the fermentation of our food, turning it from a complex form into simpler byproducts, and they can improve our health in a variety of ways.
3. They have the ability to competitively inhibit harmful microorganisms and seize control of our digestive system.
4. Numerous variables, including poor diet, ageing, alcohol consumption, and a number of others, may result in a decline in the viable count of probiotic microorganisms. Probiotics should therefore be a part of our everyday diets as a supplement.
5. They are expected to be significantly impacted in certain circumstances, such as after taking antibiotics. Thus, they should be taken orally in large quantities with food.
6. They deliver essential beneficial consequences to the organism.
7. They eliminate the negative effects of harmful microorganisms.
8. Because they make it possible for any amount of food to be properly digested and metabolised, they reduce the amount of food that the human body needs.
9. They lessen the additional stress placed on the digestive tract of humans.
10. They reduced the impact of the initial attack of hazardous substances on our cells by forming a biofilm.
11. Probiotics can often compensate for genetically determined traits and deficiencies in our genetic makeup, such as lactose intolerance.

10.10 Mechanisms Through Which Functional Foods and Nutraceuticals Exert Health Advantages

Numerous studies have demonstrated the neuroprotective effects of nutraceuticals and functional foods, which work at several biochemical and metabolic levels. In particular, they can guard against cell death, toxic levels of tau and amyloid beta, mitochondrial damage, and oxidative stress.[55] They have been shown to have a significant impact on the composition of the gut microbiota, which has helped researchers understand how the diversity of bacteria affects the production and accumulation of harmful proteins in the brain. It is more than just a diet to follow; it is a way of life that plays a significant part in the prevention and treatment of many diseases, as well as reducing the risk of developing diabetes, heart disease, cancer, and others of our most dreaded illnesses.[56] The same is true in terms of preventing Alzheimer's disease (AD). Because of its high metabolic activity and the amount of oxidisable components, the brain is especially vulnerable to oxidative damage, including the lipid milieu of the myelin membrane found in neuronal cells.[57] Therefore, diets high in antioxidants may enhance cognitive performance. According to studies, a number of berries have significant antioxidant capacities because they contain tannins, anthocyanins, and phenols, which can boost the plasticity of the hippocampus and aid in learning and memory function. Additionally, α-lipoic acid, which may be found in vegetables like spinach and broccoli, is essential for preserving the mitochondrial energy balance and enhancing cognitive abilities.[58]

The antioxidant-rich beverages green and black tea both contain epigallocatechin gallate, which have been demonstrated to indirectly lessen the development of amyloid plaques, one of AD's defining characteristics.[57] When taken in and absorbed by the gut, nutraceuticals function similarly to pharmaceuticals, providing preventative and/or adjuvant activity in the healing phase and increasing our resistance to dangerous environmental chemicals. Plants create many of them as defence mechanisms against insects, and they are mostly found in fruits and vegetables.[58] As a result, they are of considerable importance to humans due to the positive effects that may be utilised even at low dosages. The fact that a growing number of innovative studies suggest that the impact of diet on mental health may be passed down through generations, which is a fascinating idea, is an intriguing aspect of the situation, despite the fact that the molecular processes underpinning the impact of nutrition on epigenetics are still not fully understood.[59] The ability of a dietary element to influence non-genetic events—even those that nonetheless result in potentially heritable phenotypic changes—opens a therapeutic pathway that may be useful in the management, if not only the control, of the illness development.[60] The antioxidant, anti-estrogenic, anti-inflammatory, immunomodulatory, and anticarcinogenic effects of these nutrients make them significant for the human body (Figure 10.1) even though they are not necessary for survival, unlike vitamins, for example. Additionally, they are less potent as bioactive substances than pharmaceuticals, but because they are routinely ingested in large quantities as part of a meal, they can nevertheless have important long-term physiological impacts.[61] Thus, phytonutrients or phytochemicals may have an anti-chronic inflammatory effect and may aid in the prevention of age-related and chronic disorders, including atherosclerosis and arthritis. By modifying the makeup of the microbiota, the aforementioned bioactive substances have profound effects on the gut environment.[62] The biological functions of functional dietary components can also differ since the gut microbiota itself can impact their metabolites and relative effects, which can even vary from person to person. Although numerous animal and *in vitro* studies have been conducted to examine the bioactivity of various functional foods (nutraceuticals), clinical trials involving humans are currently rare and inconclusive.[63] It must be remembered that any chemical changes to the initial bioactive component that take place during

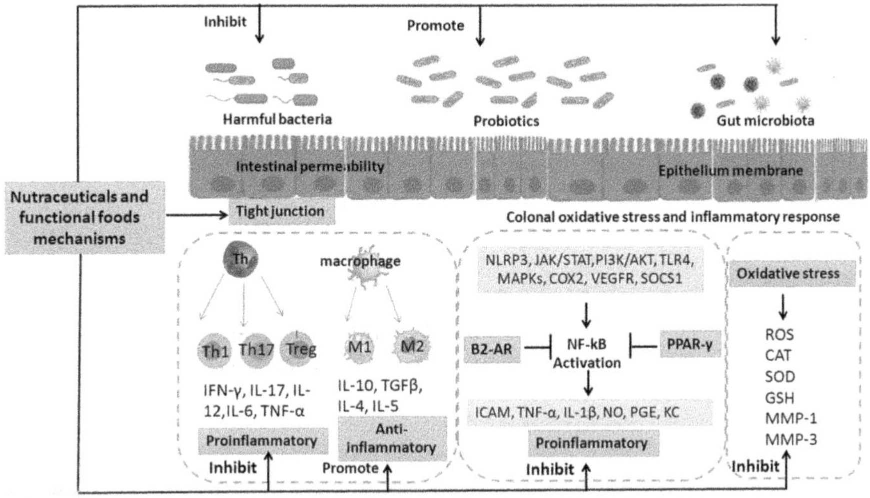

Figure 10.1 Some of the proposed mechanisms by which functional foods and nutraceuticals exert their beneficial effects.

digestion or storage could significantly alter both the compound's bioactivity and bioavailability.

To further explain this idea, it should be remembered that while a compound's antioxidant activity, as measured in a test tube, can be used to gauge a product's quality, this does not necessarily mean that it will perform well *in vivo*. Pomegranate ellagitannins (ETs) are an example of this effect.[64] Pomegranates are one of the most potent *in vitro* antioxidants because of the high antioxidant potential of ETs. In contrast, these substances are not absorbed in living organisms and are heavily degraded by the colon microbiota into urolithins, which are no longer endowed with the ETs' initial antioxidant capacity.[65] Onion, garlic, grapes, rosemary, broccoli, spinach, turmeric, parsley, and other foods have significant antioxidant properties and guard against a number of neurodegenerative illnesses, including AD. For instance, antioxidant compounds found in odourless aged garlic extracts (AGE) help to reduce oxidative damage.[55] AGE has been demonstrated to prevent cellular damage brought on by the buildup of amyloid beta peptide (Aβ) *in vitro* and to mitigate the toxicity of Aβ.

Additionally, glycogen synthase kinase-3 (GSK) activity is regulated to control tau protein hyperphosphorylation, usually in relation to the aetiology of AD.[66] Garlic's ability to penetrate through the blood–brain barrier (BBB), a characteristic that further supports its promise as a medicinal substance, has been used to investigate the neuroprotective effects of AGE on AD-related cognitive impairment. In other words, a bioactive component's distinct ability to cross, or not, the BBB is an important property deserving of attention in the AD therapy strategy.[67]

10.11 Applications of Microbial Functional Foods and Nutraceuticals

The term "nutrients" refers to a class of compounds with well-established nutritional roles, such as vitamins, minerals, amino acids, and fatty acids, which are naturally present in our food and have a wide range of possible health advantages.[68] The majority of vegetables, whole grains, dairy products, fruits, and animal products like meat and poultry all include vitamins that are helpful in the treatment of diabetes, cancer, cataracts, heart disease, stroke, and other conditions.[69] The minerals present in plants, animals and dairy products are useful in osteoporosis and anaemia, building bones, teeth and strong muscles, and improving nerve impulses and heart rhythm. The term "nutraceuticals" describes a large group of substances;[55] here, only some important nutraceuticals and functional foods and their applications are described below:

Dietary fibers: Dietary fibers are chemicals of vegetable origin that are found in meals that increase the amount of intestinal contents without being digested in the large intestine. In chemistry, "dietary fibre" is defined as carbohydrate polymers having a degree of polymerisation of at least 3 that are neither digested nor absorbed in the small intestine.[70] Examples include fruit, barley, oats, lignin, cellulose, pectin, and other items. An appropriate intake of these fibres through food is linked to a decreased risk of gastrointestinal problems, colon cancer, hypertension, diabetes, obesity, and cardiovascular disease.[71]

Probiotics: Probiotics are living bacteria found in food that have a variety of positive effects on the human body. The enhancement of the intestinal microbiota has been attributed to the benefits of probiotic-rich diets that are prepared either naturally or through commercial fermentation techniques.[72]

Polyunsaturated fatty acids: Polyunsaturated fatty acids found in oily fish, flax seeds, soybeans, and other foods include omega 3 fatty acids such as linolenic acid, eicosapentaenoic acid, and docosahexaenoic acid. Omega 6 fatty acids, such as α-linoleic acid and arachidonic acid, are also present in foods like corn, soybean, sunflower, and wheat.[73]

Prebiotics: Prebiotics are non-living substances, in contrast to probiotics. They are short-chain polysaccharides with distinctive chemical structures, particularly fructose-based oligosaccharides, that are present in food either naturally or artificially yet nonetheless have a variety of beneficial effects on our bodies while not being digested. The composition and function of microbiota fluctuate in the colon, affecting the development and activity of particular bacteria that benefit the host's health.[74] Prebiotics act as the nutrition for the probiotics and promote their growth in the gastrointestinal tract. The chicory root, bananas, tomatoes, alliums, beans, *etc.*, are a few examples of these edibles.[72]

Polyphenols: In order to defend themselves against reactive oxygen species and photosynthetic stress, plants synthesise phytochemicals called polyphenols. Others include the anti-inflammatory and antioxidant characteristics of flavonoids, anthocyanins, and phenolic acids,[75] which can be found in a range of foods.

Additionally, some nutraceuticals are found in more than one food, such as quercetin, which is found in red grapes, broccoli, onions, and other foods. This makes mentioning the foods that are higher in a substance or family of substances while discussing nutraceuticals worthwhile. The quantity of nutraceuticals in food is affected by several factors, including the soil on which plants grow and how animals are fed. As a result, eating a gorgeous red tomato that has ripened in the sun or merely a rosé tomato is not the same.[22] In actuality, adverse reactions and toxicity are frequently recorded due to the consumption of nutraceuticals and functional foods. Perhaps this contamination comes from pesticides, other hazardous plants, metals, fertilisers, *etc.*[76] The possible health advantages of functional foods are mentioned below:

a. Enhance the nutritional value of food.
b. Aid in extending life.
c. Assist in preventing specific medical conditions.
d. May appear more "natural" than conventional treatments and be associated with fewer negative side effects.
e. Improve the nutritional content of food for those with special needs, for example foods that are nutrient-rich for the elderly.
f. Be readily available and reasonably priced.

10.12 Future Perspectives

The realm of "neuro-nutraceuticals" currently offers strong research ideas, as was extensively noted in this chapter, and in the near future it may open interesting windows on effective and secure treatments, even in individuals with dementia who have already been diagnosed. We hope that several food-borne chemicals will eventually help AD therapy, because of the preclinical evidence that they may be beneficial in preventing and treating cognitive decline. To validate their positive effects on humans, additional research is required. Currently, adopting an active lifestyle along with a diet that fits the criteria of the Mediterranean diet (perhaps enhanced with walnuts and olive oil) is the strongest suggestion in terms of behaviour and nutrition—while a healthy cognitive reserve still exists. Recent research by Ravi *et al.* highlights the beneficial, mild, and preventive anti-AD properties of Mediterranean diets, which are mostly constituted of fruits, vegetables, and omega-3 fatty acids. We can battle this debilitating neurological condition by making the necessary lifestyle changes. Maintaining a healthy diet is vital for people of all ages. However, it is crucial to encourage specific nutritional practices among the elderly owing to the possibility of malnutrition, decreased absorption, appetite loss, and difficulty chewing, all of which can impair a person's nutritional status. In reality, nutritional intake and assimilation can be reduced in ageing individuals, which may help to explain the contradictory outcomes of clinical research involving dietary treatments. These techniques aim to fundamentally restore the variety of gut flora in the elderly in an effort to lessen certain negative effects of ageing.

10.13 Conclusion

Due to the high demand for value-added nutraceuticals to protect people from diseases, the metabolic engineering of microbial-based platforms has made amazing strides in manufacturing nutraceuticals over the past few decades. The preparation and industrial-scale production of complex natural compounds like carotenoids from simple carbon sources were made possible thanks to metabolic engineering in microorganisms. Novel or more complex value-added substances could be created in microbes naturally, particularly with the clarification of biosynthetic pathways and the development of synthetic biology technologies. The effectiveness of enzymes produced by plants or animals, which restrict the production of the target molecules, is one of the main drawbacks of microbial platforms. Various non-microbial manufacturing methods, such as those using algae and plant-based cell cultures, have also been developed to produce naturally occurring or synthetic value-added nutraceuticals and functional foods. Even though there is no question about the role that nutraceuticals and functional foods play in preventing, limiting, and treating numerous diseases, it is important to remember that everyone responds differently to nutraceuticals. An individual's assimilation of nutraceuticals and susceptibility to a specific disease will depend on their lifestyle, environmental exposures, and genetic makeup.

References

1. C. M. Galanakis, Functionality of food components and emerging technologies, *Foods*, 2021, **10**(1), 1–26.
2. S. Pai, A. Hebbar and S. Selvaraj, A critical look at challenges and future scopes of bioactive compounds and their incorporations in the food, energy, and pharmaceutical sector, *Environ. Sci. Pollut. Res.*, 2022, **29**(24), 35518–35541.
3. S. K. Saha, E. McHugh, P. Murray and D. J. Walsh, Microalgae as a source of nutraceuticals, *Phycotoxins: Chemistry and Biochemistry*, 2nd edn, 2015, pp. 255–291.
4. B. McNeil, D. Archer, I. Giavasis and L. Harvey, *Microbial Production of Food Ingredients, Enzymes and Nutraceuticals*, 2013, pp. 1–610.
5. J. Angelin and M. Kavitha, Exopolysaccharides from probiotic bacteria and their health potential, *Int. J. Biol. Macromol.*, 2020, **162**, 853–865.
6. X. Lyu, Y. Lyu, H. Yu, W. Chen, L. Ye and R. Yang, Biotechnological advances for improving natural pigment production: a state-of-the-art review, *Bioresour. Bioprocess.*, 2022, **9**(1), 8.

7. F. Gültekin, Food additives and microbiota, *North. Clin. Istanb.*, 2019, **7**(2), 192–200.
8. A. M. Boutanaev, T. Moses, J. Zi, D. R. Nelson, S. T. Mugford and R. J. Peters, *et al.*, Investigation of terpene diversification across multiple sequenced plant genomes, *Proc. Natl. Acad. Sci. U. S. A.*, 2015, **112**(1), E81–E88.
9. A. Masyita, R. Mustika Sari, A. Dwi Astuti, B. Yasir, N. Rahma Rumata and T. B. Emran, *et al.*, Terpenes and terpenoids as main bioactive compounds of essential oils, their roles in human health and potential application as natural food preservatives, *Food Chem.: X*, 2022, 100217.
10. G. Pizzino, N. Irrera, M. Cucinotta, G. Pallio, F. Mannino and V. Arcoraci, *et al.*, Oxidative stress: harms and benefits for human health, *Oxid. Med. Cell. Longevity*, 2017, **2017**(1), 8416763.
11. N. Panjiar, S. Mishra, A. N. Yadav and P. Verma, Functional Foods from Cyanobacteria: An Emerging Source for Functional Food Products of Pharmaceutical Importance, *Microbial Functional Foods and Nutraceuticals*, John Wiley & Sons Ltd, 2017, pp. 21–37.
12. C. M. Hasler, Functional foods: Benefits, concerns and challenges – A position paper from the American Council on Science and Health, *J. Nutr.*, 2002, **132**(12), 3772–3781.
13. T. Hasler Shimizu, Health claims on functional foods: the Japanese regulations and an international comparison, *Nutr. Res. Rev.*, 2003, **16**(2), 241–252.
14. M. Zhao, S. Chen, X. Ji, X. Shen, J. You and X. Liang, *et al.*, Current innovations in nutraceuticals and functional foods for intervention of non-alcoholic fatty liver disease, *Pharmacol. Res.*, 2021, 105517.
15. M. Mahajan and M. Rishi, Functional Foods and Nutraceuticals: An Overview, *Res. Rev. Biotechnol. Biosci.*, 2016, 1–11.
16. A. Cencic and W. Chingwaru, The role of functional foods, nutraceuticals, and food supplements in intestinal health, *Nutrients*, 2010, **2**(6), 611–625.
17. K. Jalgaonkar, M. K. Mahawar, B. Bibwe, P. Nath and S. Girjal, *Nutraceuticals Funct. Foods*, 2019, 231–250.
18. A. M. Metwaly, M. M. Ghoneim, I. H. Eissa, I. A. Elsehemy, A. E. Mostafa and M. M. Hegazy, *et al.*, Traditional ancient Egyptian medicine: A review, *Saudi J. Biol. Sci.*, 2021, **28**(10), 5823–5832.
19. M. Alali, M. Alqubaisy, M. N. Aljaafari, A. O. Alali, L. Baqais and A. Molouki, *et al.*, Nutraceuticals: Transformation of conventional foods into health promoters/disease preventers and safety considerations, *Molecules*, 2021, **26**(9), 2540.
20. G. Hoti, A. Matencio, A. R. Pedrazzo, C. Cecone, S. L. Appleton and Y. K. Monfared, *et al.*, Nutraceutical Concepts and Dextrin-Based Delivery Systems, *Int. J. Mol. Sci.*, 2022, **23**(8), 4102.
21. J. Wang, S. Guleria, M. A. G. Koffas and Y. Yan, Microbial production of value-added nutraceuticals, *Curr. Opin. Biotechnol.*, 2016, **37**, 97–104.

22. J. Wang, L. Das, E. Bhaumik, U. Raychaudhuri and R. Chakraborty, Role of nutraceuticals in human health, *J. Food Sci. Technol.*, 2012, **49**(2), 173–183.
23. A. G. Atanasov, B. Waltenberger, E. M. Pferschy-Wenzig, T. Linder, C. Wawrosch and P. Uhrin, *et al.*, Discovery and resupply of pharmacologically active plant-derived natural products: A review, *Biotechnol. Adv.*, 2015, **33**(8), 1582–1614.
24. V. F. Wendisch, Microbial production of amino acids and derived chemicals: Synthetic biology approaches to strain development, *Curr. Opin. Biotechnol.*, 2014, **30**, 51–58.
25. D. N. Carruthers and T. S. Lee, Translating advances in microbial bioproduction to sustainable biotechnology, *Front. Bioeng. Biotechnol.*, 2022, **10**, 1–22.
26. V. Geraldes and E. Pinto, Mycosporine-like amino acids (Maas): Biology, chemistry and identification features, *Pharmaceuticals*, 2021, **14**(1), 1–17.
27. A. L. Tardy, E. Pouteau, D. Marquez, C. Yilmaz and A. Scholey, Vitamins and minerals for energy, fatigue and cognition: A narrative review of the biochemical and clinical evidence, *Nutrients*, 2020, **12**(1), 228.
28. F. J. Rodrigues, M. F. Cedran, J. L. Bicas and H. H. Sato, Encapsulated probiotic cells: Relevant techniques, natural sources as encapsulating materials and food applications – A narrative review, *Food Res. Int.*, 2020, **137**, 109682.
29. R. Bonku and J. Yu, Health aspects of peanuts as an outcome of its chemical composition, *Food Sci. Hum. Wellness*, 2020, **9**(1), 21–30.
30. D. O. Kennedy, B vitamins and the brain: Mechanisms, dose and efficacy—A review, *Nutrients*, 2016, **8**(2), 68.
31. V. Kumar, R. Amrutha, J. J. Ahire and N. K. Taneja, Techno-Functional Assessment of Riboflavin-Enriched Yogurt-Based Fermented Milk Prepared by Supplementing Riboflavin-Producing Probiotic Strains of Lactiplantibacillus plantarum, *Probiotics Antimicrob. Proteins*, 2024, **16**(1), 152–162.
32. V. Capozzi, P. Russo, M. Teresa Dueñas, P. López and G. Spano, Lactic acid bacteria producing B-group vitamins: a great potential for functional cereals products, *Appl. Microbiol. Biotechnol.*, 2012, **96**, 1383–1394.
33. J. G. Leblanc, J. E. Laiño, M. J. del Valle, V. Vannini, D. van Sinderen and M. P. Taranto, *et al.*, B-Group vitamin production by lactic acid bacteria – current knowledge and potential applications, *J. Appl. Microbiol.*, 2011, **111**(6), 1297–1309.
34. K. Thakur, S. K. Tomar and S. De, Lactic acid bacteria as a cell factory for riboflavin production, *Microb. Biotechnol.*, 2016, **9**(4), 441–451.
35. V. Kumar, M. Vijendra, R. Ananya and A. Jayesh, Riboswitch – mediated regulation of riboflavin biosynthesis genes in prokaryotes, *3 Biotech*, 2022, 1–14.
36. J. Angelin and M. Kavitha, Exopolysaccharides from probiotic bacteria and their health potential, *Int. J. Biol. Macromol.*, 2020, **162**, 853–865.

37. P. Gupta and B. Diwan, Bacterial Exopolysaccharide mediated heavy metal removal: A Review on biosynthesis, mechanism and remediation strategies, *Biotechnol. Rep.*, 2017, **13**, 58–71.
38. M. Lemieszek and W. Rzeski, Anticancer properties of polysaccharides isolated from fungi of the Basidiomycetes class, *Wspolczesna Onkol.*, 2012, **16**(4), 285–289.
39. M. Mora-Pale, S. P. Sanchez-Rodriguez, R. J. Linhatrdt, J. S. Dordick and M. A. G. Koffas, Metabolic engineering and in vitro biosynthesis of phytochemicals and non-natural analogues, *Plant Sci.*, 2013, **210**, 10–24.
40. A. M. Ehrenworth, S. Sarria and P. Peralta-Yahya, Pterin-dependent mono-oxidation for the microbial synthesis of a modified monoterpene indole alkaloid, *ACS Synth. Biol.*, 2015, **4**(12), 1295–1307.
41. D. A. Bird and P. J. Facchini, Berberine bridge enzyme, a key branch-point enzyme in benzylisoquinoline alkaloid biosynthesis, contains a vacuolar sorting determinant, *Planta*, 2001, **213**(6), 888–897.
42. J. Wang, S. Guleria, M. A. G. Koffas and Y. Yan, Microbial production of value-added nutraceuticals, *Curr. Opin. Biotechnol.*, 2015, **29**, 153–159.
43. M. Mora-Pale, S. P. Sanchez-Rodriguez, R. J. Linhatrdt, J. S. Dordick and M. A. G. Koffas, Metabolic engineering and in vitro biosynthesis of phytochemicals and non-natural analogues, *Plant Sci.*, 2013, **210**, 10–24.
44. S. Rezac, C. R. Kok, M. Heermann and R. Hutkins, Fermented foods as a dietary source of live organisms, *Front. Microbiol.*, 2018, **9**(8), 1785.
45. A. Terpou, A. Papadaki, I. K. Lappa, V. Kachrimanidou, L. A. Bosnea and N. Kopsahelis, Probiotics in food systems: significance and emerging strategies towards improved viability and delivery of enhanced beneficial value, *Nutrients*, 2019, **11**(7), 1591.
46. S. V. N. Vijayendra and R. C. Gupta, Performance evaluation of bulk freeze dried starter cultures of dahi and yoghurt along with probiotic strains in standardized milk of cow and buffalo, *J. Food Sci. Technol.*, 2014, **51**(12), 4114–4119.
47. R. Mohammadi, A. M. Mortazavian, R. Khosrokhavar and A. G. Da Cruz, Probiotic ice cream: Viability of probiotic bacteria and sensory properties, *Ann. Microbiol.*, 2011, **61**(3), 411–424.
48. S. Misra, P. Pandey, C. G. Dalbhagat and H. N. Mishra, Emerging Technologies and Coating Materials for Improved Probiotication in Food Products: a Review, *Food Bioprocess Technol.*, 2022, **15**(5), 998–1039.
49. M. Aspri, P. Papademas and D. Tsaltas, Review on non-dairy probiotics and their use in non-dairy based products, *Fermentation*, 2020, **6**(1), 1–20.
50. H. S. Gill and F. Guarner, Probiotics and human health: A clinical perspective, *Postgrad. Med. J.*, 2004, **80**(947), 516–526.
51. F. Cristofori, V. N. Dargenio, C. Dargenio, V. L. Miniello, M. Barone and R. Francavilla, Anti-Inflammatory and Immunomodulatory Effects of Probiotics in Gut Inflammation: A Door to the Body, *Front. Immunol.*, 2021, **12**(2), 1–21.

52. M. Zommiti, M. G. J. Feuilloley and N. Connil, Update of probiotics in human world: A nonstop source of benefactions till the end of time, *Microorganisms*, 2020, **8**(12), 1–33.
53. A. A. Amara and A. Shibl, Role of Probiotics in health improvement, infection control and disease treatment and management, *Saudi Pharm. J.*, 2015, **23**(2), 107–114.
54. A. Chandarakesan, S. Muruhan and R. R. A. Sayanam, Morin Inhibiting Photocarcinogenesis by Targeting Ultraviolet-B-Induced Oxidative Stress and Inflammatory Cytokines Expression in Swiss Albino Mice, *Int. J. Nutr., Pharmacol., Neurol. Dis.*, 2018, **8**, 41–46.
55. A. Atlante, G. Amadoro, A. Bobba and V. Latina, Functional foods: An approach to modulate molecular mechanisms of alzheimer's disease, *Cells*, 2020, **9**(11), 1–62.
56. Y. J. Zhang, S. Li, R. Y. Gan, T. Zhou, D. P. Xu and H. B. Li, Impacts of gut bacteria on human health and diseases, *Int. J. Mol. Sci.*, 2015, **16**(4), 7493–7519.
57. A. Gella and N. Durany, Oxidative stress in Alzheimer disease, *Cell Adhes. Migr.*, 2009, **3**(1), 88–93.
58. E. Kelly, P. Vyas and J. T. Weber, Biochemical properties and neuroprotective effects of compounds in various species of berries, *Molecules*, 2017, **23**(1), 26.
59. A. J. Stevens, J. J. Rucklidge and M. A. Kennedy, Epigenetics, nutrition and mental health. Is there a relationship?, *Nutr. Neurosci.*, 2018, **21**(9), 602–613.
60. M. Franzago, D. Santurbano, E. Vitacolonna and L. Stuppia, Genes and diet in the prevention of chronic diseases in future generations, *Int. J. Mol. Sci.*, 2020, **21**(7), 2633.
61. H. Cory, S. Passarelli, J. Szeto, M. Tamez and J. Mattei, The Role of Polyphenols in Human Health and Food Systems: A Mini-Review, *Front. Nutr.*, 2018, **5**(9), 1–9.
62. J. J. Ahire, A. Rohilla, V. Kumar and A. Tiwari, Quality Management of Probiotics: Ensuring Safety and Maximizing Health Benefits, *Curr. Microbiol.*, 2024, **81**(1), 1.
63. A. M. Valdes, J. Walter, E. Segal and T. D. Spector, Role of the gut microbiota in nutrition and health, *BMJ*, 2018, **361**, 36–44.
64. N. B. Sadeer, D. Montesano, S. Albrizio, G. Zengin and M. F. Mahomoodally, The versatility of antioxidant assays in food science and safety—chemistry, applications, strengths, and limitations, *Antioxidants*, 2020, **9**(8), 1–39.
65. S. Asgary, S. Javanmard and A. Zarfeshany, Potent health effects of pomegranate, *Adv. Biomed. Res.*, 2014, **3**(1), 100.
66. D. M. D. L. Navarro, J. J. Abelilla and H. H. Stein, Structures and characteristics of carbohydrates in diets fed to pigs: A review, *J. Anim. Sci. Biotechnol.*, 2019, **10**(1), 1–17.
67. B. Mathew and R. Biju, Neuroprotective Effects of Garlic: A Review, *Libyan J. Med.*, 2008, **3**(1), 23–33.

68. Y. Chen, M. Michalak and L. B. Agellon, Importance of nutrients and nutrient metabolism on human health, *Yale J. Biol. Med.*, 2018, **91**(2), 95–103.
69. M. Asif, The prevention and control the type-2 diabetes by changing lifestyle and dietary pattern, *J. Educ. Health Promot.*, 2014, **3**(1), 1.
70. J. Lunn and J. L. Buttriss, Carbohydrates and dietary fibre, *Nutr. Bull.*, 2007, **32**(1), 21–64.
71. V. P. Nirmala Prasadi and I. J. Joye, Dietary fibre from whole grains and their benefits on metabolic health, *Nutrients*, 2020, **12**(10), 1–20.
72. P. Markowiak and K. Ślizewska, Effects of probiotics, prebiotics, and synbiotics on human health, *Nutrients*, 2017, **9**(9), 1021.
73. A. Balić, D. Vlašić, K. Žužul, B. Marinović and Z. B. Mokos, Omega-3 *versus* Omega-6 polyunsaturated fatty acids in the prevention and treatment of inflammatory skin diseases, *Int. J. Mol. Sci.*, 2020, **21**(3), 741.
74. A. Megur, E. B. M. Daliri, D. Baltriukienė and A. Burokas, Prebiotics as a Tool for the Prevention and Treatment of Obesity and Diabetes: Classification and Ability to Modulate the Gut Microbiota, *Int. J. Mol. Sci.*, 2022, **23**(11), 6097.
75. A. N. Panche, A. D. Diwan and S. R. Chandra, Flavonoids: An overview, *J. Nutr. Sci.*, 2016, **5**, e47.
76. F. Colombo, C. Di Lorenzo, S. Biella, S. Vecchio, G. Frigerio and P. Restani, Adverse effects to food supplements containing botanical ingredients, *J. Funct. Foods*, 2020, **72**(2), 103990.
77. I. Miyahisa, N. Funa, Y. Ohnishi, S. Martens, T. Moriguchi and S. Horinouchi, Combinatorial biosynthesis of flavones and flavonols in *Escherichia coli*, *Appl. Microbiol. Biotechnol.*, 2006, **71**(1), 53–58.
78. S. Wang, S. Zhang, A. Xiao, M. Rasmussen, C. Skidmore and J. Zhan, Metabolic engineering of *Escherichia coli* for the biosynthesis of various phenylpropanoid derivatives, *Metab. Eng.*, 2015, **29**, 153–159.
79. Q. Huang, Y. Lin and Y. Yan, Caffeic acid production enhancement by engineering a phenylalanine over-producing *Escherichia coli* strain, *Biotechnol. Bioeng.*, 2013, **110**(12), 3188–3196.
80. S. Zhao, J. A. Jones, D. M. Lachance, N. Bhan, O. Khalidi and S. Venkataraman, *et al.*, Improvement of catechin production in *Escherichia coli* through combinatorial metabolic engineering, *Metab. Eng.*, 2015, **28**, 43–53.
81. R. P. Pandey, P. Parajuli, N. Koirala, J. H. Lee, Y. Park Il and J. K. Sohng, Glucosylation ofisoflavonoids in engineered *Escherichia coli*, *Mol. Cells*, 2014, **37**(2), 172–177.
82. J. Li, Z. Li, C. Li, J. Gou and Y. Zhang, Molecular cloning and characterization of an isoflavone 7-O-glucosyltransferase from Pueraria lobata, *Plant Cell Rep.*, 2014, **33**(7), 1173–1185.
83. L. Yu and D. J. O'Sullivan, Production of galactooligosaccharides using a hyperthermophilic β-galactosidase in permeabilized whole cells of Lactococcus lactis, *J. Dairy Sci.*, 2014, **97**(2), 694–703.

84. C. Zhang, L. Liu, L. Teng, J. Chen, J. Liu and J. Li, *et al.*, Metabolic engineering of *Escherichia coli* BL21 for biosynthesis of heparosan, a bioengineered heparin precursor, *Metab. Eng.*, 2012, **14**(5), 521–527.
85. B. Weissmann and K. Meyer, The Structure of Hyalobiuronic Acid and of Hyaluronic Acid from Umbilical Cord, *J. Am. Chem. Soc.*, 1954, **76**(7), 1753–1757.
86. S. A. Survase, P. S. Saudagar and R. S. Singhal, Enhanced production of scleroglucan by Sclerotium rolfsii MTCC 2156 by use of metabolic precursors, *Bioresour. Technol.*, 2007, **98**(2), 410–415.
87. X. R. Li, G. Q. Tian, H. J. Shen and J. Z. Liu, Metabolic engineering of *Escherichia coli* to produce zeaxanthin, *J. Ind. Microbiol. Biotechnol.*, 2015, **42**(4), 627–636.
88. S. B. Da Silva, V. V. Cantarelli and M. A. Z. Ayub, Production and optimization of poly-γ-glutamic acid by Bacillus subtilis BL53 isolated from the Amazonian environment, *Bioprocess Biosyst. Eng.*, 2014, **37**(3), 469–479.
89. X. S. Chen, X. D. Ren, N. Dong, S. Li, F. Li and F. L. Zhao, *et al.*, Culture medium containing glucose and glycerol as a mixed carbon source improves ε-poly-l-lysine production by Streptomyces sp. M-Z18, *Bioprocess Biosyst. Eng.*, 2012, **35**(3), 469–475.

CHAPTER 11

Recent Insights in Potential Applications of Edible Packaging Materials in Food Industries

SHUBHI SINGH AND SMRITI GAUR*

Department of Biotechnology, Jaypee Institute of Information Technology, Noida, India
*Email: smritigaurjiit@gmail.com

11.1 Introduction

Food contamination is one of the leading global problems which results in onset of food insecurity, food waste and economic losses. Larger portions of food are generally contaminated through foodborne pathogens which contribute a major threat to illness, leading to increased rates of illness and deaths.[1] Several food industries have faced major challenges in food quality degradation and its loss due to microbial contamination caused by some widely spread pathogens, namely *Staphylococcus aureus*, *Salmonella* spp. and *Listeria monocytogenes*, leading to decreased economic losses faced by retailers and marketers.[2] The microbial contamination of food products can also result in decreased nutritional and sensorial quality of the food products, hence decreasing the overall acceptability by the consumers.[3] Microbial contamination can affect the quality of food products any time from food processing to food storage up to food transportation. Hence, there is a need to develop effective packaging systems which can protect food products from

Food Chemistry, Function and Analysis No. 44
Functional Foods of the Future
Edited by Vijai Kumar Gupta, Minaxi Sharma, Smriti Gaur and Ramesh Chander Kuhad

Published by the Royal Society of Chemistry, www.rsc.org

the external environment and reduce contamination without affecting the nutritional and sensorial parameters of food. There are several materials which are presently utilized as packaging materials of food items. Amongst several materials, plastics have dominated the packaging material market and have crossed 300 billion USD in year 2022.[4] Plastics are extracted from fossil fuels containing non-renewable resources and are considered to be non-biodegradable as well as non-renewable, leading to extremely serious environmental issues related to exploitation of natural resources and its degradation.[5] The better alternative is to switch to bio-based edible packaging which could prevent the environmental losses, prevent food contamination, and support enhancing the shelf life of food products.

Edible packagings are easily degradable, are renewable, reduce the environmental stress and provide a great opportunity to replace the current plastic-based packaging material in the food industry. Advancements in research have explored several biomolecules as well as bio-based composites which could be used as safer and healthier alternatives for food packaging. Several biomolecule-based coatings have been applied in fruits and vegetables to enhance the shelf life up to several weeks. Further, carbohydrate- and protein-based food packaging films have displayed enhanced mechanical and physical barrier properties in maintaining a reduced count of microbial contaminants in food products.[6] However, the addition of biopolymeric supplements during development of edible packaging has enhanced the mechanical and physical properties to a great extent. Besides this, added bio-based nanoparticles have been found to interact with the substrate material at the atomic and molecular levels, also resulting in enhancement of overall functional properties. Amongst different nanoparticles, zinc oxide nanoparticles have been explored to a great extent to develop zinc oxide nanocomposite edible packagings exhibiting antimicrobial as well as photocatalytic activities.[7,8]

At present, there are several biomolecules as well as bio-based composite materials which have been utilized in development of edible packaging, however it is important to investigate the interaction patterns between the two distinctive matrices. This chapter comprehensively discusses recent insights into different biomolecules which have been utilized as the core substrate material in development of edible packaging. Moreover, the applications of novel edible packaging in the food sector are mentioned herein too, hence providing enough information to understand the behaviour and applicability of edible packagings which might be considered as sustainable and renewable alternatives in the future.

11.2 Biomolecules in the Development of Edible Packaging Materials

Edible packagings are packaging materials which are developed using natural, renewable and plant-based biodegradable materials that can be easily

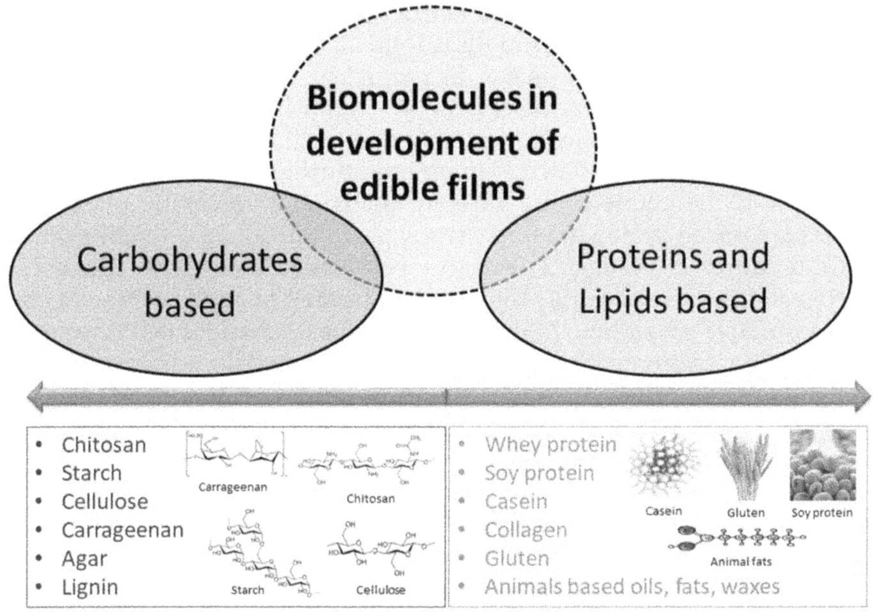

Figure 11.1 The various biomolecules utilized in development of edible films.

consumed along with food products. These edible packaging materials can be easily processed, recycled and disposed in a sustainable manner without damaging the environment.[9] Biomolecules are naturally and abundantly present in all plants and have been explored using advanced research technologies as substrate materials in the development of edible packagings. Generally, biomolecule-based edible packagings include carbohydrate-based packaging, protein-based materials, and lipid-based matrices (Figure 11.1), which are comprehensively discussed in this section.

11.2.1 Edible Packaging Developed Using Carbohydrates

Recent exploratory research studies have investigated the utilization of several carbohydrate-based sources in the development of edible packaging material for food industries. Polysaccharides like chitosan, cellulose and starch have been largely utilized due to their biodegradable and non-toxic nature.[10] Amongst these, chitosan is the most extensively used biopolymer in edible packaging development due to film-forming properties related to nano-sized crystalline sections.[11] In addition, chitosan exhibits antimicrobial properties.[12] These functional properties are often dependant on pH value and the degree of deacetylation, which causes protonation of the amino group in chitosan at lower pH values, hence leading to high antimicrobial activities.[13,14] A recent study by Pavinatto and co-workers has developed a chitosan-based edible packaging film for enhancement of the shelf life of strawberries. These films were prepared by addition of glycerol

(30%), which resulted in increased elasticity as well as hydrophobicity. The antimicrobial activity of chitosan films were observed against Gram positive and Gram negative bacteria as well as fungi. The results demonstrated that the developed films were successful in inhibiting microbial attack on strawberries, hence leading to enhanced shelf life without affecting the original taste, aroma and flavor.[15] In order to further enhance the functional properties of the edible films, different phenolic or prebiotic plant-based extracts are added during the film casting procedure along with chitosan. To illustrate, different *Prunella* L. extracts were added along with chitosan to develop chitosan-based edible films with enhanced phenolic content as well as antioxidative properties.[16] Similarly, xylooligosaccharides (XOS) were also added to the chitosan film solution which resulted in enhanced tensile strength and reduced oxygen permeability of films. The developed films also resulted in enhanced functional properties as the prebiotic content of the films was enhanced by addition of XOS.[17]

Apart from utilizing chitosan, cellulose and starch have also been extensively utilized in development of edible packaging materials. Starch contains amylose (linear) and amylopectin (branched) molecules linked through α-1,4 glycosidic linkages and α-1,6 glycosidic linkages, respectively. On the other hand, cellulose is a linear polymer of β-D-glucopyranose units connected by β-1,4 linkages. Starch and cellulose are readily available, low-cost, non-toxic, degradable and possess high mechanical properties.[14] Starch-based edible films are known to possess high resistance to oxygen or low oxygen permeability, hence resulting in delayed spoilage of food products.[18] It is known that high moisture penetration into food products results in greater microbial attack by spoilage microbes. Therefore, edible films are often developed using natural materials which can resist water as well as oxygen permeability. Recently, starch films have been developed by extraction of starch from various natural sources like rice, cassava and yam.[19–21] The starch-based edible films exhibited higher tensile strength, mechanical properties, physicochemical properties, antimicrobial properties as well as lower oxygen and water permeability.

11.2.2 Edible Packaging Developed Using Proteins and Lipids

Several protein sources have been investigated for their mechanical, physicochemical, thermal and barrier properties for utilization in the development of edible food-packaging materials. Whey proteins, soy proteins, gelatin, casein and collagen are abundantly available protein sources and have been explored in the development of edible films.[22] These protein sources are non-toxic, biodegradable and are easily fabricated with other additives. Whey proteins are known to form transparent films with higher flexibility and water-holding capacity (WHC). The water-holding capacities are higher due to the presence of hydrophilic amino acids present in whey proteins which allows for more water permeability.[23] On the other hand, gelatin is also a widely available protein-based polymer which is widely used in film formation along with other additives, in order to enhance the tensile strength as well as plasticity of the edible films.[24] Bhatia and others have utilized casein protein along with sodium alginate to develop protein-based edible films. The films were modified by addition of bitter orange oil in order to enhance their functional properties. The protein-based edible films exhibited higher transparency, water absorption capacities, moisture content as well as water solubility.[25] Similarly, casein-based films were developed along with the addition of gallic acid, which improved the overall phenolic content as well as the antioxidative behavior of the films. However, it is worth noting that addition of gallic acid to the casein-based edible films decreased the transparency and altered the color characteristics of the films.[26] As discussed, protein-based edible films are known for their great mechanical properties, however they still exhibit shortcomings in terms of important physical characteristics like water vapor permeability. In response to the limitations of protein-based films, several methods have been suggested to address these issues. These strategies encompass cross-linking techniques, combining them with other biopolymers through blending by physical, chemical, and enzymatic methods, as well as enhancing their properties by incorporating nanoparticles. Despite advancements in protein-based bio-nanocomposites, their development is still in the nascent stages, highlighting the need for further research and exploration in the future.

Lipids are considered crucial sources in the development of various edible films. Lipid sources such as animal-derived oils, fats, and waxes as well as natural resins have been widely utilized in order to enhance the surface hydrophobicity, flexibility, cohesiveness and moisture penetrability in edible films. Recently, Vu and others have developed hydroxypropylmethylcellulose/shellac-based composite films (HPMC/Sh composite) from oil/water emulsions. The results demonstrated that the edible films produced from oil/water emulsions exhibited a smoother surface and enhanced elasticity as well as flexibility.[27] Additionally, it has been observed that saturated and unsaturated fatty acids have an impact on edible films. These fatty acids can create stable layers within the film structure and enhance the moisture barrier properties. Furthermore, the incorporation of nanoparticles as reinforcement has been shown to be a successful approach for greatly enhancing the properties of lipid-based films. Moreover, the successful integration of nanoparticles as reinforcements has demonstrated a highly effective method for significantly improving the properties of lipid-based films.

11.3 Applications of Novel Edible Packaging Materials in Food Industries

Contamination and deterioration of food and beverages are ongoing concerns for the food sector. Foodstuffs require appropriate packaging during distribution, storage, and commercialization in order to prevent the growth of foodborne pathogens and spoiling microorganisms. Food is typically packaged in plastics which are derived from non-renewable fossil fuel resources. This kind of material is rarely recyclable and can be difficult to dispose. As edible films are biodegradable, renewable, and environmentally friendly bio-based materials, they are now considered to be potential alternatives in food packaging. The successful applications of edible films in food industries as active packaging materials are presented in Table 11.1.

Table 11.1 Applications of novel edible films developed using various biomolecules.

Material used for developing edible films	Food products	Beneficial claims	Ref.
Chitosan + oregano essential oil	Apricots	Antifungal activity against black spot rot	28
Chitosan + *Nigella sativa* extract	Grapes	Enhanced antioxidative properties, total phenolic content and increased shelf life of grapes	30
Chitosan + grape seed/ sea buckthorn essential oil	Apples	Maintenance of post-harvest parameters in apples	31
Chitosan + grape seed extract	Whole grain cookies	Enhanced total phenolic content and shelf life, reduced peroxide formation	32
Whey protein + chitosan	Sausages	Inhibited microbial spoilage and lipid oxidation	33
Cassava starch + carboxymethylcellulose (CMC)	Bananas	Enhanced antioxidative properties, shelf life and reduced lipid oxidation	34
Whey protein + oregano oil + garlic oil + nisin + natamycin	Kasar cheese	Reduced microbial spoilage	35
Soy protein isolates + cardamom essential oil	Iranian white cheese	Enhanced antimicrobial, antioxidative properties and shelf life	36
Corn zein + dill leaves extract + β-cyclodextrin	Carp fillets	Lower changes in pH during storage and enhanced shelf life	37
Egg proteins + vegetable oils	Cakes	Enhanced barrier against water permeability, reduced microbial spoilage	38
Chitosan + sea fennel plant extract	Fish burgers	Enhanced antioxidative properties and shelf life	39
Whey protein + rosemary essential oil	Fresh spinach	Reduced weight loss and transmission rate of oxygen in spinach leaves	40
Polyvinyl alcohol + gelatin + *Amaranthus* leaf extract	Fish and chicken	Enhanced shelf life by reducing oxidative rancidity	41
Nanocellulose	Fresh spinach	Retention of chlorophyll, color and moisture content and reduction in respiration rate	42

Currently, there is a wide range of innovative edible films being utilized in food packaging. These films include biopolymer substances, nanoparticles/nanofillers, bio-based composites, and bio-nanocomposites. To illustrate, edible films based on chitosan and oregano essential oil edible films have been developed by Guo and co-workers. The films exhibited strong antimicrobial properties as they successfully prohibited microbial attack by black spot rot, a spoilage fungus. The films were successfully utilized in enhancing the shelf life of apricots.[28] For a variety of reasons, essential oils are often used in the development of edible films. First, essential oils have inherent antibacterial properties which can help in preventing the growth of bacteria, fungi, and other pathogens, hence extending the shelf life of food products.[6] Furthermore, essential oils can contribute to the desired flavor and aroma of the edible films, therefore improving the sensory experience of consumers. In addition, essential oils are made from natural ingredients and are thought to be safer and more environmentally friendly than synthetic additions.[29] Several studies have also incorporated essential oils from a variety of natural sources, along with natural bio-based materials in the development of edible films. For instance, sea buckthorn essential oil, garlic oil, cardamom essential oil and rosemary essential oil have been successfully incorporated along with chitosan and whey protein isolates in developing edible films. The results demonstrated that novel films exhibited enhanced antimicrobial activities, hence leading to reduced microbial spoilage and enhanced shelf life. Moreover, the antioxidative properties and total phenol content was also observed to be increased upon supplementation of essential oils.[31,35,36,40]

Li and co-workers have demonstrated the utilization of carboxymethylcellulose (CMC) along with cassava starch in the development of edible films for enhancing the shelf life of bananas. CMC has been utilized due to various essential properties, namely high water-holding capacity, high mechanical and tensile strength, high flexibility as well as its ability to bind other biomolecules and additives. It also acts as a barrier against oxygen and other external contaminants, hence preserving the quality and freshness of the packaged food. The utilization of CMC along with cassava starch resulted in enhanced antioxidative properties, longer shelf life and reduced lipid oxidation in bananas.[34] Lipid oxidation is a major concern in the food industry as it can lead to the deterioration of taste, texture, and nutritional quality of the product. The application of CMC as well as polyvinyl alcohol into edible films is known to prevent the interaction of oxygen with the lipids present in the food. This barrier developed from CMC and polyvinyl alcohol reduces the exposure of lipids to oxygen, thereby minimizing the occurrence of lipid oxidation. In a recent study, polyvinyl alcohol, gelatin and *Amaranthus* leaf extract were utilized in developing edible films, with the aim of enhancing the shelf life of fresh fish and chicken. The results demonstrated that the novel films exhibited reduced lipid oxidation and greatly enhanced the shelf life as compared to unwrapped food samples.[41]

11.4 Insights into Smart and Intelligent Edible Packaging Materials

In recent times, there has been a growing interest regarding the advancements and applications of smart packaging materials. These materials are now being recognized for their ability to respond effectively to variations in the system, such as changes in the quality, safety, or maturity of packaged food. These smart or intelligent packaging materials offer other advantages too, including improved product monitoring, enhanced safety measures, and the ability to provide timely indications of any changes or issues with the packaged food products.[43] Smart/intelligent edible films are equipped with several sensor technologies in order to determine the changes occurring in food products (Figure 11.2). Table 11.2 represents insights into recent studies on the utilization and development of smart edible films in food industries.

A pH indicator has been developed to display an apparent change when the pH of the packaged food significantly changes. These pH changes in the food might be a result of enzyme activity, chemical reactions, or microbial development. Nowadays, natural colors are more preferred over synthetic colors as consumers demand items with clean/safe labels. Numerous kinds of natural pigments with distinctive colors have been explored that show variations in response to environmental pH fluctuations. Recently several studies have explored various natural pigments and developed novel smart edible packaging materials. To illustrate, anthocyanin from red barberry and saffron have been extracted and utilized in development of edible films based on κ-carrageenan and gelatin. The results demonstrated that smart films were successful in determining the shelf life of fish. Smart films that exhibited color changes in response to food freshness were also observed to enhance bacteriostatic activities, hence leading to improved shelf life.[44] Similarly, anthocyanin from other natural sources has also been extracted. Roselle anthocyanin along with curcumin has been incorporated in glycerol- and gelatin-based smart films which detected changes pH as well as release of amines from shrimps during the storage period. As the amounts of amine gases increased during spoilage of shrimps, the pH of the films changed and hence a notable color change was observed. Hence, these smart films were proved to have successful applications in food packaging.[49] Several other

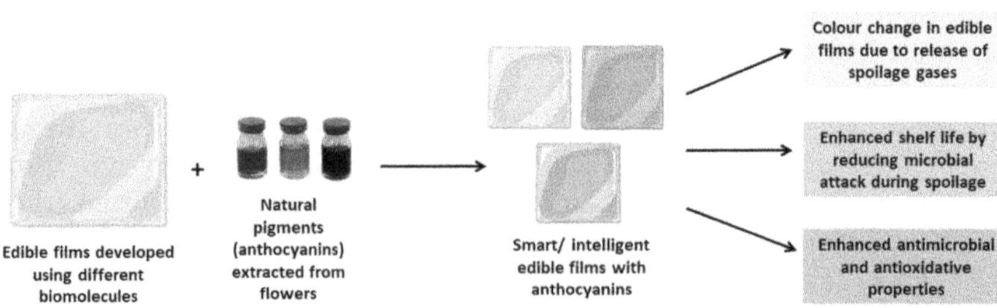

Figure 11.2 Development and characteristics of smart/intelligent edible films.

Table 11.2 Recent studies on development of novel smart/intelligent edible films.

Edible film matrix	Coloring agents	Indicators	Food products	Functions	Ref.
κ-Carrageenan + gelatin +	Saffron and red barberry anthocyanin	pH indicators	Fish	Smart films exhibited color changes in response to food freshness, enhanced bacteriostatic activities	44
Cassava starch	Pomegranate peel powder	Gas indicators	Lamb meat	Film color changed from red to green during storage and was related with TVBN	45
Starch + agar	Shikonin	pH indicators	Shrimp	Shikonin enhanced mechanical and antioxidative properties, and displayed color changes from red to blue during storage	46
Chitosan + gelatin + starch	Raspberry anthocyanin + curcumin	pH indicators	Meat	Enhanced opacity, water solubility, thickness and color change during storage	47
Starch + gelatin	Purple sweet potato anthocyanins	pH indicators	Mushrooms	Film color changed from green to purplish gray during storage	48
Gelatin + glycerol	Roselle anthocyanin + curcumin	pH and gas indicators	Shrimp	Release of volatile amines detected by smart films indicated by color changes	49

studies have also developed biomolecule-based films supplemented with natural pigments as pH indicators in quality assessments of meat and mushrooms.[47,48] It is known that anthocyanins have different characteristic colours under acidic, neutral, and alkaline conditions, hence they can be readily used as sensors over a wide pH range (Figure 11.2).

Apart from pH indicators, gas indicators are also considered crucial in estimating the freshness of food. Food often produces gases as a result of natural respiration processes or during the attack of spoilage microbes. Additionally, gases such as oxygen have the potential to enter or exit food packaging, thereby affecting the vulnerability of food products to chemical degradation and lipid oxidation. Recently, smart edible films have been developed which contain sensors that are sensitive to different kinds of gases and other volatile compounds and help in providing an indication of food quality. Gas sensors are often developed by involving the capacity of certain natural pigments to undergo chemical degradation in response to certain gases. To illustrate, anthocyanins are known to undergo degradation when exposed to ammonia gas which is released during food spoilage, hence leading to a notable change in colour. Recently, Mohseni-Shahri and Moeinpour have developed a film based on glycerol and gelatin for quality evaluation of shrimps, supplemented with roselle anthocyanin and curcumin. The release of volatile amines during storage of shrimps resulted in degradation of anthocyanins and color changes in films.[49] Similarly, an edible film based on cassava starch and pomegranate peel powder was also developed for quality assessment of lamb meat. During the storage period, the color of the film changed from red to green due to the chemical reaction between anthocyanins and total volatile basic nitrogen (TVBN), which was released from the lamb meat during storage.[45]

11.5 Conclusion and Future Prospects

Recently, there has been a growing interest in finding alternative solutions for food packaging materials. Traditional packaging materials are often derived from non-renewable fossil fuels and generally face challenges in terms of recycling and disposal. However, researchers have turned their attention to bio-based edible films as potential options for packaging applications. These films possess unique chemical and physical properties, including natural antibacterial properties and antioxidative properties. Advanced coatings, films, and biological additives are being explored as potential methods to enhance food preservation and address sustainability concerns in packaging. As discussed, smart edible films have the capability to enhance food quality, improve safety measures, extend the shelf life of products, and effectively reduce food waste.

While edible films have shown significant potential for development, there are several challenges that need to be addressed. Future studies must be focused on development of edible films with enhanced stability, optical, mechanical and barrier properties. Furthermore, there is a need to create edible films cost-effectively on a larger scale as well as exploring new natural materials and implementing innovative structural designs and techniques in the development of edible films.

References

1. D. G. Nyachuba, Foodborne illness: is it on the rise?, *Nutr. Rev.*, 2010, **68**(5), 257–269.
2. A. Chatterjee and J. Abraham, Microbial contamination, prevention, and early detection in food industry, in *Microbial contamination and food degradation*, Academic Press, 2018, pp. 21–47.
3. S. Singh and S. Gaur, Development of rapid and non-destructive Electric nose (E-nose) system for shelf life evaluation of different edible seeds, *Food Chem.*, 2023, 136562.
4. M. Mittal, D. Mittal and N. K. Aggarwal, Plastic accumulation during COVID-19: call for another pandemic; bioplastic a step towards this challenge?, *Environ. Sci. Pollut. Res.*, 2022, 1–15.
5. V. Girgenti, C. Peano, C. Baudino and N. Tecco, From "farm to fork" strawberry system: Current realities and potential innovative scenarios from life cycle assessment of non-renewable energy use and green house gas emissions, *Sci. Total Environ.*, 2014, **473**, 48–53.
6. N. A. Al-Tayyar, A. M. Youssef and R. Al-Hindi, Antimicrobial food packaging based on sustainable Bio-based materials for reducing food-borne Pathogens: A review, *Food Chem.*, 2020, **310**, 125915.
7. S. M. Mousavi-Kouhi, A. Beyk-Khormizi, M. S. Amiri, M. Mashreghi and M. E. T. Yazdi, Silver-zinc oxide nanocomposite: From synthesis to antimicrobial and anticancer properties, *Ceram. Int.*, 2021, **47**(15), 21490–21497.
8. M. Divya, B. Vaseeharan, M. Abinaya, S. Vijayakumar, M. Govindarajan, N. S. Alharbi, S. Kadaikunnan, J. M. Khaled and G. Benelli, Biopolymer gelatin-coated zinc oxide nanoparticles showed high antibacterial, antibiofilm and anti-angiogenic activity, *J. Photochem. Photobiol., B*, 2018, **178**, 211–218.
9. L. K. Ncube, A. U. Ude, E. N. Ogunmuyiwa, R. Zulkifli and I. N. Beas, Environmental impact of food packaging materials: A review of contemporary development from conventional plastics to polylactic acid based materials, *Materials*, 2020, **13**(21), 4994.
10. H. Namazi and M. Mosadegh, Bio-nanocomposites based on naturally occurring common polysaccharides chitosan, cellulose and starch with their biomedical applications, *Recent developments in bio-nanocomposites for biomedical applications*, 2011, pp. 379–397.
11. M. Azmana, S. Mahmood, A. R. Hilles, A. Rahman, M. A. B. Arifin and S. Ahmed, A review on chitosan and chitosan-based bionanocomposites: Promising material for combatting global issues and its applications, *Int. J. Biol. Macromol.*, 2021, **185**, 832–848.
12. P. K. Dutta, S. Tripathi, G. K. Mehrotra and J. Dutta, Perspectives for chitosan based antimicrobial films in food applications, *Food Chem.*, 2009, **114**(4), 1173–1182.
13. M. Hosseinnejad and S. M. Jafari, Evaluation of different factors affecting antimicrobial properties of chitosan, *Int. J. Biol. Macromol.*, 2016, **85**, 467–475.

14. S. Chaudhary, V. P. Jain and G. Jaiswar, The composition of polysaccharides: monosaccharides and binding, group decorating, polysaccharides chains, in *Innovation in Nano-Polysaccharides for Eco-sustainability*, Elsevier, 2022, pp. 83–118.
15. A. Pavinatto, A. V. de Almeida Mattos, A. C. G. Malpass, M. H. Okura, D. T. Balogh and R. C. Sanfelice, Coating with chitosan-based edible films for mechanical/biological protection of strawberries, *Int. J. Biol. Macromol.*, 2020, **151**, 1004–1011.
16. İ. Erken, S. Şahin, B. Karkar, B. Akça and C. Özakın, Chitosan based edible film incorporating different Prunella L. extracts, characterization and their antioxidant properties, *J. Food Process. Preserv.*, 2022, **46**(7), e16658.
17. J. Xu, R. Xia, T. Yuan and R. Sun, Use of xylooligosaccharides (XOS) in hemicelluloses/chitosan-based films reinforced by cellulose nanofiber: Effect on physicochemical properties, *Food Chem.*, 2019, **298**, 125041.
18. M. A. García, A. Pinotti, M. N. Martino and N. E. Zaritzky, Characterization of starch and composite edible films and coatings, *Edible films and coatings for food applications*, 2009, pp. 169–209.
19. J. Jeevahan and M. Chandrasekaran, Influence of nanocellulose additive on the film properties of native rice starch-based edible films for food packaging, *Recent Pat. Nanotechnol.*, 2019, **13**(3), 222–233.
20. Y. Zhou, X. Wu, J. Chen and J. He, Effects of cinnamon essential oil on the physical, mechanical, structural and thermal properties of cassava starch-based edible films, *Int. J. Biol. Macromol.*, 2021, **184**, 574–583.
21. M. Nagar, V. S. Sharanagat, Y. Kumar and L. Singh, Development and characterization of elephant foot yam starch–hydrocolloids based edible packaging film: physical, optical, thermal and barrier properties, *J. Food Sci. Technol.*, 2020, **57**, 1331–1341.
22. C. Tan, F. Han, S. Zhang, P. Li and N. Shang, Novel bio-based materials and applications in antimicrobial food packaging: Recent advances and future trends, *Int. J. Mol. Sci.*, 2021, **22**(18), 9663.
23. S. Gunasekaran, S. Ko and L. Xiao, Use of whey proteins for encapsulation and controlled delivery applications, *J. Food Eng.*, 2007, **83**(1), 31–40.
24. M. G. A. Vieira, M. A. Da Silva, L. O. Dos Santos and M. M. Beppu, Natural-based plasticizers and biopolymer films: A review, *Eur. Polym. J.*, 2011, **47**(3), 254–263.
25. S. Bhatia, A. Al-Harrasi, M. S. Al-Azri, S. Ullah, A. E. D. A. Bekhit, A. Pratap-Singh, M. K. Chatli, M. K. Anwer and M. F. Aldawsari, Preparation and physiochemical characterization of bitter Orange oil loaded sodium alginate and casein based edible films, *Polymers*, 2022, **14**(18), 3855.
26. S. Bhatia, A. Al-Harrasi, M. S. Al-Azri, S. Ullah, H. A. Makeen, A. M. Meraya, M. Albratty, A. Najmi and M. K. Anwer, Gallic Acid Crosslinked Gelatin and Casein Based Composite Films for Food Packaging Applications, *Polymers*, 2022, **14**(19), 4065.

27. T. H. Vu, T. H. T. Le, H. T. Nguyen, T. T. Nguyen, V. K. Nguyen, T. T. H. Pham, N. Y. Truong and T. L. Nguyen, Effects of emulsifiers on engineering and microstructural properties of hpmc/shellac composite films, in *Materials Science Forum*, Trans Tech Publications Ltd, 2022, vol. 1051, pp. 181–188.
28. S. Guo, T. Li, M. Chen, C. Wu, X. Ge, G. Fan, X. Li, D. Zhou, X. Zhao and T. Yang, Sustainable and effective Chitosan-based edible films incorporated with OEO nanoemulsion against apricots' black spot, *Food Control*, 2022, **138**, 108965.
29. L. S. Nerio, J. Olivero-Verbel and E. Stashenko, Repellent activity of essential oils: a review, *Bioresour. Technol.*, 2010, **101**(1), 372–378.
30. B. Karkar, İ. Patır, S. Eyüboğlu and S. Şahin, Development of an edible active chitosan film loaded with Nigella sativa L. extract to extend the shelf life of grapes, *Biocatal. Agric. Biotechnol.*, 2023, **50**, 102708.
31. P. A. Popescu, L. M. Palade, I. C. Nicolae, E. E. Popa, A. C. Miteluţ, M. C. Drăghici, F. Matei and M. E. Popa, Chitosan-Based Edible Coatings Containing Essential Oils to Preserve the Shelf Life and Postharvest Quality Parameters of Organic Strawberries and Apples during Cold Storage, *Foods*, 2022, **11**(21), 3317.
32. D. Molnar, D. Novotni, M. Kurek, D. Iveković, H. Bionda and M. Ščetar, Characteristics of edible films enriched with fruit by-products and their application on cookies, *Food Hydrocolloids*, 2023, **135**, 108191.
33. H. Esmaeili, N. Cheraghi, A. Khanjari, M. Rezaeigolestani, A. A. Basti, A. Kamkar and E. M. Aghaee, Incorporation of nanoencapsulated garlic essential oil into edible films: A novel approach for extending shelf life of vacuum-packed sausages, *Meat Sci.*, 2020, **166**, 108135.
34. S. Li, Y. Ma, T. Ji, D. E. Sameen, S. Ahmed, W. Qin, J. Dai, S. Li and Y. Liu, Cassava starch/carboxymethylcellulose edible films embedded with lactic acid bacteria to extend the shelf life of banana, *Carbohydr. Polym.*, 2020, **248**, 116805.
35. A. C. Seydim, G. Sarikus-Tutal and E. Sogut, Effect of whey protein edible films containing plant essential oils on microbial inactivation of sliced Kasar cheese, *Food Packag. Shelf Life*, 2020, **26**, 100567.
36. B. Hajirostamloo, M. Molaveisi, P. Jafarian Asl and M. M. Rahman, Novel soy protein isolate film containing cardamom essential oil microcapsules: study of physicochemical properties and its application in Iranian white cheese packaging, *J. Food Meas. Charact.*, 2023, **17**(1), 324–336.
37. A. Zolfaghari, B. Bazargani-Gilani and N. Aghajani, Edible film based on corn zein containing dill extract and essential oil/β-cyclodextrin inclusion complex: Shelf life enhancement of common carp fillet, *Food Sci. Nutr.*, 2023, **11**, 4275–4288.
38. T. De Pilli, Development of a vegetable oil and egg proteins edible film to replace preservatives and primary packaging of sweet baked goods, *Food Control*, 2020, **114**, 107273.
39. D. Rico, I. Albertos, O. Martinez-Alvarez, M. E. Lopez-Caballero and A. B. Martin-Diana, Use of sea fennel as a natural ingredient of edible films

for extending the shelf life of fresh fish burgers, *Molecules*, 2020, **25**(22), 5260.

40. A. Abedi, L. Lakzadeh and M. Amouheydari, Effect of an edible coating composed of whey protein concentrate and rosemary essential oil on the shelf life of fresh spinach, *J. Food Process. Preserv.*, 2021, **45**(4), e15284.
41. S. R. Kanatt, Development of active/intelligent food packaging film containing Amaranthus leaf extract for shelf life extension of chicken/fish during chilled storage, *Food Packag. Shelf Life*, 2020, **24**, 100506.
42. K. Pacaphol, K. Seraypheap and D. Aht-Ong, Development and application of nanofibrillated cellulose coating for shelf life extension of fresh-cut vegetable during postharvest storage, *Carbohydr. Polym.*, 2019, **224**, 115167.
43. M. Vanderroost, P. Ragaert, F. Devlieghere and B. De Meulenaer, Intelligent food packaging: The next generation, *Trends Food Sci. Technol.*, 2014, **39**(1), 47–62.
44. M. A. Sani, M. Tavassoli, S. A. Salim, M. Azizi-lalabadi and D. J. McClements, Development of green halochromic smart and active packaging materials: TiO_2 nanoparticle-and anthocyanin-loaded gelatin/κ-carrageenan films, *Food Hydrocolloids*, 2022, **124**, 107324.
45. A. Esfahani, A. Mohammadi Nafchi, H. Baghaei and L. Nouri, Fabrication and characterization of a smart film based on cassava starch and pomegranate peel powder for monitoring lamb meat freshness, *Food Sci. Nutr.*, 2022, **10**(10), 3293–3301.
46. P. Ezati and J. W. Rhim, Starch and agar-based color-indicator films integrated with shikonin for smart packaging application of shrimp, *ACS Food Sci. Technol.*, 2021, **1**(10), 1963–1969.
47. A. Duan, J. Yang, L. Wu, T. Wang, Q. Liu and Y. Liu, Preparation, physicochemical and application evaluation of raspberry anthocyanin and curcumin based on chitosan/starch/gelatin film, *Int. J. Biol. Macromol.*, 2022, **220**, 147–158.
48. Z. Zong, M. Liu, H. Chen, M. A. Farag, W. Wu, X. Fang, B. Niu and H. Gao, Preparation and characterization of a novel intelligent starch/gelatin binary film containing purple sweet potato anthocyanins for Flammulina velutipes mushroom freshness monitoring, *Food Chem.*, 2023, **405**, 134839.
49. F. S. Mohseni-Shahri and F. Moeinpour, Development of a pH-sensing indicator for shrimp freshness monitoring: Curcumin and anthocyanin-loaded gelatin films, *Food Sci. Nutr.*, 2023, **11**, 3898–3910.

CHAPTER 12

Functional Energy Drinks: Today's Choice?

SHRUTIKA,[a] SHIKHA MAHAJAN,*[a] KANDI SRIDHAR[b] AND MINAXI SHARMA*[c]

[a] Department of Food & Nutrition Punjab Agricultural University, Ludhiana 141004, India; [b] Department of Food Technology, Karpagam Academy of Higher Education (Deemed to be University), Coimbatore 641021, India; [c] Research Centre for Life Science and Healthcare, Nottingham Ningbo China Beacons of Excellence Research and Innovation Institute (CBI), University of Nottingham Ningbo China, Ningbo-315000, China
*Emails: shikha_bathla@yahoo.com; minaxi86sharma@gmail.com

12.1 Introduction

Functional energy drinks are beverages that have become increasingly popular in recent years due to their unique blend of ingredients that focus not only on providing an energy boost but also offer various health benefits. Functional energy drinks are tailored to contain ingredients that promote mental and physical performance, such as caffeine, amino acids, vitamins, and other herbal extracts, which distinguish them from normal beverages and makes them a better option for health-conscious individuals who not only drink to quench their thirst but want to achieve certain health goals.[1]

Consumption of energy drinks has become popular among people which has resulted in significant growth in their market, offering a variety of brands and flavors that are easily available in supermarkets and convenience stores.[2] In the early 20th century soft drinks containing sugar and caffeine were first marketed as "energy boosters", which is when functional energy

Food Chemistry, Function and Analysis No. 44
Functional Foods of the Future
Edited by Vijai Kumar Gupta, Minaxi Sharma, Smriti Gaur and Ramesh Chander Kuhad

Published by the Royal Society of Chemistry, www.rsc.org

drinks first gained popularity. The U.S. launched the first energy drink in 1949, known as "Dr. Enuf".[2] However, energy drinks as we know them today did not start to appear until the 1960s. The first widely available energy drink was a Japanese product called Lipovitan, which contained taurine, B vitamins and an amino acid that was known to improve athletic performance. In the 1980s, energy drinks became popular in Europe, with brands like Red Bull and Monster Energy consumed almost in every house. These drinks were rich in caffeine, as well as additional ingredients like ginseng and guarana, and which provided many benefits to the consumers.[3]

Functional energy drinks continue to be an appealing choice for individuals who are looking for a quick boost of energy and enhanced performance. They are designed in a such way that they offer specific formulations to meet different requirements of people for their pre-workout energy as well as post-workout recovery. Many brands now supplement their energy drinks with amino acids like tyrosine and theanine, which are believed to improve mental sharpness and reduce stress, as well as vitamins and antioxidants which support overall health by improving the immune system and by scavenging reactive oxygen species (ROS). Additionally, it is believed that the consumption of energy drinks in higher doses can lead to potential health risks, particularly among young people.[4] Still, for them functional energy drinks are a convenient and effective way to stay alert and focused throughout the day. It becomes important for parents, educators, and healthcare professionals to provide young people with accurate information about the potential risks and benefits of consuming these energy drinks in terms of the right dose and the nutritional information claimed.

12.2 Formulation of Energy Drinks

An array of ingredients is used to formulate functional energy drinks, and each of these ingredients is claimed to possess a particular health benefit or stimulant action. As the main focus of the energy drink is on providing quick energy, to fulfil these criteria the basic ingredients added to the drink is glucose/sugars and caffeine.[5] But, with increased access to information and a growing awareness of the importance of nutrition, people are taking a more proactive approach to their dietary habits which make them seek out foods and beverages that are nutrient-dense. So other ingredients like vitamin B, green tea extract, guarana, *Ginkgo biloba*, carnitine, glucuronolactone, ginseng, antioxidants and inositol are added in order to provide additional health benefits.[6] According to Alford *et al.*, functional energy drinks provide beneficial effects to the consumer like improving their psychomotor performance (reaction time, concentration and memory), mental alertness (choice reaction, concentration and memory) and physical endurance. It was found that their consumption enhanced aerobic endurance (maintaining 65–75% max. heart rate) and aerobic performance (maintaining max. speed) on cycle ergometers.[7] Energy drink's effect on the time to exhaustion during treadmill exercise was studied by Walsh.[8] It was observed that a significant increase in time to exhaustion during a moderate intensity endurance run was seen as well as improvement in perceived feelings of focus, energy and delayed fatigue.

The different types of ingredients used for formulation of energy drink are discussed as below:

- Caffeine: Caffeine is a key ingredient in most energy drinks and plays a vital role in providing the desired functional benefits. It is a central nervous system stimulant that can help to improve mental alertness, focus, and physical performance. It works by blocking the action of adenosine, a neurotransmitter that causes drowsiness, and promotes the release of other neurotransmitters such as dopamine and norepinephrine, which can improve mood and increase heart rate and blood pressure.[9,10] Most energy drinks contain 70–200 mg of caffeine per 16 oz serving.[3] The caffeine content in energy drinks is often higher than that in other caffeinated beverages such as coffee or tea, making it important to monitor and limit consumption to avoid potential health risks. Caffeine mobilizes fat stores and stimulates working muscles to use fat as a fuel, which delays muscle glycogen depletion and allows for prolonged exercise.[11,12]
- Glucose or sugar: Sugars play a major role in energy drinks by providing a quick source of energy for the body. Many energy drinks contain added sugars, such as high fructose corn syrup, to help mask the bitter taste of some of the other ingredients, including caffeine and other stimulants. Administration of glucose or other carbohydrates before, during, and after prolonged exercise (>1 hour) has been shown to postpone fatigue, conserve muscle glycogen, and improve performance.[13]
- L-Carnitine: The body utilizes lysine and methionine to form carnitine, responsible for the transport of fatty acids for β-oxidation and preventing accumulation of acetyl CoA. Carnitine, specifically the isomer propionyl-L-carnitine (PLC), exhibits a high affinity for both skeletal and cardiac muscle and is rapidly converted to L-carnitine when given exogenously. PLC supplementation offers great health effects including decreased body weight and abdominal adiposity, vascular inflammation, triglycerides, low-density lipoprotein cholesterol, atherosclerotic lesions and lipid peroxidation, improved peripheral arterial disease symptoms, and possible improvement in myocardial function after ischemia.[14]
- Guarana: Guarana, a plant from Brazil, contains a caffeine concentration in 2–15 percent of its dry weight (about twice the amount of coffee beans), possesses antioxidant effects, and also can decrease platelet aggregation as reported in ref. 15.
- Glucuronolactone: Glucuronolactone is a naturally occurring metabolite in the liver derived from glucose that serves as a precursor for synthesis of ascorbic acid, an antioxidant, and as a structural component of connective tissues.[15]

- Taurine: Taurine is one of the non-essential amino acids that is found in higher concentrations in the heart, brain, and skeletal muscle. It is involved in the process of conjugating bile acids with chenodeoxycholic acid and cholic acid. Taurine has a potential effect on insulin and insulin receptors which may benefit diabetic patients' blood glucose levels.[16]
- Ginseng: Consumption of Korean red ginseng (KRG) results in vasodilation as it can increase nitric oxide levels in healthy individuals. This effect is seen mainly due to ginsenosides, which are a class of steroid glycosides found exclusively in the ginseng plant.[17]
- B vitamins: Commonly added B vitamins are riboflavin (B2), niacin (B3), pyridoxine (B6), inositol (B8), folate (B9), and cobalamin (B12). These vitamins are added by the manufacturers as an energy enhancer for their energy drinks.[18]
- Inositol: Inositol, which is a component of phosphatidylinositol lipids, also serves as a secondary messenger, where it triggers the release of calcium in cells and the transmission of messages between neural cells, and facilitates the transport of fat within cells.[19]

Apart from providing abundant health benefits, overconsumption of energy drinks can be responsible for causing the following adverse health effects. The situation becomes alarming as this overconsumption is largely seen in younger people including adolescents. The following are a few of the adverse effects commonly seen due to the overconsumption of energy drinks:

- Caffeine overdose: Energy drinks are high in caffeine, which can cause caffeine overdose when consumed in excess. This can lead to symptoms such as rapid heart rate, high blood pressure, nervousness, anxiety, and insomnia.[3] A spectrum of metabolic, physiological and hormonal effects has been seen in athletes as reported by Sokmen *et al.*[20]
- Weight gain and obesity: Increased intake of energy drink not only puts a person at a risk of getting obese but also increases the chance of developing type 2 diabetes. This is because of the excessive content of sugars in energy drinks.[21]
- Dehydration: Energy drinks can cause dehydration due to their high caffeine and sugar content. This can lead to headaches, dizziness, and fatigue.
- Cardiac problems: The high levels of caffeine and other stimulants in energy drinks can lead to cardiac problems, including increased heart rate, abnormal heart rhythms, and even heart attacks.[22]
- Addiction: The caffeine and sugar in energy drinks can lead to addiction, making it difficult to quit or reduce consumption.
- Insomnia: The high caffeine content in energy drinks can also cause insomnia or disrupted sleep patterns, which can have negative impacts on overall health.
- Neurological and psychological effect: Overconsumption is often linked with muscle twitching, restlessness, and periods of inexhaustibility.[21]

- Interactions with other substances: Energy drinks can interact negatively with other substances, including prescription medications and alcohol.
- Dental problems: Energy drinks are high in sugar and acidic ingredients, which can lead to tooth decay and other dental problems over time.[23]

In moderation, energy drinks can offer a brief boost in alertness and energy levels, which can be beneficial in certain situations. But it is important to understand that regular and excessive consumption of these drinks can cause a variety of serious health problems. Added stimulants and high doses of sugar can lead to weight gain, heart palpitations, elevated blood pressure *etc.* Long-term use may interfere with sleep cycles and cause insomnia as well as mental health problems. In conclusion, while occasional and responsible consumption of energy drinks may have limited benefits, their chronic intake should be approached with caution due to the substantial health risks associated with excessive use. So, it is important to make a judicial choice in order to prevent or decrease the chance of health concerns.

12.3 Popular Brands and Products

The market for functional energy drinks is diverse and dynamic, with several popular brands working for consumer attention. Each brand has its own unique selling proposition, targeting specific demographics and catering to varying consumer preferences. Consumers are drawn to popular brands not only for the quality of their products but also for the status and lifestyle they represent. The appeal of popular brands often lies in their ability to consistently deliver products that meet or exceed consumer expectations. In this competitive market, several popular brands have emerged, each offering unique formulations and marketing strategies to capture the attention of consumers. There are many functional energy drink brands on the market, each with its own unique product lines. Here is an overview of some of the most popular brands:

- Red Bull: It is one of the most well-known energy drink brands, with a line of products that includes Red Bull Energy Drink, Red Bull Sugar-free, and Red Bull Total Zero. These drinks contain caffeine, taurine, B vitamins, and other ingredients designed to boost energy levels.
- Monster energy: It offers a wide range of energy drinks, including Monster Energy, Monster Energy Ultra, and Monster Energy Rehab. These drinks contain caffeine, taurine, and other ingredients designed to enhance mental and physical performance.
- Rockstar: It has a line of energy drinks that includes Original, Zero Carb, and Recovery. These drinks contain caffeine, taurine, and other ingredients designed to improve focus, alertness, and energy levels.
- Bang energy: Bang energy offers a line of energy drinks that contain caffeine, amino acids, and other ingredients designed to boost energy

and mental focus. They also offer a range of other functional beverages, including protein shakes and ready-to-drink coffees.

- Celsius: Celsius offers a range of energy drinks that are marketed as fitness drinks. These drinks contain caffeine, green tea extract, and other ingredients designed to enhance physical performance and burn calories.
- 5-hour energy: 5-hour energy is a brand that specializes in small, concentrated energy shots. These shots contain caffeine, B vitamins, and other ingredients designed to provide a quick energy boost.

12.3.1 Comparison of Ingredients and Nutritional Value Among Different Brands/Products

Energy drink companies uses a blend of ingredients in the formulation of their product with the main motive being to entice a broader audience and enhance flavour profiles. These ingredients can vary from basic ingredients like caffeine and sugar to taste and flavour developers like fruit extracts and preservatives. With the unique combination of ingredients present in the drink the nutritional value also changes. Nowadays a critical focus is being made to improve the functional value of drinks, which can be achieved by adding components like vitamin B, taurine, *etc.*, as already discussed. Table 12.1 compares the different ingredients and their nutritional value.

As per the table, the differentiation of diverse brands of energy drink on the basis of their ingredient has been done. The comparison clearly shows that

Table 12.1 Comparison of ingredients in energy beverages.[a]

Ingredient	Red Bull	Rockstar	Monster
Calories	220	280	200
Carbohydrates	54 g sucrose, glucose	62 g sucrose, glucose	54 g sucrose, glucose, sucralose, maltodextrin
Sodium	Only listed as sodium citrate	80 mg sodium citrate	360 mg (16% RDA) sodium citrate, sodium chloride
Caffeine	160 mg	160 mg Part of a 1.35 g "energy blend"	Only listed as part of a 5000 mg "energy blend"
Taurine	2000 mg	2000 mg	2000 mg
Glucuronolactone	Only listed (1200 mg)	None listed	Only listed as part of a 5000 mg "energy blend"
Niacin (B_2)	200% RDA niacinamide (40 mg)	40 mg (200% RDA) niacinamide	40 mg (200% RDA) niacinamide
Inositol (B_2)	Only listed	50 mg Part of a 1.35 g "energy blend"	Only listed as part of a 5000 mg "energy blend"

Table 12.1 (*Continued*)

Ingredient	Red Bull	Rockstar	Monster
Pyridoxine hydrochloride (B)	500% RDA (10 mg)	4 mg (200% RDA)	4 mg (200% RDA)
Cyanocobalamin (B_{12})	160% RDA, listed as vitamin B (10 μg)	12 μg (200% RDA)	12 μg (200% RDA)
Riboflavin (B_2)	None listed	6.8 mg (400% RDA)	3.4 mg (200% RDA)
Pantothenic acid (B_2)	100% RDA calcium pantothenate (10 mg)	20 mg (200% RDA) calcium pantothenate	None listed
Ginseng extract	None listed	50 mg Part of a 1.35 g "energy blend"	400 mg
Guarana extract	None listed	50 mg Part of a 1.35 g "energy blend".	Only listed as part of a 5000 mg "energy blend"
Ginkgo biloba leaf extract	None listed	300 mg Part of a 1.35 g "energy blend"	None listed
Milk thistle extract	None listed.	40 mg Part of a 1.35 g "energy blend"	None listed
L-Carnitine	None listed	50 mg Part of a 1.35 g "energy blend"	Only listed as part of a 5000 mg "energy blend"
Sorbic acid	None listed	Yes	Yes
Sodium benzoate	None listed	Yes, benzoic acid	Yes
Citric acid	None listed	Yes	Yes
Natural flavours	Yes	Yes	Yes
Artificial flavours	Yes	Yes	None listed
Colouring	Colours	Caramel	Colour added

[a] Adapted from ref. 6 with permission from Elsevier, Copyright 2010.

energy drink Rockstar contains the maximum amount of calories due to its high sugar content. Sugars used in the drinks are either in the form of glucose or sucrose, while almost every drink contains artificial flavour and colours.

12.4 Marketing Strategies Used by Functional Energy Drink Industries

Functional energy drink companies employ various marketing strategies to effectively promote their products and capture the attention of their target audience. With the growing demand for energy-boosting beverages, these companies utilize a combination of innovative techniques and brand positioning to stand out in a highly competitive market. Companies employ a range of strategies to create brand awareness, establish a strong brand identity, and ultimately drive consumer demand. These marketing efforts not only aim to highlight the unique benefits and ingredients of their products but also create an aspirational lifestyle that resonates with their target consumers.

Strategies used by brands to advertise their products are as follows:[24]

1. Sponsorship and endorsements: Many energy drink companies sponsor sports teams, athletes, and high-profile events to promote their products. They may also use celebrity endorsements or partnerships to build brand awareness and credibility.
2. Social media: Social media platforms are a key marketing tool for energy drink companies, with many using social media influencers and user-generated content to reach younger consumers.
3. Product sampling: Energy drink companies often provide free samples of their products at events or in stores to encourage trials and build brand loyalty.
4. Packaging and branding: The packaging and branding of energy drinks are designed to be eye-catching and visually appealing, with bright colours and bold graphics.
5. Health claims: Many energy drink companies make health claims about the benefits of their products, such as improved energy, focus, and athletic performance. These claims may be based on scientific research or anecdotal evidence.
6. Limited edition and seasonal products: Energy drink companies may release limited edition or seasonal products to create excitement and encourage consumers to try their products.
7. Targeted advertising: Energy drink companies often target their advertising to specific demographics, such as young adults, athletes or gamers, through channels such as social media, television, and online gaming platforms.

12.4.1 Advertising Regulations and Controversies Surrounding Energy Drinks

By using the above-mentioned tactics and advertising methods, brands can easily attract the attention of consumers, especially the young population, towards them. Problems arise when they use false claims just to get a profit. So, there is a need for proper and strict regulations that can have a proper check on their authenticity. In India, energy drinks are regulated as a food product by the Food Safety and Standards Authority of India (FSSAI). The FSSAI has issued several guidelines for the labelling and advertising of energy drinks in India according to which the use of misleading or false claims on energy drink packaging or in advertising is prohibited. Also, according to these guidelines, energy drink labels must include information on the caffeine content (Food Products Standards and Food Additives, FSSAI, 2011) and the presence of any other stimulants or ingredients that may have a physiological effect on the body, and a warning statement indicating that the product is not recommended for children, pregnant or lactating women, or individuals sensitive to caffeine must be mentioned in clear words.[39] In addition, the FSSAI has banned the sale of energy drinks containing more than 320 mg of caffeine per litre. While there are regulations in place, some experts have raised concerns about the potential health risks associated with energy drinks and the need for stricter regulations.[26] In 2017, the Indian Medical Association called for a ban on the sale of energy drinks to minors and for stricter regulations on their sale and distribution. However, despite all of these, there have been controversies surrounding energy drink advertisements. One of the main concerns is the potential impact of these advertisements on young people, who may be more susceptible to the marketing messages used by energy drink companies. There have also been concerns about the safety of energy drinks, particularly in relation to their high caffeine content.[27] Some studies have linked energy drink consumption to adverse health effects, such as heart palpitations, seizures, and even death in some cases.[22] As a result, there have been calls for stricter regulation of energy drinks, including restrictions on advertising and marketing to young people and warning labels that highlight the potential health risks associated with their consumption. In response to these concerns, some energy drink companies have voluntarily implemented changes to their advertising and marketing practices. For example, some companies have committed to not marketing their products to children under the age of 12 and to providing more transparent labelling and information about the ingredients and potential health risks associated with their products.

12.4.2 Impact of Marketing and Advertising on Consumer Behaviour

Marketing and advertising have a significant impact on consumer behaviour. By using a variety of strategies and techniques, companies can influence the way consumers think about their products and ultimately affect their buying decisions. One of the most important ways that marketing and advertising influence consumer behaviour is by creating brand awareness and recognition. By creating a strong brand identity and building brand loyalty, companies can encourage consumers to choose their products over those of their competitors. This can be achieved through targeted advertising campaigns, sponsorships, and endorsements that promote the brand and create a sense of familiarity and trust.[24]

It can also affect consumer behaviour by appealing to their emotions and desires. By using persuasive language, emotional appeals, and visual imagery, companies can create an emotional connection with consumers and influence their decision making. For example, advertisements that emphasize the health benefits of a product may appeal to consumers who are health conscious, while advertisements that highlight the convenience of a product may appeal to busy consumers who value efficiency and convenience.

Another way that marketing and advertising can influence consumer behaviour is by creating a sense of urgency or scarcity. By using time-limited offers, limited-edition products, or creating a sense of exclusivity, companies can create a sense of urgency and encourage consumers to make a purchase. This can be particularly effective in industries such as fashion and luxury goods, where exclusivity and scarcity are highly valued.[28]

12.5 Health and Safety Concerns

The consumption of functional energy drinks has raised significant health and safety concerns in recent years, prompting scrutiny and debate. While these beverages are marketed as providing a quick boost of energy, there are potential risks associated with their ingredients and consumption patterns. High levels of caffeine, often found in functional energy drinks, can lead to adverse effects such as increased heart rate, elevated blood pressure, and disrupted sleep patterns. Additionally, the high sugar content in some energy drinks raises concerns about obesity, dental health, and the risk of developing chronic conditions like diabetes. The combination of stimulants and other additives present in these beverages also raises questions about potential interactions with medications, pre-existing health conditions, and the overall impact on long-term health. As a result, consumers, healthcare professionals, and regulatory bodies are increasingly focusing on understanding and mitigating the health and safety implications of consuming functional energy drinks.

Excessive consumption of functional energy drinks can cause a broad spectrum of health issues:

- Cardiovascular effect: The ergogenic effect of caffeine can cause a rise in arterial blood pressure and heart rate. Overconsumption can cause many heart issues like ventricular arrhythmias, ST-segment elevation, and QT prolongation.[22] Myocardial infarction has been reported in healthy 17 and 19 year old boys, with the reason behind this being the consumption of energy drinks.[29]
- Neurological and psychological effects: These effects can be seen due to caffeine intoxication which takes place at a dose equal to or above 200 mg. Symptoms include insomnia, anxiety, restlessness, muscle twitching, gastrointestinal upset, and periods of inexhaustibility.[21] Higher caffeine intake can cause acute or chronic headaches on a daily basis by stimulating a pronociceptive state of cortical hyperexcitability.[30] The *Diagnostic and Statistical Manual of Mental Disorders*, 4th edition, has recognized four caffeine-induced psychiatric disorders which include caffeine-induced anxiety, caffeine intoxication, caffeine-related disorder and caffeine-induced sleep disorder.[27]
- Metabolic disorders: High energy drink intake increases the risk of obesity and type 2 diabetes. Acute caffeine intake decreases insulin sensitivity, which could explain the rise in blood glucose levels after

energy drink consumption documented in some studies.[31,32] Beaudoin *et al.* demonstrated that caffeine intake reduces insulin sensitivity in a dose-dependent manner, with a 5.8% increase in insulin for each mg per kg body-weight increase in caffeine.[33]

- Renal effects: Caffeine has been shown to increase diuresis.[34] Therefore, energy drinks should be avoided in case of prolonged exercise in hot weather conditions because they can cause dehydration. Studies have shown that dehydration at a level of 1.5 per cent during prolonged exercise can result in an increase in body temperature, heart rate and perceived rate of exertion.[35]
- Dental effects: A study done in Sweden has shown a strong relationship between energy drinks and dental erosion.[23] Similarly, Marshall *et al.* demonstrated a similar observation in American children.[36] There was a 2.4-fold increase in dental erosion with the consumption of energy drinks as they have low pH and high sugar content.

12.6 Safety Regulations and Guidelines for Energy Drink Manufacturers

Energy drink manufacturers are subject to safety regulations and guidelines established by the Food Safety and Standards Authority of India (FSSAI, 2006). Safety regulations and guidelines for energy drink manufacturers in India are:[37]

1. Labelling requirements: Energy drink manufacturers are required to provide accurate and clear labelling on their products in compliance with the FSSAI's regulations on food labelling. This includes information on the ingredients, nutritional content, and any potential allergens, as well as a warning statement indicating that the product is not recommended for children, pregnant or lactating women, or individuals sensitive to caffeine (packaging and labelling, FSSAI 2011).[38]
2. Ingredient restrictions: The FSSAI has set maximum limits for caffeine content and requires that energy drinks contain no more than 320 mg of caffeine per litre. The use of certain ingredients, such as taurine, is also subject to restrictions under the FSSAI regulations.[25]
3. Manufacturing standards: Energy drink manufacturers must adhere to good manufacturing practices (GMP) and maintain high standards of cleanliness, sanitation, and quality control in their production facilities.
4. Safety testing: Energy drink manufacturers are required to test their products for safety and quality, and to ensure that they are free from contaminants such as bacteria, toxins, or heavy metals, in compliance with the FSSAI regulations.
5. Advertising guidelines: Energy drink manufacturers must comply with the FSSAI guidelines related to their advertising and marketing

practices. Ads for energy drinks must be truthful, accurate, and not misleading, and must not make any claims that are not supported by scientific evidence (Advertising and Claims, FSSAI 2018).[40]

Energy drink manufacturers in India are subject to a range of safety regulations and guidelines aimed at ensuring the safety and quality of their products. By adhering to these regulations and guidelines, manufacturers can help to protect consumers from potential health risks associated with excessive energy drink consumption.

12.6.1 Public Health Campaigns and Initiatives Aimed at Reducing Energy Drink Consumption

In recent years, there have been several public health campaigns and initiatives aimed at reducing energy drink consumption and raising awareness about the potential health risks associated with these products. Examples of these campaigns are:

1. Campaigns targeting children and adolescents: Many public health campaigns have focused on educating children and adolescents about the potential risks associated with energy drink consumption, and encouraging them to make healthier beverage choices. These campaigns often involve school-based educational programs and social media campaigns to reach younger audiences. The American Academy of Pediatrics Committee on Nutrition, Council on Sports Medicine and Fitness released a policy statement discouraging energy drink use and encouraging water as the preferred beverage for children and adolescents.
2. Warning labels and advertising restrictions: Some countries have introduced warning labels on energy drink packaging, highlighting the potential health risks associated with excessive consumption. In addition, some countries have placed restrictions on advertising and marketing practices for energy drinks, particularly those targeting children and adolescents.[41]
3. Health professional education: Many public health campaigns have focused on educating health professionals about the potential health risks associated with energy drink consumption, and providing guidance on how to manage patients who present with energy drink-related health problems.[42]
4. Collaborative initiatives: Some public health campaigns have involved collaboration between government agencies, industry groups, and non-governmental organizations (NGOs) to raise awareness about energy drink consumption and promote healthier beverage choices.[42]

Overall, public health campaigns and initiatives aimed at reducing energy drink consumption have been focused on educating consumers, particularly children and adolescents, about the potential health risks associated with these products, and encouraging them to make healthier beverage choices. By raising awareness about the potential health risks associated with energy drink consumption and promoting healthier alternatives, these campaigns can help to protect public health and reduce the prevalence of energy drink-related health problems.

12.7 Future Trends and Directions

Continuous innovations are made to meet the consumer's requirements. These innovations in functional energy drink ingredients and product formulations are aimed at providing consumers with healthier and more convenient options. This can be achieved by using the following ingredients:

1. Natural and plant-based ingredients: Many manufacturers have started to use natural and plant-based ingredients in their energy drinks, such as green tea extract, guarana, and yerba mate. These ingredients contain natural sources of caffeine and other nutrients, and are seen as a healthier alternative to synthetic caffeine.[43]
2. Sugar-free and low-calorie: Many energy drink manufacturers have introduced sugar-free and low-calorie options in response to concerns about the high sugar content of traditional energy drinks. These drinks use alternative sweeteners such as stevia to provide a sweet taste without the added sugar.[44]
3. Functional ingredients: Some energy drinks now contain additional functional ingredients, such as vitamins, minerals, and amino acids, that are believed to provide additional health benefits, such as improved cognitive function, increased energy, and reduced fatigue.[43]
4. Non-caffeinated options: In addition to traditional caffeinated energy drinks, some manufacturers have introduced non-caffeinated options, such as energy drinks that contain only B vitamins or other non-stimulant ingredients.
5. Novel delivery systems: Some energy drinks now come in novel delivery systems, such as powders, shots, or effervescent tablets, that are designed to provide a quick and convenient way to consume the product.

12.7.1 Emerging Trends in Packaging, Branding, and Marketing of Energy Drinks

After innovation in ingredients, other emerging trends in packaging, branding, and marketing of energy drinks are focused on attracting the attention of consumers, providing more personalized options, and addressing growing concerns about the health and environmental impact of these products. As the market for energy drinks continues to evolve, it is likely that manufacturers will continue to explore new and innovative ways to reach consumers and differentiate their products in the following ways:

1. Sustainable packaging: As consumers become more environmentally conscious, many energy drink manufacturers are exploring sustainable packaging options, such as biodegradable or compostable packaging, or using recycled materials for their packaging.[45]
2. Bold and colourful designs: Energy drink manufacturers often use bold and colourful designs in their branding and packaging to attract the attention of consumers. This trend is likely to continue as manufacturers look for ways to stand out in a crowded market.[46]
3. Personalization: Some energy drink manufacturers are exploring ways to personalize their products for individual consumers. This may include offering customized blends of ingredients or personalized packaging with a consumer's name or favourite colour.
4. Influencer marketing: Social media influencers are playing an increasingly important role in the marketing of energy drinks. Manufacturers are partnering with popular influencers to promote their products and reach new audiences.[47]
5. Health and wellness messaging: With the growing concern about the health risks associated with energy drinks, some manufacturers are shifting their marketing messaging to emphasize the health and wellness benefits of their products. This may include highlighting the natural and organic ingredients used in the product or promoting the functional benefits of the ingredients.

12.7.2 Implications for the Beverage Industry and Public Health Policies

The implications for the beverage industry and public health policies with regard to energy drinks are significant. For the beverage industry, there is a need to respond to changing consumer preferences and concerns about health and sustainability. As consumers become more health conscious, the industry will need to continue innovating in product formulation and packaging to meet the demand for healthier and more sustainable products. Additionally, the industry will need to consider the potential health risks associated with energy drinks and take steps to mitigate these risks, such as providing clear labelling and educational materials for consumers.

In terms of public health policies, there is a need for increased regulation and oversight of energy drinks. This may include setting limits on the amount of caffeine and other stimulants allowed in these products, requiring warning labels about potential health risks, and restricting marketing to minors. Governments may also consider implementing public health campaigns to raise awareness about the potential health risks associated with excessive energy drink consumption and promote healthier alternatives.

Overall, there is a need for collaboration between the beverage industry and public health policymakers to ensure that consumers have access to safe and healthy beverage options, while also protecting public health. By working together, the industry and policymakers can promote responsible consumption of energy drinks and ensure that consumers have the information and resources they need to make informed choices about their health.

12.8 Conclusion

Functional energy drinks are different from ordinary drinks in respect of their ingredients and nutritional content, which includes caffeine and sugars along with other ingredients like taurine, L-carnitine, glucuronolactone, vitamin B, guarana, ginseng, inositol, *etc.* These provide additional short time benefits to the consumer like mental alertness, improved endurance, prolonged stamina, boost in energy. But, their consumption for a long period or in higher doses can result in such serious health issues that it surely overcomes the short-term benefits. So, it is very important for the consumer to choose their drink wisely. Manufacturers on their part should have a wholesome health approach such that both consumers and sellers can benefit from it. On the other hand, the government should formulate strict regulations for the manufacturing and selling of beverages in order to ensure food safety.

Public health campaigns should be developed to educate consumers about the potential health risks of energy drinks and to promote healthier alternatives. Strict banning of harmful chemicals and additives which can have an adverse effect on the body should be done. Regulations can be made regarding marketing restrictions, which prevent the targeting of young people and vulnerable populations for energy drink consumption.

Finally, the consumer should choose wisely when it comes to their food because it not only influences their health but also affects the trends in the market. Avoid purchasing drinks that contain more artificial ingredients or high caffeine and sugar content, this will automatically motivate manufacturers to come up with more healthy and safe functional energy drinks. Apart from these implications on the consumer and public health level, there is a need for ongoing research and policy development to address the potential health risks associated with these products. Further research should be done to investigate the long-term detrimental effects of energy drink consumption, especially among vulnerable populations such as children, pregnant women, and those with pre-existing medical conditions and metabolic disorders. Studies should be done to test the efficacy, claims, and safety of alternative ingredients used in energy drinks to attract consumers.

By implementing these policy recommendations and continuing to conduct research on the safety and efficacy of energy drinks, we can work to ensure that consumers have access to safe and healthy beverage options while protecting public health.

Declaration of Competing Interest

The authors declare that they have no known competing financial interests or personal relationships.

References

1. W. W. Ishak, C. Ugochukwu, K. Bagot, D. Khalili and C. Zaky, Energy drinks: psychological effects and impact on well-being and quality of life-a literature review, *Innov. Clin. Neurosci.*, 2012, **9**, 25–34.
2. A. A. Alsunni, Energy Drink Consumption: Beneficial and Adverse Health Effects, *Int. J. Health Sci.*, 2015, **9**, 468–474.
3. K. A. Clauson, K. M. Shields, C. E. McQueen and N. Persad, Safety issues associated with commercially available energy drinks, *J. Am. Pharm. Assoc.*, 2008, **48**, e55–e67.
4. A. Chen and G. Eriksson, The making of healthy and moral snacks: A multimodal critical discourse analysis of corporate storytelling, *Discourse, Context Media*, 2019, **22**, 423–445.
5. K. M. Hashem, F. J. He and G. A. MacGregor, Cross-sectional surveys of the amount of sugar, energy and caffeine in sugar-sweetened drinks marketed and consumed as energy drinks in the UK between 2015 and 2017: monitoring reformulation progress, *BMJ Open*, 2017, **7**, e018136.
6. J. P. Higgins, T. D. Tuttle and C. L. Higgins, Energy beverages: content and safety, *Mayo Clin. Proc.*, 2010, **85**, 1033–1041.
7. C. Alford, H. Cox and R. Wescott, The effects of red bull energy drink on human performance and mood, *Amino Acids*, 2001, **21**, 139–150.
8. A. L. Walsh, A. M. Gonzalez, N. A. Ratamess, J. Kang and J. R. Hoffman, Improved time to exhaustion following ingestion of the energy drink Amino Impact, *J. Int. Soc. Sports Nutr.*, 2010, **15**, 7–14.
9. L. F. Pettenuzzo, C. Noschang, E. V. P. Toigo, A. Fachin, D. Vendite and C. Dalmaz, Effects of chronic administration of caffeine and stress on feeding behavior of rats, *Physiol. Behav.*, 2008, **95**, 295–301.
10. M. Lorist and M. M. Tops, Caffeine, fatigue, and cognition, *Brain Cogn.*, 2003, **53**, 82–94.
11. E. Lopez-Garcia, R. M. V. Dam, S. Rajpathak, W. C. Willett, J. E. Manson and F. B. Hu, Major dietary patterns are related to plasma concentrations of markers of inflammation and endothelial dysfunction, *Am. J. Clin. Nutr.*, 2006, **83**, 674–680.
12. D. Laurent, K. E. Schneider, W. K. Prusaczyk, C. Franklin, S. M. Vogel, M. Krssak, K. F. Petersen, H. W. Goforth and G. I. Shulman, Effects of caffeine on muscle glycogen utilization and the neuroendocrine axis during exercise, *J. Clin. Endocrinol. Metab.*, 2000, **85**, 2170–2175.
13. M. S. El-Sayed, D. MacLaren and A. J. Rattu, Exogenous carbohydrate utilisation: effects on metabolism and exercise

performance, *Comp. Biochem. Physiol., Part A: Mol. Integr. Physiol.*, 1997, **118**, 789–803.
14. S. D. Marchi, S. Zecchetto, A. Rigoni, M. Prior and L. Fondrieschi, Propionyl-L-carnitine improves endothelial function, microcirculation and pain management in critical limb ischemia, *Cardiovasc. Drugs Ther.*, 2012, **26**, 401–408.
15. T. M. McLellan and H. R. Lieberman, Do energy drinks contain active components other than caffeine?, *Nutr. Rev.*, 2012, **70**, 730–744.
16. M. A. Moloney, R. G. Casey, D. H. O'Donnell, P. Fitzgerald, C. Thompson and D. J. Bouchier-Hayes, Two weeks taurine supplementation reverses endothelial dysfunction in young male type 1 diabetics, *Diab. Vasc. Dis. Res.*, 2010, **7**, 300–310.
17. E. Jovanovski, A. Jenkins, A. G. Dias, V. Peeva, J. Sievenpiper, J. T. Arnason, D. Rahelic, R. G. Josse and V. Vuksan, Effects of Korean red ginseng (Panax ginseng C. A. Mayer) and its isolated ginsenosides and polysaccharides on arterial stiffness in healthy individuals, *Am. J. Hypertens.*, 2010, **23**, 469–472.
18. G. M. Wardlaw and A. M. Smith, *Contemporary nutrition a functional approach*, McGraw-Hill, New York, 2009.
19. T. Wundenberg and G. W. Mayr, Synthesis and biological actions of diphosphoinositol phosphates (inositol pyrophosphates), regulators of cell homeostasis, *Biol. Chem.*, 2012, **393**, 979–998.
20. B. Sokmen, L. E. Armstrong, W. J. Kraemer, D. J. Casa, J. C. Dias, D. A. Judelson and C. M. Maresh, Caffeine use in sports: considerations for the athlete, *J. Strength Cond. Res.*, 2008, **22**, 978–986.
21. N. Bedi, P. Dewan and P. Gupta, Energy drinks: potions of illusion, *Indian Pediatr.*, 2014, **51**, 529–533.
22. M. Goldfarb, C. Tellier and G. Thanassoulis, Review of published cases of adverse cardiovascular events after ingestion of energy drinks, *Am. J. Cardiol.*, 2014, **113**, 168–172.
23. A. Hasselkvist, A. Johansson and A. K. Johansson, Dental erosion and soft drink consumption in Swedish children and adolescents and the development of a simplified erosion partial recording system, *Swed. Dent. J.*, 2009, **34**, 187–195.
24. L. Buchanan, H. Yeatman, B. Kelly and K. Kariippanon, Digital Promotion of Energy Drinks to Young Adults Is More Strongly Linked to Consumption Than Other Media, *J. Nutr. Educ. Behav.*, 2018, **50**(9), 888–895.
25. Food Safety and Standards (Food Products Standards and Food Additives) Regulations, 2011.
26. N. Gunja and J. A. Brown, Energy drinks: health risks and toxicity, *Med. J. Aust.*, 2012, **196**, 46–49.
27. L. M. Juliano, D. P. Evatt, B. D. Richards and R. R. Griffiths, Characterization of individuals seeking treatment for caffeine dependence, *Psychol. Addict. Behav.*, 2012, **26**, 948.

28. M. E. Malik, M. M. Ghafoor, H. K. Iqbal, Q. Ali, H. Hunbal, M. Noman and B. Ahmad, Impact of Brand Image and Advertisement on Consumer Buying Behavior, *World Appl. Sci. J.*, 2013, **23**, 117–122.
29. M. J. Scott, M. El-Hassan and A. A. Khan, Myocardial infarction in a young adult following the consumption of a caffeinated energy drink, *BMJ Case Rep.*, 2011, bcr0220113854.
30. J. C. Espinosa and M. F. Sobrino, Caffeine and headache: specific remarks, *Neurologia*, 2017, **32**(6), 394–398.
31. S. Lee, R. Hudson, K. Kilpatrick, T. E. Graham and R. Ross, affeine ingestion is associated with reductions in glucose uptake independent of obesity and type 2 diabetes before and after exercise training, *Diabetes Care*, 2005, **28**, 566–572.
32. F. R. Ragsdale, T. D. Gronli, N. Batool, N. Haight, A. Mehaffey, E. C. McMahon, T. W. Nalli, C. M. Mannello, C. J. Sell, P. J. McCann, G. M. Kastello, T. Hooks and T. Wilson, Effect of Red Bull energy drink on cardiovascular and renal function, *Amino Acids*, 2010, **38**, 1193–1200.
33. M.-S. Beaudoin, B. Allen, G. Mazzetti, P. J. Sullivan and T. E. Graham, Caffeine ingestion impairs insulin sensitivity in a dose-dependent manner in both men and women, *Appl. Physiol., Nutr., Metab.*, 2012, **38**, 140–147.
34. A. Riesenhuber, M. Boehm, M. Posch and C. Aufricht, Diuretic potential of energy drinks, *Amino Acids*, 2006, **31**, 81–83.
35. S. J. Montain and E. F. Coyle, Influence of graded dehydration on hyperthermia and cardiovascular drift during exercise, *J. Appl. Physiol.*, 1992, **73**, 1340–1350.
36. T. A. Marshall, S. M. Levy, B. Broffitt, J. J. Warren, J. M. Eichenberger-Gilmore, T. L. Burns and P. J. Stumbo, Dental caries and beverage consumption in young children, *Pediatrics*, 2003, **112**, e184–e191.
37. The Food Safety and Standards Act, 2006 (FSSAI Act).
38. Food Safety and standards (packaging and labeling) Regulations, 2011.
39. Food Safety and Standards (Food Products Standards and Food Additives) Regulations, 2011.
40. Food Safety and Standards (Advertising and Claims) Regulations, 2018.
41. M. B. Schneider and H. J. Benjamin, Sports drinks and energy drinks for children and adolescents: are they appropriate?, *Pediatrics*, 2011, **127**, 1182–1189.
42. I. Kraak, M. Davy, M. S. Rockwell, S. Kostelnik and E. Hedrick, Policy Recommendations to Address Energy Drink Marketing and Consumption by Vulnerable Populations in the United States, *J. Acad. Nutr. Diet.*, 2020, **120**(5), 767–777.
43. B. Donna, Clean energy' driving beverage innovation, Food Business News, 2019.
44. Red Bull, Red Bull Energy Drink, 2023. Retrieved from https://www.redbull.com/in-en/energydrink/red-bull-sugarfree.

45. C. E. Realini and B. Marcos, Active and intelligent packaging systems for a modern society, *Meat Sci.*, 2014, **98**, 404–419.
46. P. Anca and N. N. Andreea, Navigating the path from eco-friendly production to consumer Choices in cosmetics, *Annals of'Constantin Brancusi'University of Targu-Jiu*, Economy Series, Analele Universităţ ii'Constantin Brâncuşi'din Târgu-Jiu Seria Economie, 2014, vol. 1.
47. IvyPanda, New Energy Drink Marketing Strategy Research Paper, 2022, https://ivypanda.com/essays/strategic-marketing-plan-for-a-new-energy-drink/.

CHAPTER 13

Lactobacilli as Gut Modulators: Established Applications

NISHU DEVI,[a] MEENAKSHI CHITKARA,[b] HARMEET SINGH DHILLON,[a] PRADIP VISHNU BEHARE[a] AND ANIL KUMAR PUNIYA*[a]

[a] Dairy Microbiology Division, ICAR-National Dairy Research Institute, Karnal – 132001, India; [b] Animal Biotechnology Center, ICAR-National Dairy Research Institute, Karnal – 132001, India
*Email: akpuniya@gmail.com

13.1 Introduction

A significant group of lactic acid bacteria (LAB) known as *Lactobacillus* spp. are rod-shaped, non-sporulating, facultative anaerobic, Gram-positive bacteria. Being both oxidase and catalase negative, these can ferment carbohydrates and produce lactic acid as the major end-product of fermentation. Lactobacillaceae species can be divided into the following three groups: obligate homo-fermentative (group I) – *Lactobacillus acidophilus, Lacticaseibacillus delbrueckii, Lacticaseibacillus helveticus, Ligilactobacillus salivarius*; facultative hetero-fermentative (group II) – *Lacticaseibacillus casei, Latilactobacillus curvatus, Lactiplantibacillus plantarum, Latilactobacillus sakei*; and obligate hetero-fermentative (group III) – *Levilactobacillus brevis, Lentilactobacillus buchneri, Limosilactobacillus fermentum, Limosilactobacillus reuteri.* In addition, *Lactobacillus* spp. is common flora in the female vaginal tract and the gastrointestinal tract, where they help to partially block

Food Chemistry, Function and Analysis No. 44
Functional Foods of the Future
Edited by Vijai Kumar Gupta, Minaxi Sharma, Smriti Gaur and Ramesh Chander Kuhad

Published by the Royal Society of Chemistry, www.rsc.org

harmful pathogen entry by producing lactic acid.[1] The original genus *Lactobacillus* contained around 260 known species as of 2020. These organisms exhibit enormous morphological, ecological, and genetic diversity. As a result, the Lactobacillaceae family has been given a new taxonomy, and the former genus *Lactobacillus* has been divided into 25 genera with the addition of 23 unique genera. It is highly likely that there are many *Lactobacillus* strains with the same characteristics as well-known probiotics that are active in the human microbiome but go unnoticed or uncharacterized because historically probiotic lactobacilli were randomly isolated from different niches of the healthy human microbiome and characterized. Evidence suggests that the gut microbiome, namely through its impact on the immune system's development, plays a key role in early-life imprinting. This is true for the *Lactobacillus*, which dominates the vaginal microbiota in the final stages of pregnancy and is also prevalent in large quantities in human milk.

Many *Lactobacillus* species hold occupational positions in the market area of food microbiology and human nutrition, where they are used significantly in the preparation of fermented products and as probiotics. Probiotics are live microorganisms in different formulations providing health benefits when consumed, basically by improving or restoring the gut microbiota. On a phylogenetic level, they are intermixed with *Pediococcus* but are members of Lactobacillaceae. Generally found in fermented foods, like decomposing plants, milk, and milk products, these are widely accepted and are now recognized as 'Generally Recognized as Safe'.[2] Thus, they can be used as probiotics for humans and especially in the food industry to make fermented milk products such as yogurt, cheese, curd, and sour milk, and they play an important role in the manufacturing of fermented vegetable products, like pickles and sauerkraut, sourdough bread, sausages, and beverages. Some species of *Lactobacillus* are considered as probiotics to help maintain health and improve gut-associated microbes. *L. acidophilus, L. casei, L. johnsonii, L. fermentum, Lacticaseibacillus rhamnosus, L. plantarum, L. reuteri, L. salivarius, Lacticaseibacillus paracasei, L. delbrueckii* subsp. *bulgaricus* are the species involved in the maintenance of health attributes.[3] The mode of action of lactobacilli is explained in Figure 13.1.

Probiotics are believed to work in two ways: either by preventing or reducing the colonization of the intestinal tract with pathogenic microorganisms or by interacting with the gut-associated lymphoid tissue to lessen inflammatory reactions and promote tolerance to both the probiotics themselves and possibly foods. The incidence of necrotizing enterocolitis in premature neonates is decreased if *Lactobacillus* and *Bifidobacterium* colonize the gut or if breast milk is used instead of formula. By attaching to specific receptors on immune cells and other tissues, lactobacilli enhance the innate and adaptive immune response in the host, increasing the generation of cytokines and chemokines, as well as inducing the activation of dendritic cells and macrophages. Mechanisms of the interaction of lactobacilli with immune cells are explained in Figure 13.2.

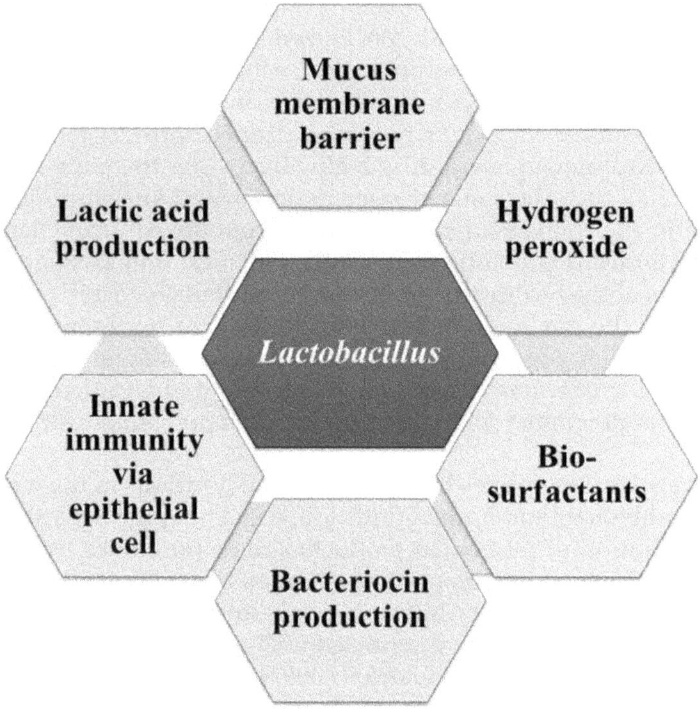

Figure 13.1 Mode of action of various *Lactobacillus* species.

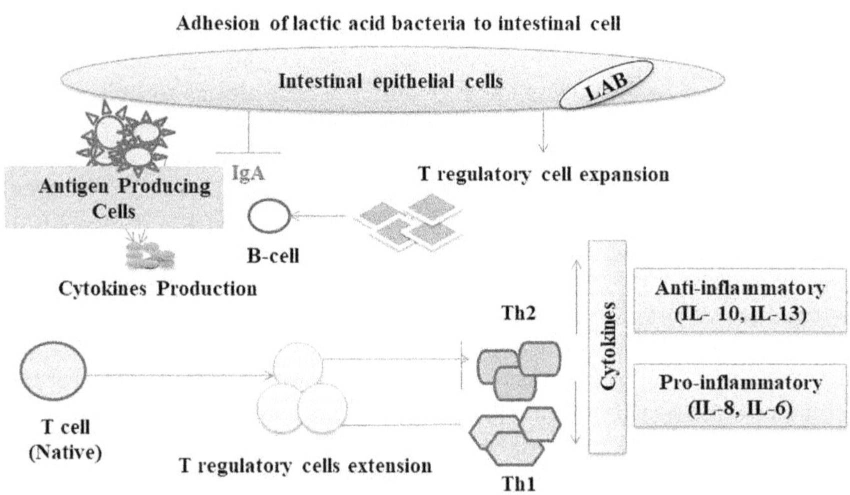

Figure 13.2 Lactobacilli interact with immune cells to regulate cytokine production and improve immune response in host cell.

The kind of strain and species of lactobacilli determine how immune cells respond. The prevention of atopic and dermatitis infection has been demonstrated in numerous probiotic research using various strains of *Lactobacillus* in both animals and humans. Additionally, the probiotic lactobacilli strains boost the immunogenicity of vaccines given orally to prevent diseases including polio, cholera, and influenza.[4] The greatest beneficial results have been associated with the two probiotic strains *L. rhamnosus* GG and *Bifidobacterium lactis* BB-12 for prevention and *L. reuteri* SD-22 for therapy.

13.2 Microbes and Gut Microbiota

Gut microbiota includes a diverse group of bacteria, such as symbiotic bacteria, which exist in a beneficial relationship with each other. Found to reside in the human intestinal tract, gut microflora accounts for more than 100 trillion bacteria. Disruption of the balance amongst the microflora species disrupts the metabolic system and prompts gastrointestinal, hepatic, respiratory, cardiovascular, and other disorders. Different genetic and environmental factors influence gut-microbiome composition. For instance, infants born naturally inherit roughly 40% of their mother's intestinal microflora, whereas the gut-microbiome composition changes significantly following surgical delivery. Nutrition plays a major role in determining gut-microbiome composition during the first two years of life, influenced by many factors including age, diet, drugs, and environmental conditions. The gut microbiome is linked to the pathogenesis of numerous diseases, including cancer, autism, depression, inflammatory bowel disease (IBD), colorectal carcinoma, infectious and non-infectious chronic liver diseases, obesity, and others, according to studies published in the past ten years.[5] A recent study revealed that the gut microbiome influences human brain function *via* the gut–brain axis. The gut microbiota is well established to provide numerous benefits to human health, including protection from disease-causing pathobiont bacteria, and an imbalance of gut microbiota can significantly contribute to depression. A complex relationship exists between the bacterial species that make up the 1–2 kg bulk of the human gut microbiota. The majority of gut bacteria form close bonds with their hosts, improving immunity, regulating the release of bile juice and secondary bile acids into the liver, and affecting the synthesis of metabolites and vitamins. New therapeutic strategies for managing dysbiosis, such as the administration of probiotics, prebiotics, synbiotics, pharmabiotics, or fecal microbiota transplantation, are promising in recent studies. The strains of *Lactobacillus bulgaricus*, *L. casei*, *L. helveticus*, and *Pediococcus pentosaceus* KID7 are known to help treat several conditions, such as inflammation and hypertension, and these also reduce the liver/body weight ratio and raise the non-alcoholic fatty liver disease activity score in mice models.[6]

The metabolic inheritance of the gut microbiome thus improves our biochemical adaptability by delivering important dosages of enzymes not produced by human DNA and active in processes like the synthesis of vitamins, polyphones, and polysaccharide digestion. The pressure of evolution, which led to the emergence of symbiotic traits in microbes, is assumed to be the reason for this privilege.[7] In addition to being highly influenced by internal, external, biotic, and abiotic elements, the makeup of the gut microbiota fluctuates in proportion to the pH, oxygen concentration, digestive flow rate, and secretions along the digestive system.[7] In the case of SARS-CoV-2 infections, a healthy gut microbiota has demonstrated a synergistic effect in the operation of vaccines, underscoring its significance for a particular person's health. Dysregulation of the gut microbiota (gut dysbiosis) is linked to hyperinflammation caused by risk factors in severe cases of COVID-19. This causes both acute and chronic inflammation, which weakens the body's natural anti-inflammatory systems. It makes things worse for the patients by intensifying the cytokine storm and pro-inflammatory tone in the gut. As a result, modifying the gut microbiota and supplementing it with commensal bacterial metabolites may be able to reduce the severity of COVID-19's hyperinflammatory response.[8]

The gut microbiota contributes to carcinogenesis through a variety of species-specific processes, including initiating inflammation, activating carcinogens and tumorigenic pathways, and disrupting host DNA. The gut microbiota promotes the maturation of gut-associated lymphoid tissues, the development of intestinal epithelial barrier function, and the support of immunological and intestinal epithelial cells toward commensal organisms and dietary antigens. The gut microbiota supports a range of body activities, including energy recovery from food digestion, pathogen defense, immune response regulation, and strengthening the biochemical barriers of the stomach and intestine. Changes in microbial composition could affect these processes. While healthy bacteria exist in the gut, certain dangerous bacteria can infiltrate the gastrointestinal system and cause infection. From birth to death, there is always a relationship between the gut microbiome and the host, which is formed during childhood. Furthermore, the gut microbiome produces substances that alter epigenetic change, resulting in long-term effects on individuals. Probiotic lactobacilli are important in maintaining the intestinal environment in both humans and animals.[9] Moreover, because they contribute to the formation of the mucosal immune system in the gastrointestinal tract, gut bacteria are vital for gastrointestinal immune surveillance. *Lactobacillus* spp. are a minor but crucial element of the human gut microbiota that are regularly implicated in numerous instances intimately related to human health.[10] The probiotic *L. reuteri* reduced the course of experimental autoimmune encephalomyelitis in mice, a widely used animal model of multiple sclerosis. In experimental autoimmune encephalomyelitis mice, *L. reuteri* treatment decreased TH1/TH17 cell proliferation and the production of their associated cytokines, interferon-gamma (IFN-γ)/IL-17. *L. reuteri*, a probiotic that alters the gut microbiota to modulate immunological responses in experimental autoimmune encephalomyelitis[10] has a unique possibility for reducing the severity of multiple sclerosis. Chronic stress affects the makeup of the gut microbiota while

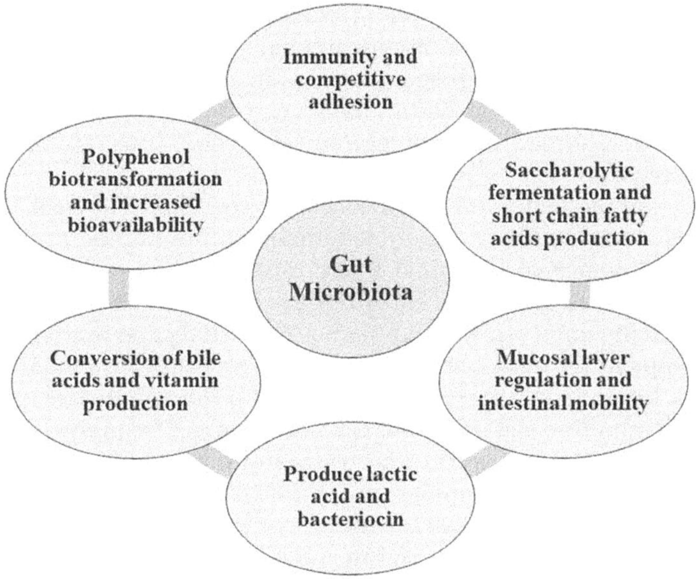

Figure 13.3 Different functions of the gut microbiota.

also disrupting the intestinal barrier, making it permeable and increasing blood levels of immunomodulatory bacterial cell wall components such as lipopolysaccharide.[11] The gut microbiota can ferment non-digestible sugars known as prebiotics, such as fructooligosaccharide, oligofructose, inulin, galactose, and xylose, which contain oligosaccharides, to meet energy needs. Furthermore, changing the gut microbiota can be harmful if the gut ecology undergoes severe abnormal alterations. Several probiotic species have been reported to prevent many degenerative disorders. Probiotics have been demonstrated to improve gut microbiota composition by enhancing intestinal microflora, epithelial barrier function, and cytokine release.[7,12] Figure 13.3 discusses various functions of gut microbiota.

13.3 Lactobacilli as Gut Modulators

Human gut LAB promote homeostasis and lower the possibility of metabolic depression. Probiotics that are frequently utilized include *Bifidobacterium* and lactobacilli.[13] The gut microbiota, which includes *Lactobacillus* spp., is crucial in regulating oxidative stress in host cells and tissues, particularly in the gastrointestinal tract. Excessive reactive oxygen species, which can be produced by a small number of NADPH oxidase enzymes, are associated with oxidative stress. More and more *Lactobacillus* species have been reliably linked to the human gastrointestinal tract in recent years. A few studies have emphasized how beneficial *Lactobacillus* species are for the balance these can bring about in the gut microbiome. For instance, it has been found that introducing various mixtures of *L. plantarum* and *L. fermentum* affects the diversity and functionality of gut microbiota in mice. According to 16S rRNA (ribonucleic acid) gene compositional data, a combination of two *L. fermentum* strains may be responsible for the improved anti-inflammatory activity of gut microbiota. Additionally, *L. plantarum* and *L. fermentum* increased the number of *Lactobacillus* spp. and lowered the amount of Enterobacteriaceae.[1]

Probiotic supplementation has also been shown to help treat the symptoms of enteritis. It functions by enhancing immunomodulation, which controls the balance of intestinal microorganisms, raises levels of advantageous metabolites in the gut, and stops the growth of pathogenic bacteria, thereby reducing enteritis. Both human clinical trials and animal models of IBD have repeatedly demonstrated the safety and efficacy of LAB. The gut microbiota fundamentally serves to protect the intestinal mucosa, and intestinal inflammation caused by gut dysbiosis is most frequently observed in people with IBD. When *Lactobacillus* was administered to colitis-prone animals, the gut's microbial population stabilized and essentially returned to normal levels. Changes in gut microbial composition also have a natural impact on the levels of several significant metabolites, some of which operate as a "bridge" for interactions between the microorganisms and the host cells. *Lactobacillus* strains considerably aid in reducing the signs of enteritis in the animal model by enhancing intestinal barrier integrity, reducing inflammation, and restoring intestinal bacteria and their metabolites to close to normal levels.[14] These changes in the microbial composition lead to an increase in the synthesis of beneficial metabolites and a decrease in the production of pro-inflammatory compounds.[15] Rheumatoid arthritis, a chronic autoimmune-mediated inflammatory illness, is characterized by synovitis and joint degradation. Evidence that the dysbiosis of the gut microbiota, in particular *Prevotella copri,* is significantly linked with the disease and that the disturbance was restored following methotrexate therapy provides compelling evidence that the abnormal gut microbiota is involved in the progression of rheumatoid arthritis in addition to the genetic and environmental factors that contribute to its onset. Studies have discovered significant alterations in the gut microbiota of rheumatoid arthritis patients that are related to clinically significant serologic traits. Probiotics may therefore have a big impact. This finding is further supported by the oral administration of *L. casei* which also significantly reduces oxidative stress, restores gut flora, and prevents the development of arthritis. Further research is required to determine how the gut microbiota affects metabolites in autoimmune illnesses like rheumatoid arthritis. Probiotics may be able to prevent or reduce the progression of arthritis, although the underlying mechanisms are yet unclear. According to certain studies, consuming lactobacilli may impact how effectively people respond to rheumatoid arthritis therapy. In one study, rats with collagen-induced arthritis were used to test the effects of two *L. casei* strains on immune responses, gut microbiota, and plasma metabolites. This was done using an integrated cross-omics approach that included plasma metabolomics and fecal 16S rRNA high-throughput sequencing.[4] Therefore, targeting the gut microbiota can be a strategy to improve the therapeutic efficacy in the treatment of rheumatoid arthritis and arthritis. To optimize treatment medication, it is essential to understand how *Lactobacillus* modulates rheumatoid arthritis. By lowering Th1-associated immune responses and modulating gut microbiota, the *L. casei* cocktail lessens inflammation in arthritic patients.[16]

Gliomas, which are tumors of the brain's glial cells, are among the most malignant cancers. There is evidence that some probiotics can inhibit tumor growth and that the gut microbiota plays a part in the development of cancers. Combining *B. lactis* with *L. plantarum* reduced tumor volume, prolonged longevity, and repaired damage to the intestinal barrier in an orthotropic mouse model of glioma. The combination of *B. lactis* and *L. plantarum* suppresses the phosphoinositide 3-kinase (PI3K)/AKT pathway and lowers the production of Ki-67 and N-cadherin, according to additional studies. The probiotic mixture also altered the fatty acids and oxygen-containing organic molecules in feces. It is important to highlight that prolonged antibiotic use modifies the gut microbiota, which affects the way mice's brain immune systems react and encourages the growth of gliomas. By altering the PI3K/AKT pathway, gut microbiota, and metabolites, *B. lactis* and *L. plantarum* delivered together prevent the formation of gliomas, at least in part.[17,18] It follows that the inhibitory effect of a combination of probiotics on the progression of glioma disease is associated with the alteration of gut microbiota composition, which is characterized by an increase in the relative abundance of *Lactobacillus* and a decrease in other potentially pathogenic bacteria. The gut microbiota of healthy mice and mice with glioma was different, and the proportion of *Lactobacillus* initially decreased and then gradually increased, according to a recent study. Serotonin and dopamine levels in the striatum of germ-free mice can be increased by administering *Lactobacillus plantarum*. Additionally, selective serotonin reuptake inhibitors can inhibit the progression of glioma cell lines by inducing apoptosis in proliferating cells. The results demonstrated that *B. lactis* and *L. plantarum* altered the PI3K/AKT pathway, gut microbiota, and metabolites to partially prevent the growth of gliomas.[18]

13.4 Maintenance of Gut Ecosystem and Disease Prevention

The central nervous system and the gut microbiota are linked by an intricate network of bidirectional communication known as the gut–brain axis. Probiotic lactobacilli contain retained microbial-associated molecular patterns, such as particular cell wall polysaccharides, peptidoglycan, and lipoprotein, that are tethered in cytoplasmic membranes, in contrast to other Gram-positive bacteria. However, the immune system still detects them as foreign substances. The pattern recognition receptors, which include toll-like receptors, C-type lectin receptors, and nucleotide oligomerization domain-like receptors, are expressed on several cell types, including epithelial and immune cells. These canonical microbial-associated molecular patterns can interact with these pattern recognition receptors. The innate immune response, which includes membrane activation, generation of antimicrobial factors, and presentation of antigens, is facilitated by pattern recognition receptor signaling. The chemical composition, length, and substitutions of microbial-associated molecular patterns, which are produced by a variety of lactobacilli, vary between strains and species, resulting in probiotic strain specificity.[19] Both human and animal cells respond well to *Lactobacillus* supplementation. Most piglets eat less after weaning as their diets switch from liquid milk to solid feed. Reduced feed and water intake result in numerous issues with the small intestine, which in turn results in reduced ability to digest and absorb nutrients and slows growth. The gut is a crucial defense mechanism against hazardous environmental factors. Piglets with weak intestinal defenses are more vulnerable to weaning stress, which can lead to infection and illness. According to Guo *et al.*,[20] the tight junction protein's lower expression during weaning results in the pathogen penetrating the barrier and blocking claudin-1, which then allows the bacteriocin to enter the body. Weaned piglets have increased proinflammatory cytokine production, which is linked to the health of the intestinal epithelium and inflammatory diseases.[21] Consuming probiotic bacteria thereby aids in intestinal function by preserving paracellular permeability, boosting mucous layer and immune system activity, and modifying the makeup and activity of the local gut microbiota. Adding *L. plantarum* to the diet enhances antioxidant activity, growth rate, and intestinal barrier function in post-weaning piglets as well as the interaction between them.[22]

Colorectal cancer, the third most frequently diagnosed cancer, is caused by an accumulation of genetic, epigenetic, and environmental stressors. The main environmental stressors are dietary factors like food consumption, physical activity, and smoking. Numerous gut bacteria, such as *Fusobacterium nucleatum*, *Peptostreptococcus anaerobius,* and the enterotoxigenic *Bacteroides fragilis*, promote the growth of tumors, increase inflammation, and damage DNA, all of which contribute to colorectal carcinogenesis. On the other side, colorectal cancer patients also have bacteria, primarily *Lachnospiraceae* species and *Bifidobacterium animalis*, which have a preventive impact against the disease. Several strategies employed for gut microbiota modulation include antibiotics, fecal microbiota transplantation, probiotics, prebiotics, and postbiotics.[23] *L. casei* improves the liver/body weight ratio and non-alcoholic fatty liver disease activity score. Treatment with *Lactobacillus*, *Lactococcus*, *Bifidobacterium,* and acetic acid bacteria causes a reduction of inflammatory cytokines such as IL-1, IL-1β, TNF-α, IL-6 and IL-8 in patients suffering from non-alcoholic fatty liver disease. Treatment with *L. bulgaricus*, *L. helveticus,* and *P. pentosaceus* improved steatohepatitis through modulation of gut microbiota composition and inflammatory pathways in the gut–liver axis of non-alcoholic fatty liver disease patients.[6]

The protective effect of gut microbiota modulation against diabetes has been confirmed in recent observations. Probiotic bacteria (*Lactobacillus, Bifidobacterium,* and *Lactococcus*) reduce hyperglycemia, hyperinsulinemia, and oxidative stress in high fructose-induced diabetic rats. The treated animals showed a significant improvement in fasting glycemia and oral glucose tolerance. Thus, altering the gut microbiota improved rats' inability to tolerate glucose and altered the inflammatory and metabolic state of the host cell.[6,24] The anticarcinogenic properties of *L. rhamnosus* and *Streptococcus thermophilus* regulate the gut microbiota and are crucial in the prevention of cancer. By encouraging cancer cell apoptosis, *Lactobacillus gallinarum* inhibits the progression of colorectal cancer in mouse models. A metabolite produced by *L. gallinarum*, such as indole lactic acid, may also have an impact on cancer cells. By encouraging apoptosis in human and animal colorectal cancer cells by raising the production of different pro-apoptotic proteins including Bax, casp3, and p53, a study on mice treated with both heat-killed and live *L. rhamnosus* GG has proven the anticancer impact. Regular use of *Lactobacillus* as a probiotic may help to balance the microbiota in the gut and lessen the likelihood of chronic inflammation and the generation of substances that can cause cancer. By metabolizing histamine, *L. reuteri* prevents colon cancer linked to inflammation. Probiotics may modify the gut microbiota's makeup to slow the spread of cancer, according to certain research. For example, *L. salivarius* decreases tumor growth *via* modulating intestinal microbiota composition. According to a recent study, the intestinal microbiota's production of L-tryptophan plays a key role in maintaining intestinal homeostasis. By triggering apoptosis, L-tryptophan produced by *L. gallinarum* is thought to reduce colorectal cancer.[25]

To enhance the bifidobacteria and lactobacilli that already exist in the gut microbiota, several dietary strategies have been devised. When more bifidobacteria and/or lactobacilli bind to harmful or carcinogenic substances, this makes it easier for the body to eliminate them through feces.[15] In Japan and around the world, consumption of a specific strain of probiotic bacteria called *L. paracasei* strain Shirota averaged between 8.9 and 41 million bottles per day in 2019. *L. paracasei* strain Shirota enters the bowel alive, ensuring a healthy balance of gut flora. Frequent *L. paracasei* strain Shirota use has also been linked to improved bowel movements, immunological regulation, and infection prevention. These findings suggest that a person's gut microbiota is continuously stabilized and maintained by their lifestyle choices. Additionally, a study discovered for the first time that routine ingestion of probiotics, notably *L. paracasei* strain Shirota, may aid in stabilizing the gut flora of the elderly. *L. paracasei* strain Shirota is resistant to bile and stomach acids and enters the gut alive, suggesting that the production of organic acids such as lactic acid is enhanced and the pH in the intestine is decreased. *L. paracasei* strain Shirota contributes to the maintenance of a healthy balance of the gut microbiota by increasing the abundance of advantageous bacteria like bifidobacteria and lactobacilli and decreasing the number of dangerous bacteria. The stabilization of gut microbiota in individuals consuming fermented milk products containing *L. paracasei* strain Shirota three days per week may be directly attributed to these benefits in gut microbiota and habitats by *L. paracasei* strain Shirota.[26]

L. gallinarum inhibits colorectal cancer carcinogenesis in both male and female mice, as evidenced by the fact that oral administration of *L. gallinarum* reduces the frequency and size of intestinal tumors in ApcMin/+ mice. This effect was confirmed in an AOM/DSS-induced colorectal cancer animal model. Further demonstrating the effect of *L. gallinarum* on the gut microbiota is the 16S rDNA gene sequencing on ApcMin/+ mouse feces samples after gavage of *L. gallinarum* for 8 weeks. Numerous studies have shown that taking probiotics regularly improves the imbalance in the intestinal microbiota, reducing the risk of chronic inflammation and the creation of substances that cause cancer when the intestinal tract is dysbiotic. Gut dysbiosis alters the gut microbiota and raises the risk of tumor development. Probiotics may be able to help the body reduce this alteration and return to normal health.[25] Figure 13.4 discusses the mechanism by which lactobacilli compete with pathogenic microorganisms for adhesion.

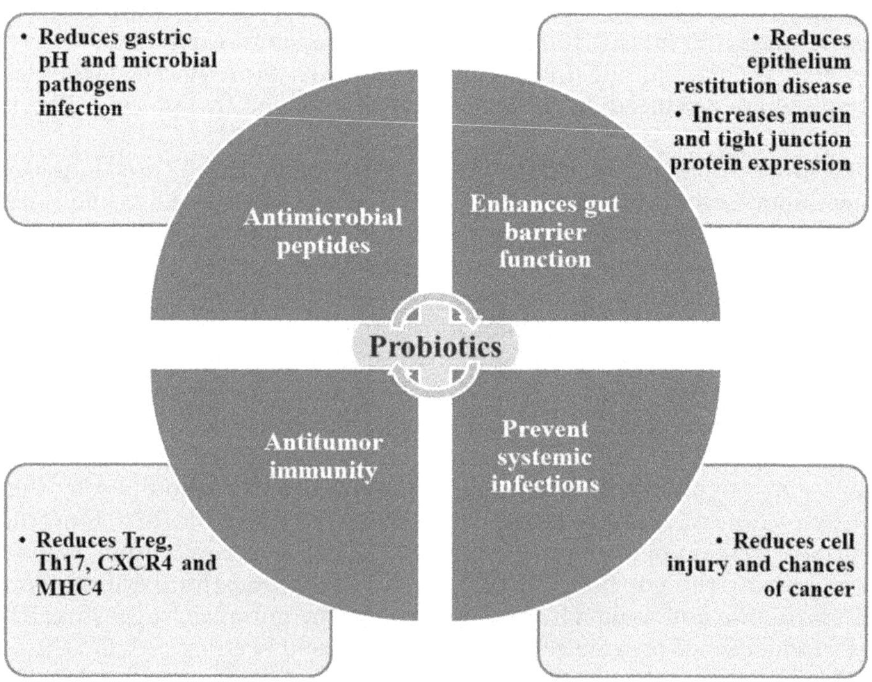

Figure 13.4 Mechanism of action of probiotic lactobacilli to replace gut pathogens to prevent colorectal cancer disease.

13.5 Lactobacilli and Its Metabolites Influence Gut Microbiota Growth

Through a variety of effector molecules, probiotic supplementation aids in the gut flora's regeneration and repair, which in turn lessens disease symptoms and fortifies the immune system. *L. brevis* B50 reportedly stimulates the production of genes related to effector molecules. Table 13.1 discusses the role of lactobacilli in treating various gastrointestinal disorders. Several essential flora, including Firmicutes, *Lactobacillus,* and *Bifidobacterium*, are universally important for the prevention and treatment of metabolic disorders,[27,28] even though the gut flora varies based on factors including age, environment, and food. Exopolysaccharide (EPS) controls the commensal bacteria growth in the gut, just as in fermentable substrates. In D-gal-induced mice, EPS produced by *L. helveticus* increased Bacteroidetes and decreased Firmicutes, indicating that EPS may reduce obesity. Because non-alcoholic fatty liver disease has higher concentrations of Proteobacteria and *Enterobacter* and lower concentrations of *Ruminococcus* and Firmicutes, it is assumed that the regulation of the gut microbiota affects the condition. According to 16s rRNA research, *L. plantarum* LPL-1 decreases the number of proteobacteria *via* secretion of bacteriocins. By using EPS produced by *L. plantarum* and *L. rhamnosus*, *Ruminococcus* was increased both *in vivo* and *in vitro*. These results suggest that bacteriocins and EPS can regulate metabolic diseases by elevating short-chain fatty acid (SCFA) levels. SCFAs may be used as a therapy in the treatment of non-alcoholic fatty liver disease, since they have the potential to prevent *Enterobacter* proliferation.[29] Low relative abundances of *Faecalibacterium* or *Romboutsia* may limit the growth of Enterobacteriaceae by inhibiting the synthesis of SCFAs. The Pamer group found that the growth of *E. coli* and *K. pneumoniae* was inhibited *in vivo* by a combination of SCFAs and an acidic pH and that this was because SCFAs promote intracellular acidification.[30] The propionic, butyric, and acetic acids produced by microorganisms act as indirect neuromodulators, and the enteric nervous system releases bioactive chemicals that can change the microbial environment directly or indirectly, such as gastrointestinal motility, permeability, pH level, and mucus secretion. For instance, enterochromaffin cells in the colon are activated by SCFAs and release serotonin. Indeed, studies on germ-free animals show increased plasma Trp and hippocampal 5-hydroxytryptamine levels following intestinal bacterial colonization, suggesting that the intestinal microbiota regulates the peripheral availability of Trp and, as a result, the synthesis of serotonin at the level of the central nervous system. Sometimes SCFA can work in conjunction with G-protein-coupled receptors. Bifidobacteria-produced SCFA acetate decreases intestinal inflammation by activating the GPR43 receptor. It is especially obvious when acetate binds to the GPR43 receptor and inhibits the production of the pro-inflammatory IL-18, the levels of which are elevated in cases of colon cancer.[31] Experiments with vagotomized mice highlighted the importance of the vagus nerve in communication between

Table 13.1 *Lactobacillus* strains studied in animal and human clinical settings for their ability to treat inflammatory disorders.

Microbes	Infection	Animal model	Clinical outcome	Immune response
Apilactobacillus kunkeei	Respiratory inflammation	Female-specific pathogen-free mice	Improvement in pulmonary infection	↑Th1 cytokine and IgA
L. murinus	Respiratory inflammation	6 to 8 week old female mice	↑Lung Th17 cells	↓Pulmonary infection
L. acidophilus	Diabetes	High fat and streptozotocin-induced mice	↓Inflammation in the colon and liver, reshape gut microbiome	↓IL-8, TNFα and IL-1β in liver and colon ↓Expression of glycogen synthase kinase
L. gasseri	Obesity	Mice	↓Body weight and fat tissue mass	↑Proinflammatory genes CCL2 and CCR2
L. salivarius	GIT inflammation	DSS-induced colitis in mice	↓Gastric inflammation	↓IL-1β, IL-12 and TGF-β level
L. casei	Respiratory inflammation	Immune-deficient 8 week old adult Swiss albino mice	↓Lung inflammation ↑Weight recovery	↓Lung bacterial composition
L. casei	GIT inflammation	TNBS-induced colitis in mice	↓Gastric inflammation	↑Treg level
L. paracasei	Cardiovascular diseases	High-fat diet mice	↑Cholesterol metabolism	LXR/inflammatory axis of LPS-stimulated alveolar macrophages, shift in gut microbiome
L. paracasei	Diabetes	High-fat and streptozotocin-induced rats	↓Insulin resistance	↓HbA1c, glucagon, and leptin level gut-microbiome modulator
L. rhamnosus	GIT inflammation	TNBS-induced colitis in mice	↓Gastric inflammation	↑IL-10/TNF-α level
L. plantarum	GIT inflammation	DSS-induced colitis in mice	↓Gastric inflammation	↑TLR2/TLR6 heterodimer
L. brevis	Cardiovascular diseases	Six week old rats	Antihypertensive effect by blocking LPS-induced NO production	↓COX-1, COX2, iNOS, and TNF-α level
L. fermentum	Cardiovascular diseases	Human subjects	↓Serum total cholesterol HDL, LDL triglyceride, HbA1c level	↓IL-6 level
L. reuteri	GIT inflammation	*C. difficile* induced colitis in mice	↓Gastric inflammation	↑IL-10 level

the gastrointestinal tract and the central nervous system. The treatment of mice with *L. rhamnosus* resulted in anxiolytic and antidepressant effects that faded following vagotomy by reducing the expression of the γ-aminobutyric acid receptor in the brain. Similarly to how *Bifidobacterium longum* had no behavioral or anxiolytic effect in vagotomized mice with chronic colitis, there was a reduction in psychological colitis comorbidities after *B. longum* administration in mice with an undamaged vagus nerve.[7]

13.6 Role of Lactobacilli in Immunobiosis

The interaction between the enterocyte and the underlying mucosal gut-associated lymphoid tissues and a eubiotic microbial ecosystem sets off a complex network in which the innate and adaptive arms of immunity play a crucial role. The immune system must distinguish between "useful" antigens (like food and commensals) and pathogens (*i.e.* pathobionts) to either induce tolerance or activate the immune response, as appropriate, given that the gut microbial biomass has undeniable benefits and the potential to be aggressive. The host must rely on general host defense systems (*i.e.* anatomic, physiologic, phagocytic/endocytic, and inflammatory) for barriers to innate immunity because the intestinal epithelium is a front created to ward off pathogens. The physical barrier is composed of two layers: a mucus layer on the outside that is home to bacteria and an inside layer that is fortified by interepithelial junctions. The bacterial communication process known as quorum sensing drives the formation of biofilms, the production of secondary metabolites, and the development of a bacterial competitive system in both Gram-positive and Gram-negative bacteria. During intrauterine life, the embryo, which has a genetic composition that is partially inherited from the father, represents an antigenic non-self to the mother's immune system and may be prone to rejection (*i.e.* abortion) as a result.[32,33] Scientific studies revealed that the diversity of the gut microbiota influences immune system development from birth. To develop gut-associated lymphoid tissue, contact with bacteria is essential for the establishment of efficient immunity, both at the peripheral and intestinal levels. Gut microbial metabolites play a significant role in inflammatory signaling, interacting both directly and indirectly with the host's immune cells. SCFAs perform a variety of tasks by interacting with G-protein-coupled receptors. Acetate, a *Bifidobacterium*-derived SCFA, controls intestinal inflammation by activating the GPR43 receptor. When acetate binds to this receptor, it appears to inhibit the release of the pro-inflammatory IL-18, which is abundant in colon cancer.[7,34,35]

13.7 Conclusions

Lactic acid bacteria, commonly known as gut modifiers, offer a wide range of health benefits. These beneficial bacteria act as probiotics, supporting a healthy gut microbiome and aiding in digestion. They also play a crucial role in preventing life-threatening diseases like cancer, thanks to their antioxidant properties that protect cells from damage. LAB has been used in food production for decades, contributing to the fermentation process that creates dairy products like yogurt, cheese, kefir, and dahi. In the industrial setting, these are used as starter cultures in combination with probiotics to enhance gut health, fight diseases, and prevent constipation. By lowering the gut's pH level, LAB create an environment that hinders the growth of harmful bacteria while promoting the growth of beneficial ones. Consuming foods or supplements rich in LAB can therefore lead to improved digestion, reduced inflammation, and a stronger immune system. Understanding how bifidobacteria and lactobacilli interact with host cells and other gut microflora components is a key to developing effective probiotics and prebiotics. To support the use of prebiotics in promoting a healthy gut microbiome, we need to analyze the molecular interactions between beneficial bacteria, harmful bacteria, and human mucosal cells. Specifically, we must investigate the link between increased levels of beneficial bacteria like bifidobacteria and lactobacilli and various health indicators. These include immune stimulation, impact on colon cancer biomarkers (*i.e.* fecal water genotoxicity, butyrate production, and microbial enzyme activities), relief from inflammatory bowel disease symptoms, and their role in resisting colonization by gastrointestinal pathogens.

References

1. Y. Kong, K. J. Olejar, S. L. On and V. Chelikani, The potential of *Lactobacillus* spp. for modulating oxidative stress in the gastrointestinal tract, *Antioxidants*, 2020, **9**(7), 610.
2. P. Heczko, Ł Kozień and M. Strus, Special Issue "An Update on *Lactobacillus*", *Microorganisms*, 2023, **11**(6), 1400.
3. R. D. C. S. Ranadheera, S. K. Baines and M. C. Adams, Importance of food in probiotic efficacy, *Food Res. Int.*, 2010, **43**(1), 1–7.
4. E. J. Quinto, P. Jiménez, I. Caro, J. Tejero, J. Mateo and T. Girbés, Probiotic lactic acid bacteria: a review, *Food Nutr. Sci.*, 2014, **5**(18), 1765.
5. I. Milosevic, A. Vujovic, A. Barac, M. Djelic, M. Korac, A. Radovanovic Spurnic and A. Amedei, Gut-liver axis, gut microbiota, and its modulation in the management of liver diseases: a review of the literature, *Int. J. Mol. Sci.*, 2019, **20**(2), 395.
6. N. Y. Lee, S. J. Yoon, D. H. Han, H. Gupta, G. S. Youn, M. J. Shin and K. T. Suk, *Lactobacillus* and *Pediococcus* ameliorate progression of non-alcoholic fatty liver disease through modulation of the gut microbiome, *Gut Microbes*, 2020, **11**(4), 882–899.
7. C. Altieri, B. Speranza, M. R. Corbo, M. Sinigaglia and A. Bevilacqua, Gut-microbiota, and multiple sclerosis: background, evidence, and perspectives, *Nutrients*, 2023, **15**(4), 942.
8. J. Chen and L. Vitetta, Modulation of Gut Microbiota for the Prevention and Treatment of COVID-19, *J. Clin. Med.*, 2021, **10**(13), 2903.
9. G. Bora, A. Gunna, M. M. Kumar, S. Morya, C. G. Awuchi and F. Menaa, The Gut microbiota and chronic diseases: Role of probiotics: Gut Microbiota, *J. Appl. Nat. Sci.*, 2023, **15**(2), 692–703.

10. B. He, T. K. Hoang, X. Tian, C. M. Taylor, E. Blanchard, M. Luo and Y. Liu, *Limosilactobacillus reuteri* reduces the severity of experimental autoimmune encephalomyelitis in mice by modulating gut microbiota, *Front. Immunol.*, 2019, **10**, 385.
11. J. F. Cryan and T. G. Dinan, Mind-altering microorganisms: the impact of the gut microbiota on brain and behaviour, *Nat. Rev. Neurosci.*, 2012, **13**(10), 701–712.
12. M. A. K. Azad, M. Sarker, T. Li and J. Yin, Probiotic species in the modulation of gut microbiota: an overview, *BioMed Res. Int.*, 2018, **2018**, 9478630.
13. K. Suda and K. Matsuda, How microbes affect depression: underlying mechanisms *via* the Gut–brain axis and the modulating role of probiotics, *Int. J. Mol. Sci.*, 2022, **23**(3), 1172.
14. B. Pot, G. E. Felis, K. D. Bruyne, E. Tsakalidou, K. Papadimitriou, J. Leisner and P. Vandamme, The genus *Lactobacillus*, *Lactic Acid Bact.: Biodiversity Taxon.*, 2014, 249–353.
15. Y. Liu, H. Wang, S. Gui, B. Zeng, J. Pu, P. Zheng and P. Xie, Proteomics analysis of the gut–brain axis in a gut microbiota-dysbiosis model of depression, *Transl. Psychiatry*, 2021, **11**(1), 568.
16. Z. Fan, R. P. Ross, C. Stanton, B. Hou, J. Zhao, H. Zhang and W. Chen, *Lacticaseibacillus casei,* CCFM1074 alleviates collagen-induced arthritis in rats via balancing Treg/Th17 and modulating the metabolites and gut microbiota, *Front. Immunol.*, 2021, **12**, 680073.
17. N. Sudo, Y. Chida, Y. Aiba, J. Sonoda, N. Oyama, X. N. Yu and Y. Koga, Postnatal microbial colonization programs the hypothalamic–pituitary–adrenal system for stress response in mice, *J. Physiol.*, 2004, **558**(1), 263–275.
18. L. Wang, S. Li, H. Fan, M. Han, J. Xie, J. Du and F. Peng, *Bifidobacterium lactis* combined with *Lactiplantibacillus plantarum* inhibit glioma growth in mice through modulating PI3K/AKT pathway and gut microbiota, *Front. Microbiol.*, 2022, **13**, 986837.
19. P. Baarlen, J. M. Wells and M. Kleerebezem, Regulation of intestinal homeostasis and immunity with probiotic lactobacilli, *Trends Immunol.*, 2013, **34**(5), 208–215.
20. L. Guo, M. Meng, Y. Wei, F. Lin, Y. Jiang, X. Cui, G. Wang, C. Wang and X. Guo, Protective effects of live combined *B. subtilis* and *E. faecium* in polymicrobial sepsis through modulating activation and transformation of macrophages and mast cells, *Front. Pharmacol.*, 2019, **9**, 1506.
21. G. Chevalier, E. Siopi, L. Guenin-Macé, M. Pascal, T. Laval, A. Rifflet and P. M. Lledo, Effect of gut microbiota on depressive-like behaviors in mice is mediated by the endocannabinoid system, *Nat. Commun.*, 2020, **11**(1), 6363.
22. J. Wang, H. Ji, S. Wang, H. Liu, W. Zhang, D. Zhang and Y. Wang, Probiotic *Lactiplantibacillus plantarum* promotes intestinal barrier function by strengthening the epithelium and modulating gut microbiota, *Front. Microbiol.*, 2018, **9**, 1953.

23. W. Fong, Q. Li and J. Yu, Gut microbiota modulation: a novel strategy for prevention and treatment of colorectal cancer, *Oncogene*, 2020, **39**(26), 4925–4943.
24. J. Han, H. Lin and W. Huang, Modulating gut microbiota as an anti-diabetic mechanism of berberine, *Med. Sci. Monit.: Int. Med. J. Exp. Clin. Res.*, 2011, **17**(7), RA164.
25. N. Sugimura, Q. Li, E. S. H. Chu, H. C. H. Lau, W. Fong, W. Liu and J. Yu, *Lactobacillus gallinarum* modulates the gut microbiota and produces anti-cancer metabolites to protect against colorectal tumourigenesis, *Gut*, 2022, **71**(10), 2011–2021.
26. R. Amamoto, K. Shimamoto, S. Park, H. Matsumoto, K. Shimizu, M. Katto and Y. Aoyagi, Yearly changes in the composition of gut microbiota in the elderly, and the effect of lactobacilli intake on these changes, *Sci. Rep.*, 2021, **11**(1), 12765.
27. S. Rastogi and A. Singh, Gut microbiome and human health: Exploring how the probiotic genus *Lactobacillus* modulate immune responses, *Front. Pharmacol.*, 2022, **13**, 1042189.
28. S. Stojanov, A. Berlec and B. Štrukelj, The influence of probiotics on the firmicutes/bacteroidetes ratio in the treatment of obesity and inflammatory bowel disease, *Microorganisms*, 2020, **8**(11), 1715.
29. X. Gao, J. Zhao, H. Zhang, W. Chen and Q. Zhai, Modulation of gut health using probiotics: the role of probiotic effector molecules, *J. Future Foods*, 2022, **2**(1), 1–12.
30. M. Magruder, E. Edusei, L. Zhang, S. Albakry, M. J. Satlin, L. F. Westblade and J. R. Lee, Gut commensal microbiota and decreased risk for *Enterobacteriaceae bacteriuria* and urinary tract infection, *Gut Microbes*, 2020, **12**(1), 1805281.
31. P. Zheng, B. Zeng, C. Zhou, M. Liu, Z. Fang, X. Xu and P. Xie, Gut microbiome remodeling induces depressive-like behaviors through a pathway mediated by the host's metabolism, *Mol. Psychiatry*, 2016, **21**(6), 786–796.
32. T. Hori, K. Matsuda and K. Oishi, Probiotics: A dietary factor to modulate the gut microbiome, host immune system, and gut–brain interaction, *Microorganisms*, 2020, **8**(9), 1401.
33. K. M. Tuohy, G. C. M. Rouzaud, W. M. Bruck and G. R. Gibson, Modulation of the human gut microflora towards improved health using prebiotics-assessment of efficacy, *Curr. Pharm. Des.*, 2005, **11**(1), 75–90.
34. Y. Liu, H. Zhang, A. Xie, J. Sun, H. Yang, J. Li and Y. Liang, *Lacticaseibacillus rhamnosus* and *Lactiplantibacillus plantarum* combination treatment ameliorated colitis symptoms in a mouse model by altering intestinal microbial composition and suppressing inflammatory response, *Mol. Nutr. Food Res.*, 2023, 2200340.
35. E. Aizawa, H. Tsuji, T. Asahara, T. Takahashi, T. Teraishi, S. Yoshida and H. Kunugi, Possible association of *Bifidobacterium* and *Lactobacillus* in the gut microbiota of patients with major depressive disorder, *J. Affective Disord.*, 2016, **202**, 254–257.

CHAPTER 14

Religious and Cultural Perspectives on Food Selection

ASHISH DAHIYA,*[a] SHELLEY DUGGAL*[b] AND NIRAJ KUMAR*[a]

[a] Institute of Hotel & Tourism Management, M.D University, Rohtak – Haryana, India; [b] School of Hospitality, Amity University, Noida, India
*Emails: ashish@mdu.ac.in; sduggal@amity.edu; niraj.mdu@gmail.com

14.1 Introduction

Ethnic or cultural affiliations can be conceived through dietary habits. People frequently use food to uphold their cultural identities. Individuals with varying cultural origins consume distinct meals. Families' dietary preferences and dislikes are influenced by the places they call home and the ancestors they are descended from. One of the most significant aspects of religious rituals is eating, and food plays a significant role in religious culture as a means of demonstrating respect within communities. Since many of religions uphold religious laws, food preparation varies throughout them. Food is an exploration of culture, and that is its meaning. There is a rich cultural foundation to what we eat, how we get it, who makes it, who sits at the table, and who eats first. There is no relationship that is closer than that of a family, and food has a significant influence on the roles, norms, and traditions that define these relationships. It illuminates our most fundamental ideas about ourselves and other people, and aids in the discovery of attitudes, customs, and rituals related to food. Food and culture are related to one another. Our identities, cultures, and surroundings are negotiated in this way.

Food Chemistry, Function and Analysis No. 44
Functional Foods of the Future
Edited by Vijai Kumar Gupta, Minaxi Sharma, Smriti Gaur and Ramesh Chander Kuhad

Published by the Royal Society of Chemistry, www.rsc.org

14.1.1 Food as Identity – Cultural and Religious Significance

Culture is defined as the beliefs, values, and attitudes that members of a group or community practise and acknowledge. Culture is the entirety of the intricate system of customary behavior that the human race has evolved and that is passed down from generation to generation.[1] Individuals within a society are the carriers of culture, since culture is a learned concept that is not inherited and is connected to the social structure.[1] It might be argued that knowledge of culture is passed down orally from one generation to the next and that it manifests itself in stories, cultural values, rituals, and beliefs, among other forms.

Considering all of these factors, it is possible to state that food can be regarded as one of the easiest points of entry into a community for understanding and evaluating cultures, subcultures, and traditions.[2] This is because, while food is universal, different cultural groups and societies have different dietary preferences because of differences in religion, ethnic behavior, and beliefs. Thus, it may be said that a culture's dietary preferences, customs, and habits can provide insights into that society. What impact does culture have on dietary preferences? Food preferences are primarily mediated by exposure.[3] There is typically a positive correlation between exposure and liking, with culture playing a major role in how foods are introduced.[4] Certain cuisines are typically linked to family and memorable occasions, and they can serve as comfort foods during stressful or depressing times. Research indicates that flavour conditioning influences dietary decisions as well.[5] Family food regulations during childhood and other social factors can have an impact on this flavour conditioning.[6,7] Food choices may be influenced, for instance, by some of the food that the mother puts on the table on important occasions and shares with the group. Therefore, it can be said that the next generation inherits family food customs and culture, such as Ramadan, Seker Bayrami, and Kurban Bayrami.[2] Sharing food is the primary means of fostering social bonds, values, and the celebration of specific holidays and occasions in practically every culture. This is primarily evident in the case of customary celebration meals and certain religious holiday meals where participants engage in ritualistic behavior.[3] According to Heiman *et al.*,[8] religion also has a significant influence on culture, dietary preferences, lifestyle, and patterns of food intake. Communities' dietary choices are determined by their respect for tradition, adherence to religious prohibitions (such as Muslims' abstinence from pork), and their personal preferences (Table 14.1).[2]

Food holds allegorical significance in religious practices and rituals. There is no denying the fact that religions often impose dietary restrictions like *kosher* in Judaism and *halal* in Islam.

Not only this, there are religious calendars where fasting or feasting is incorporated by various religions pertaining to specific food habits observed during these practices. Religious rites or food as offerings or sacrifices are practiced by various religions. Food is also served as a cultural identifier,

Table 14.1 Main religions of the world and their food beliefs.

Religion	Population	Food beliefs
Christianity	2.32 billion (approx.)	• Bread and wine became a symbol of the religion • Diverse sectarian conceptions of the Eucharist
Islam	1.90 billion (approx.)	• The concepts of *halal* (allowed) and *haram* (prohibited) • All marine animals are allowed and do not need to be killed • Examples of *haram* animals include predators and pork • Alcohol and other intoxicants are prohibited
Hinduism	1.16 billion (approx.)	• The healthiest diet is vegetarianism since it provides enough nutrients and lowers the chance of infection • Fried food, or *pukka*, is preferred over "watery food" because of its purity and safety • The status of food in accordance with the concepts of *sattva*, *rajas*, and *tamas* • Cows are considered sacred and giving animals because of their milk • Alcoholic beverages are avoided but not completely forbidden
Buddhism	0.50 billion (approx.)	• The *Pancasila* (Five Precepts) outlines the following food laws: • The *Ahimsa* principle prohibits harming any creature to avoid negative *karma* • There are two types of meat: *uddissakatamasa* (blameful) and *pavattamasa* (blameless) • Fish is a sacred animal and a symbol of prosperity • Intoxicating substances, such as alcohol, are prohibited
Sikhism	0.026 billion (approx.)	• Many followers follow a strict vegetarian diet, abstaining from consuming any fish, meat or eggs • Those who are not vegetarian may choose to abstain from eating pork or beef. • Alcohol, tobacco and other intoxicants are prohibited • One should not fast for any religious reasons
Judaism	0.014 billion (approx.)	• *Kosher* and *treifah* concepts • The prohibition of blood and *heleb* (separated solid fats) • The acceptance of terrestrial animals with hooves and chewing gum • The exclusion of marine animals with fins and scales and the proscription of scavenging and predatory birds • A devout man's *shechitah,* or method of animal slaughter, is required for all *kosher* animal products • Meat and dairy items cannot be combined; separate utensils must be used to handle meat and dairy products

since recipes, flavours, and cooking methods are deeply rooted through generations. There is an array of eating habits, table manners, and serving methods that emerge from different unique cultures, but still food acts as a unifier and brings people together, instilling a sense of community and shared identity. Human civilizations have witnessed the amalgam of culinary traditions owing to migration and displacement of communities from one place to another. This chapter develops insights into the historical evolution of food practices within regions, religions, and cultures. It probes the way food practices exercise an influence on community dynamics, social strata, and gender roles. This chapter also helps in understanding the impact of globalization on traditional food practices and the preservation of culinary heritage. It offers a kaleidoscopic canvas for research and cultural studies, culinary innovations, and sociological analysis. It bridges the gaps between the interwoven connections of food, identity, religion, culture, and society.

14.2 Religious Dietary Practices

14.2.1 Hinduism

The two biggest countries with a predominantly Hindu population are India and Nepal, with over 1 billion people living there as of 2010, according to Hackett *et al.*[9] in the Global Religious Futures Report. Hindus hold that meal preparation requires meticulous attention to detail, as food is essential to preventing "spiritual pollution". The degree of food purity is determined by a number of factors, including who prepares the food, who serves it, and who it is consumed with.[10] This type of commandment does not simply concentrate on the food procedures. Surprisingly, Hindus believe that the kitchen is the purest area in the house, and they have no trouble placing their god's picture, the *murti*, there.[10]

Hindus also have the belief that as food is ingested by humans, it will have an impact on their spiritual elements. In addition, some Hindus hold the belief that food has an impact on both human emotions and health. Being a vegetarian is the greatest way to be healthy because it reduces your risk of contracting diseases and offers enough nutrients.[11]

The *Chandogya Upanishad* makes this commandment very clear: "When curd is churned, its subtle part rises upward and that becomes clarified butter". In this same manner, the subtle aspect of food rises to the surface and becomes mind when consumed.[10]

The fascinating idea that fried food (*pukka*) is safer and purer than food cooked in water is also held by Hindus. They contend that food that is considered "watery" is more susceptible to the effects of contaminants.[10] In Hinduism, food also has ceremonial roles, mostly in the service of the gods. A dish is safe from any dangerous substances if it was given to a Hindu deity, who neutralised it. According to Hindu tradition, food can be divided into three categories: *sattva*, *rajas*, and *tamas*.[10,11] This is because different types of food are said to have distinct reactions when consumed. The first category is of sattvic food, which is considered the best among the other categories by Hindus.

Food for the second group, known as *rajas*, typically has a sour, bitter, or salty flavour that contributes to the formation of bad character and poor health. In contrast, food for the first category is typically juicy, oily, and pleasant, which formulates a good mood and character, and preserves positive spiritual characteristics. Not to be overlooked is tamasic food, which consists of food that has been overcooked, partially cooked, spoiled, partially consumed, and so on. Due to its inherent negative qualities, this type of food should be avoided.[11] For instance, meat was categorised as tamasic food, which Hindus should avoid as it results in ill karma, as the *Law of Manu* makes abundantly clear (ref. 10, p. 211):

> *Let him avoid meat because it can never be obtained without harming living things, and harming sentient beings is inimical to heavenly happiness.*

Most Hindus probably know that eating beef and pork is forbidden. Hindus see cows as sacred animals, and they have a special place in their hearts since they are thought to contribute much, especially with their milk, which is very helpful in making dairy products.[10] Because of this, the cow is considered the most giving animal in the Hindu religion, because its milk is utilised to make a variety of goods that are useful to humans. One further Hindu commandment is to refrain from alcohol, even though this is not as rigorous as in Buddhism.[10] This is based on the following passage from Chapter 11, verse 94, of the book *Manu Smriti*:[12]

> *Since dirt is considered evil, and spirits are the defiling filth ejected from rice, drinking spirits is not appropriate for priests, rulers, or commoners.*

This commandment makes sense because alcohol has the inherent ability to change a person's state of mind to one of wickedness, which then results in poor character. While not all Hindus are subject to this restriction, the majority of them abstain from alcohol.[10] In conclusion, there are justifications for some of the unique food-related regulations found in Hinduism, such as the prohibition on alcoholic beverages and the promotion of vegetable consumption, which are both reasonable in terms of encouraging a healthy lifestyle.

14.2.2 Buddhism

Aside from Jews and Hinduism, Buddhism is another major religion in the globe. In general, Buddha's teachings encourage moderation in all facets of life. Food regulations also incorporate the principle of moderation, as

Buddhists believed in finding the middle path to enlightenment.[10] It takes this kind of approach to reach enlightenment.

Buddhism's food law is constructed based on the moral precepts of *Pancasila*, often known as the Five Precepts.[10] The first commandment of *Pancasila* is generally known as *Ahimsa*, which means that one should not harm any creature in order to avoid accumulating negative karma in the afterlife. Later, this concept also influenced the teachings of Jainism and Hinduism. However, when it comes to implementation, Jainism takes a more literal interpretation of what *Ahimsa* means. The Jains hold that all living things, including plants as well as animals, cannot be destroyed.

Some Buddhists, meanwhile, continue to maintain that the only creatures that are forbidden from being killed are animals. It should be mentioned that there is disagreement among the Buddhist community regarding how the food law should be understood and applied. This is due to the fact that not all Buddhists follow a strict vegetarian diet and abstain from eating meat.

While there are numerous Buddhist holy texts that provide evidence of Buddha's rejection of meat, the *Anguttara Nikaya* and *Vinaya* texts also demonstrate that the Buddha himself enjoyed eating meat. Chapter 5, code 187 of the *Anguttara Nikaya* describes a man who gave his servant instructions to locate meat.[13] Later, Buddha was offered the opportunity to consume the meat as well. On the other hand, Chapter 1, code 239 of *Vinaya* describes a scene in which a group of men killed a cow, prepared the meat, and gave the nun the tastiest pieces. However, she decided to give it to Buddha. The Buddha only forbade eating particular creatures, such as lions, snakes, and hyenas, according to the same source.[13] This rule seems to imply that eating other creatures is genuinely acceptable for Buddhists. *Vinaya* contains a passage that claims some animals or animal products, such as crocodiles, pigs, bears, fat, oil, and so on, might be useful for therapeutic purposes.[13] The Tibetan Buddhist community is one example of a Buddhist group that currently consumes meat due to poor soil quality that makes it difficult to grow healthy vegetables.[10] According to Kwon and Tamang,[14] Tibetan Buddhism considers eating small animals like chickens, ducks, and pigs to be a sinful act and only permits consuming large animals like cows. However, they do not consume fish because they regard it as a sacred animal; rather, they revere it as a representation of wealth and long life.[14]

The two sorts of meat, according to Theravada Buddhist teachings, are *pavattamasa* (blameless) and *uddissakatamasa* (blameful).[10] The only meat that monks are allowed to eat is in the second category. To prevent negative karmic repercussions, however, the monks are not allowed to be directly involved in its preparation. Theravada Buddhism is also well-known for taking everything that is presented in the alms bowl, which is held by the Buddhist monk (ref. 10, p. 208). This type of monk is known as a *bhikkhu*, which is still used today and literally means "one who begs". The senior monks will share their bowl with the junior monks.

Not to mention, the Buddhist tradition forbids the use of any intoxicating substances, including alcoholic beverages, as they have the potential to alter one's mental state and inspire immoral behavior.[10]

A Buddhist must be free from all intoxicating influences in order to achieve nirvana. In addition, the Buddha reiterated this stringent prohibition in his final remarks before passing away. He declared:

Here is my final piece of advice to you monks. Everything that makes up the world is malleable. They don't last. Make a conscious effort to achieve your freedom.

The teachings of the Buddha are not universally applicable, but they do impart certain unique moral values. Among these are the teachings of mutual cooperation between monks and laypeople, as the former rely on the latter for their material survival and the latter on the monks for future good deeds.

14.2.3 Christianity

When we talk about food in Christianity, the first items that come to mind are the bread and wine, which have come to symbolise the religion. The bread is cooked to remember a significant event that occurred prior to Jesus' crucifixion and as a gesture of gratitude, since Jesus shared these two holy foods with his devoted disciples by saying: "My body is this bread. My blood is this wine. I shall be with you each time you partake of this bread and sip this wine in my honour".[15] As a result, the Eucharist has been recognised as a sacred communion that can build bonds between Christians and subtly promote unity.

Several significant Christian authorities held the view that members of their religion should practise vegetarianism. This is because, aside from maintaining health, they thought that vegetarianism was the best way to encourage a tranquil existence. As a result, many of the church fathers, including John Chrysostom, Boniface, and Origen, led rigorous vegetarian lives throughout their lives.[16] For instance, we can look at the case of one of the major church fathers, Basilius the Great, who stressed that if people stopped eating meat, vegetarianism would bring prosperity and a pleasant existence. Another example from a more recent era is the Methodist Church's founder, John Wesley, who advocated vegetarianism as a way to maintain good health. "Thanks be to God: I have been delivered from all physical ills since I gave up flesh and wine", he declared.[16]

Ellen White, who founded the Seventh-day Adventist Church and followed a strict vegetarian diet, is another notable person. She also denounced the idea that God erred greatly in not providing meat to Adam and Eve in the Garden of Eden. She thought that eating fruits, nuts, and vegetables would be plenty to provide the body with healthy nutrients. She went on to say:

> *If meat were necessary for strength or health, the Lord would have erred in sending Adam and Eve with nourishment. It is incorrect to believe that eating animal products increases muscle strength. Without it, the system's needs can be more effectively met and a more active lifestyle can be lived. Together with fruits, nuts, and vegetables, grains have all the nutritional qualities needed to create healthy blood. Those who consume meat are merely consuming leftover grains and vegetables. How much better to receive it directly through consuming the nourishment that God made available to us.*

14.2.4 Islam

Islam is currently the second largest religion in the world after Christianity, with 1.8 billion adherents as of 2015.[17] Islam is thought to be the religion with the fastest growth rate in the world and this number is still rising. The religion is anticipated to overtake Christianity in the next twenty years.[18] Islam is a way of life rather than just a religion. In addition to their required daily prayers, Muslims observe their religion as a means of praising their God in all spheres of their lives, including relationships, commerce, and business. Food consumption is a part of this.

Muslims adhere to strict dietary laws that are primarily found in their holy book, the *Quran*, as well as the Hadith and Sunnah, which are reliable accounts of the sayings and deeds of their last prophet, Prophet Muhammad (Peace Be Upon Him). According to authors,[19] these two sources are regarded as their principal sources or guidelines. The holy *Quran* (2 : 172), one of the primary sources of Islamic law, states, in general, that everything is permitted for human consumption unless proven contrary, a belief shared by most academics. This is where the Arabic word *halal*, which literally means "permissible" or "lawful", came from. Although the term "*halal*" is frequently used to refer to a wide range of facets of Muslim life, including banking and money, amenities and hotels, and so on, the word itself is mostly appropriate for food and beverages.

The Islamic text makes reference to *haram* food, or forbidden food, in multiple places. These scriptures state that all clean, pure food is allowed to be consumed, with the exception of a few categories. The following are a few of them:[20]

- Pork or any of its byproducts
- Carrion
- Alcoholic beverages
- Flowing or congealed blood
- Animals killed outside of the law of Islam.

Thus, Muslims are mostly prohibited from consuming pork, alcohol, blood, carrion, and animals that have been improperly slaughtered.

Furthermore, food that would be otherwise permitted for consumption is no longer be allowed if it contains any of these *haram* substances.

Muslims are also mindful of the food's ingredients, distribution and serving methods, logistical considerations like storage and transportation, and the preparation procedures like using utensils and machinery. What makes all of Islam's laws and practices special and distinct is that, in addition to encouraging *halal* eating, the religion also promotes the idea of *tayyib*, which is Arabic for "pure" or "good".[21] Islamic law mandates that food be *halal*, and it also promotes a healthy, nutrient-dense diet with the right amount of food safety. This suggests that Islam chose to prioritise high quality from farm to table.[22]

Muslims have traditionally practiced *tayyib* (or *halalan tayyiban*) in relation to food consumption, which has an impact on the menu choices made by Muslims.[20,21] Muslims are still bound by Islamic law when they travel, albeit there are some exceptions depending on the situation, such as when eating non-*halal* food is required to prevent a risk to one's life. Muslims can find ease in this flexibility within Islam since, according to the *Quran*, the religion does not intend to stress its adherents (2:286). Muslims make decisions about their menus based on their responsibilities as Muslims. Muslims prefer to go to places where there are more options for *halal* food; some may even check if items supplied in stores or restaurants have received *halal* certification.[23]

14.2.5 Judaism

According to 2015 Pew Research Centre research, there were approximately 14 million Jews worldwide in 2010. It is anticipated that this number will rise over the ensuing years to approximately 16 million in 2050. Similar to Muslim concepts of *halal* and *tayyib*, Jews have their own dietary laws and food preparation standards known as *kosher*, which translates to "fit" in Biblical Hebrew.[24] Like Islam, the food laws have their roots in biblical law and are found in their holy scriptures. The original five books of the *Torah* are the primary source of the laws, which have been expanded upon by the rabbis to protect Jews from breaking any of the basic laws and to take into account new issues and technological advancements.[25]

Given the historical connections between these two religions, there are certain similarities between *halal* and *kosher*. Foods that are suitable for ingestion by adherents of Judaism are determined by the *kosher* dietary regulations, also known as *kashrut*.[26] Apart from the fact that the Islamic notion of *halal* and *kosher* are comparable, there are fewer categories in *kosher* than in Islam's *halal* laws. Judaism's dietary regulations, which centre on the animal kingdom, primarily addressed four difficulties:[25–27]

- Permitted animals
- Prohibition of blood
- Prohibition of combining meat and milk
- Parts not allowed for ingestion.

Pigs are among the non-*kosher* animals;[25] these animals are comparable to the non-*halal* animals in Islam. In addition, *kosher* emphasises the importance of the slaughtering or butchering procedures, specific equipment, guaranteeing total separation of milk and dairy products, and supply chain traceability.[26,28] These characteristics are similar to the Islamic concepts of *halal* and *tayyib* and have the same effect on Jewish menu choices.

Because they must abide by the Jewish food commandments, practicing Jews are prone to base their meal choices on these requirements, which frequently leads them to request particular options, particularly when dining out. To fill this market niche, industry participants or service providers must make sure that the menu options they offer are in line with Jewish preferences.

14.3 Food as a Symbol of Identity

Food is a powerful symbol of cultural heritage, showcasing regional, social, national, and religious significance. Food recipes and cooking methods are proudly passed down from one generation to another, thus enriching cultural heritage.[29] Regional dishes made from region-specific ingredients become an emblem of that specific area. Community lunch, sharing of food, and communal cooking marks the social identity of a society. National symbols are represented in food, strengthening the bond between food and the nation. Some specific dishes of a nation become representative of the country on a global platter. Religious dietary laws, rituals, and practices associated with specific foods give them religious significance.[30]

14.3.1 Food in Religious and Cultural Festivals

Specific foods play an important role in religious celebrations and rituals with great symbolic and cultural significance. In Christianity, bread and wine during Holy Communion symbolize the Last Supper and the sacrifice of Jesus. In Judaism, the four cups of wine are considered as promises of redemption, whereas the bitter herbs correspond to the bitterness of slavery. Dates and water are consumed as the first items to break the fast in meal of *iftar* during the Islamic month of Ramadan as they are good sources of energy and hydration. In Hinduism, sweets are offered to deities during festivals or religious ceremonies, signifying devotion and gratitude. During special occasions, *karah prasad* is offered to everyone in Sikhism. In many Hindu festivals and in Buddhism, vegetarian meals are prepared as offerings to symbolize non-violence and compassion towards humanity. Chinese New Year is celebrated with dishes like dumplings, fish, and noodles that symbolize good luck, wealth, and longevity.[31] Pumpkin pie, cranberry sauce, and turkey are considered as traditional Thanksgiving foods in the United States.

From the above discussion, it is clearly discernible that foods are not merely culinary items but denote tradition, culture, and religious essence.

They show the vignettes of history, values, faith, and spiritual beliefs that blend the customs and traditions of diverse communities and foster a sense of unity among participants.[32]

14.4 Challenges and Adaptations – Intersecting Identities, Modernization and Globalization

When religious and cultural identities intersect, there emerge conflicts and tensions within or outside communities giving birth to social divisions and misunderstandings. Such misinterpretations of certain religious or cultural practices may lead to stereotypes and biases against specific communities or marginalization of certain groups. There are specific dietary restrictions practiced by different religions and cultures that can create challenges in communal settings with diverse groups.[33] Not only this, certain foods are considered taboo or forbidden in specific religious or cultural contexts. Even misunderstandings or misinterpretations related to the significance of certain dietary foods might lead to unintended contempt for those with different beliefs.

Contrary to this, intersectionality provides an opportunity to acknowledge varied cultures and religious beliefs, nurturing a more inclusive society with greater tolerance and empathy through dialogue and exchange between different groups. The global cultural diversity is enriched by the contributions of numerous intersecting religious and cultural identities that put in efforts to preserve their heritage and traditions. Intersectionality also promotes social cohesion, solidarity, and harmony among distinct groups standing firm on the strong pillars of the values of inclusivity and cooperation.[34] It includes appreciating the complicated nature and nuances of these identities in the making of a society that is built upon values, mutual respect, and cooperation amongst various members. It provides a symphony of the rich diversity of culinary traditions and practices where people are given opportunities to explore and acknowledge distinct cuisines and cooking techniques by sharing meals and traditional dishes. Food is often considered the centre-piece of various cultural and religious practices that unifies people having similar belief systems. Such a junction of cultural and religious identities helps preserve and hand down the traditional recipes and culinary heritage from one generation to another. It is essential to promote inclusivity and embrace culinary diversity so that individuals from different religious and cultures feel respected, valued and understood.

Modernization and globalization has exercised great influence on traditional dietary practices. Processed and convenient foods are examples of modernization. There has been a shift in food choices from traditional nutrient-rich foods to fast foods that are high in refined sugars, unhealthy fats, additives and preservatives, leading to numerous health issues like obesity, diabetes and cardiovascular diseases. With modernization, there is a plethora of food choices available from different cultures that can enrich culinary experiences but also result in a shift from local traditional foods. Inactive work habits, fewer physical activities, and altered meal patterns adversely affect traditional practices of community lunch, communal sharing, *etc.* Modernization may lead to the potential loss of cultural heritage as the younger generation may not acquire knowledge and skills of traditional cooking methods and recipes. Not only this, modernization may dilute cultural and religious identities as more globalized options may replace traditional foods.

The impact of modernization on traditional and cultural dietary practices is not universally negative, as it may also induce some positive changes like better access to nutritious foods, advancement of technology in food production, and more knowledge to people about the health benefits of traditional foods. The fusion of traditional wisdom and modern knowledge and technology can lead to healthier eating habits.

14.5 Cultural and Religious Dynamics – Evolving Perspectives over Time

Globalization, cultural deflection, and re-interpretation of religious practices have resulted in changing dynamics of cultural and religious practices. There is a struggle for religious dietary practices to fit in contemporary societies, like the challenges posed by certain religious dietary restrictions in urban settings, whereas they were quite feasible in agrarian societies.[35] The changing situations with advanced technology and scientific knowledge about health has led some religious communities to reinterpret ancient dietary laws to fit in modern systems. The wide array of foods available nowadays due to globalization has impacted the dietary practices of those religious communities that were previously geographically isolated. There are adaptations within cultural and religious dietary practices with a fusion of culinary traditions due to migration and cultural exchange. The commercialization of food has led to debates within religious communities on what is permissible or impermissible to consume, as processed foods might unintentionally contain ingredients conflicting with cultural and religious dietary laws.

Kosher guidelines have been modified due to industrialization and globalization, as now certification systems have been introduced to ensure food meets *kosher* standards.[36] There has been a rise in *kosher*-certified processed foods and innovative dietary products to suit modern lifestyles and in line with traditional systems too. AI-based food production and processing techniques have given birth to a discussion on the concept of what is considered *halal* and what are permissible food additives or methods of animal slaughter. Some Hindus have altered their dietary practices to the availability of more modern foods while maintaining vegetarian principles.

These alterations in cultural and religious practices demonstrate the blending of tradition with modernity while still maintaining cultural and religious practices. This evolution brings a balance between traditional practices and modern stances.

14.6 Future of Functional Foods – Cultural and Religious Perspectives

Cultural and religious perspectives play a crucial role in deciding the future of functional foods. Constancy to the guidelines of *halal* or *kosher* in ingredient sourcing will bolster the market for functional foods among Muslim and Jewish populations, respectively. Functional foods can be tailored to certain communities based on understanding the cultural and religious dietary patterns and restrictions.[37] Ethical treatment of animals and sustainable food production are the basis of many religious and cultural belief systems. Future functional foods can be adjusted to these principles, promoting ethical sourcing and sustainable practices. Most of the traditional and cultural foods are nutrient-based. Innovative functional foods can be prepared by integrating traditional recipes and modern research. Since the cultural and traditional value systems stress holistic health and wellness, future functional foods might also focus on not only health but also mental and spiritual wellbeing. The latest innovative and inclusive marketing and promotional strategies must be devised to promote awareness and acceptance amongst the masses for functional foods that cater to a diverse consumer base.

So, in the end it can be said that functional food development from the perspectives of religion and culture is not confined to specific dietary considerations, but also expands to giving respect to traditions, culture, values, and beliefs.

14.7 Conclusion: Nourishing Diversity – A Culinary Tapestry of Culture and Faith

This exploration into religious and cultural perspectives on food selection unveils the profound interplay between what we eat, our identities, and the traditions that bind us. The chapter underscores the pivotal role of food in a culture, shaping and reflecting the beliefs, values, and practices of diverse communities worldwide.

From the familial transmission of cultural knowledge through food to the allegorical significance of dietary practices in religious rituals, the intricate tapestry of culinary traditions weaves together a shared human experience. The symbolic representation of identity in specific dishes during religious festivals exemplifies how food becomes a language through which communities express devotion, gratitude, and historical narratives.

However, this culinary journey is not without its challenges. The collision of religious and cultural identities in the modern era gives rise to conflicts, stereotypes, and biases, potentially eroding the rich tapestry of diversity. Yet, the chapter remains optimistic about the potential for intersectionality to foster inclusivity, understanding, and tolerance among diverse communities. It suggests that dialogue and exchange can transform these challenges into opportunities for appreciating the multifaceted nature of cultural and religious identities.

The impact of modernization on traditional dietary practices is acknowledged, emphasizing the need to balance the advantages and disadvantages. The evolving dynamics of cultural and religious practices over time, shaped by globalization and technological advancements, demonstrate the adaptability and resilience of these traditions in the face of change.

Looking ahead, the chapter envisions the future of functional foods, where cultural and religious perspectives play a pivotal role in shaping dietary choices. It advocates for the development of innovative functional foods that respect and integrate traditional recipes while meeting modern health standards. The importance of ethical sourcing, sustainability, and holistic wellbeing is emphasized, suggesting a path forward that combines cultural heritage with contemporary nutritional science.

In essence, this exploration into the realms of food, identity, religion, and culture serves as an invaluable resource for those seeking to understand and appreciate the intricate connections that nourish our diverse global society. It calls for a celebration of culinary diversity and the preservation of cultural and religious heritage through the language of food.

References

1. *Concepts of culture: Art, politics, and society*, ed. A. Muller, University of Calgary Press, 2005.
2. B. Monin and L. M. Szczurek, *Food cultures*, 2014.
3. P. Rozin, Food and eating, *Handbook of Cultural Psychology*, 2007, vol. 2, pp. 447–477.
4. R. B. Zajonc, Attitudinal effects of mere exposure, *J. Personality Soc. Psychol.*, 1968, **9**(2p2), 1.
5. A. Eertmans, F. Baeyens and O. Van Den Bergh, Food likes and their relative importance in human eating behavior: review and preliminary suggestions for health promotion, *Health Educ. Res.*, 2001, **16**(4), 443–456.
6. J. M. de Castro, Weekly rhythms of spontaneous nutrient intake and meal pattern of humans, *Physiol. Behav.*, 1991, **50**(4), 729–738.
7. I. De Bourdeaudhuij, Perceived family members' influence on introducing healthy food into the family, *Health Educ. Res.*, 1997, **12**(1), 77–90.
8. A. Heiman, D. R. Just, B. McWilliams and D. Zilberman, Incorporating family interactions and socioeconomic variables into family production functions: The case of demand for meats, *Agribusiness: Int. J.*, 2001, **17**(4), 455–468.

9. C. Hackett, M. Stonawski, M. Potančoková, B. J. Grim and V. Skirbekk, The future size of religiously affiliated and unaffiliated populations, *Demographic Res.*, 2015, **32**, 829–842.
10. P. Gwynne, *World religions in practice: A comparative introduction*, John Wiley & Sons, 2011.
11. J. Maharaj, [cited 2024 Jul 14]. Available from: https://lakshminarayanlenasia.com/articles/WhyHindusNoMeat.pdf.
12. Z. Naik, Similarities between Islam and Hinduism – Part 9 Dr. Zakir Naik, [cited 2024 Jul 14]. Available from: https://www.islamawareness.net/Hinduism/ZakirNaik/part9.html.
13. S. Dhammika and Ven. S. Dhammika: To eat or not to eat meat – introduction. 2011 [cited 2024 Jul 14]. Available from: https://dharma-records.buddhasasana.net/ecology/ven-s-dhammika-to-eat-or-not-to-eat-meat-an-introduction.
14. D. Y. Kwon and J. P. Tamang, *Religious ethnic foods*, 2015.
15. What is holy eucharist and why is it so significant? – explore faith, [cited 2024 Jul 14]. Available from: http://www.explorefaith.org/eucharist.html.
16. J. A. Davidson, World religions and the vegetarian diet, *J. Adventist Theological Soc.*, 2003, **14**(2), 8.
17. D. Desilver and D. Masci, *World's Muslim population more widespread than you might think. Pew Research Center.*
18. M. Lipka and C. Hackett, *Why Muslims are the world's fastest-growing religious group.*
19. A. Quraishi, *Principles of Islamic Jurisprudence*. By Mohammad Hashim Kamali. Cambridge: Cambridge Islamic Texts Society1991. Pp. xxi, 417. Price not available. ISBN: 0-946-62123-3. Paper. ISBN: 0-946-62124-1, *J. Law Relig.*, 2001, **15**, 385–387.
20. *Handbook of halal food production*, ed. M. N. Riaz and M. M. Chaudry, CRC Press, 2018.
21. M. A. Ramli and M. A. Jamaludin, *Kriteria makanan yang baik dan buruk menurut perspektif kepenggunaan Islam. dalam Halal Haram dalam Penggunaan Islam*, ed. S. Ab Rahman and M. Aizat Jamaludin, Penerbit Universiti Putra Malaysia, Serdang, 2011, pp. 65–83.
22. L. H. Toh, *Straight talk: Tapping the huge halal business*, 2009.
23. J. Jafari and N. Scott, Muslim world and its tourisms, *Ann. Tourism Res.*, 2014, **44**, 1–9.
24. J. R. Eliasi and J. T. Dwyer, Kosher and Halal: religious observances affecting dietary intakes.(The Business of Dietetics), *J. Am. Diet. Assoc.*, 2002, **102**(7), 911–914.
25. J. M. Regenstein, M. M. Chaudry and C. E. Regenstein, The kosher and halal food laws, *Compr. Rev. Food Sci. Food Saf.*, 2003, **2**(3), 111–127.
26. M. Tieman and F. H. Hassan, Convergence of food systems: Kosher, Christian and Halal, *Br. Food J.*, 2015, **117**(9), 2313–2327.
27. E. Clay, G. Marks, M. M. Chaudry, M. Riaz, H. Siddiqui and J. M. Regenstein, Kosher and halal meals, in *Meals in science and practice*, Woodhead Publishing, 2009, pp. 594–637.

28. H. Campbell, A. Murcott and A. MacKenzie, Kosher in New York City, halal in Aquitaine: challenging the relationship between neoliberalism and food auditing, *Agric. Human Values*, 2011, **28**, 67–79.
29. A. Dahiya and S. Duggal, Exploring the food tourism effectiveness of official websites of BRICS nations, in *Destination Management and Marketing: Breakthroughs in Research and Practice*, IGI Global, 2020, pp. 1188–1205.
30. K. Napompech, Understanding the Perceptions' of Local Thai Cuisine of Chinese and European Tourists in Thailand., *Int. J. Hospitality Tourism Syst.*, 2023, **16**(2).
31. L. Zhang, *The changing path of Chinese New Year*, Doctoral dissertation, University of Iceland, 2017.
32. J. Baby, Influence of Festival Attractiveness, Novelty, and Experience on Attendees' Satisfaction: Moderating Role of Risk Awareness in a Cultural Festival., *Int. J. Hospitality Tourism Syst.*, 2022, **15**(2).
33. V. Gupta and S. Duggal, How do the tourists' behavioral intentions influenced by their perceived food authenticity: a case of Delhi., *J. Culinary Sci. Technol.*, 2021, **19**(4), 294–314.
34. P. H. Collins and S. Bilge, *Intersectionality*, John Wiley & Sons, 2020.
35. M. Montanari, *Food is culture*, Columbia University Press, 2006.
36. J. Fischer, Kosher Biotech: Between religion, regulation, and globalization, *Relig. Soc.*, 2018, **9**(1), 52–67.
37. S. Hasnah Hassan, Consumption of functional food model for Malay Muslims in Malaysia, *J. Islam. Mark.*, 2011, **2**(2), 104–124.

CHAPTER 15

Therapeutic Role of Natural Pigments in Functional Food Applications

PRAVEEN KUMAR DIKKALA,[a,†] MONICA NABIL GAYED IBRAHIM,[b,†] KANDI SRIDHAR,[c] VIJAI KUMAR GUPTA[d] AND MINAXI SHARMA*[e]

[a] Department of Food Technology, Koneru Lakshmaiah Education Foundation, Vaddeswaram, Andhra Pradesh 522303, India; [b] Chair of Food Science and Technology, Institute of Veterinary Medicine and Animal Sciences, Estonian University of Life Sciences, Kreutzwaldi 56/5, 51006 Tartu, Estonia; [c] Department of Food Technology, Karpagam Academy of Higher Education Deemed to be University, Coimbatore 641021, India; [d] School of Biotechnology, Dublin City University, Glasnevin, Dublin D09 K20V, Ireland; [e] Research Centre for Life Science and Healthcare, Nottingham Ningbo China Beacons of Excellence Research and Innovation Institute (CBI), University of Nottingham Ningbo China, Ningbo-315000, China
*Email: minaxi86sharma@gmail.com

15.1 Introduction

Functional and nutraceutical foods are fast entering in the corporate mainstream and gaining huge attention and consumer acceptance, as customers want healthier and safer foods. According to a report, the global

† Equal contribution.

Food Chemistry, Function and Analysis No. 44
Functional Foods of the Future
Edited by Vijai Kumar Gupta, Minaxi Sharma, Smriti Gaur and Ramesh Chander Kuhad

Published by the Royal Society of Chemistry, www.rsc.org

functional food market is estimated to reach USD 212.85 billion by 2028, with a CAGR of 2.71% during the forecast period (2023–2028).[1] Functional foods provide unique health benefits that go beyond typical daily nutritional consumption, such as increased bone health, cholesterol management, heart health, and other benefits related to eye health and eyesight, *etc.*

Colour is a vital attribute in enhancing the aesthetic appeal of a food commodity and attracting consumers towards it, which causes inflation in the market demand for coloured-food products. In previous decades, various functional and nutraceutical foods were developed using synthetic colourants, due to their high stability, cheap production costs, strong tinctorial strength, and simplicity of use in the food chain. Moreover, it is important to note that using synthetic pigments has long been associated with a number of negative effects, including toxicity, allergic reactions, and neurological and behavioral issues including hyperactivity.[2,3]

Alternatively, demand for natural food colours is rapidly increasing with increasing customer awareness to seek clean-labelled, safe, and natural origin products. According a recent report, the global market for natural food colours was valued at USD 1.8 billion in 2022 and is expected to grow at 8.6% CAGR from 2023–2032.[4] With a clean-labelled image, generally recognized as safe (GRAS) status for human consumption and improved aesthetic appeal of the food products, the natural colourants obtained from fruit and vegetable resources by natural means exhibit many pharmacological actions, such as anti-inflammatory, anti-tumour, anti-cholesterol, and anti-diabetic effects, when added to food formulations.[5,6]

Nature provides a wide range of hues to fruits and vegetables, resulting in a pigment-rich dietary supply. Pigments are chemical substances that can absorb light over the visible wavelength range and provide a wide array of colours to the food commodities. Different plant parts are employed to synthesize natural pigments with varying qualities.[7] These pigments are mostly found as secondary plant metabolites in fruits and vegetables and are commonly referred to as natural pigments.[8] Generally, plants (fruits and vegetables) include four primary types of natural pigments, chlorophylls, carotenoids, flavonoids (anthocyanins), and betalains, all of which have numerous therapeutic activities. Due to these therapeutic properties, natural pigments are now widely used in the food sector.[9] However, during food applications and processing, the natural pigments are susceptible to environmental conditions such as light, pH, temperature, UV, oxygen, and heat, resulting in colour degradation and reduction food product lifespan.[10]

This chapter provides a general overview of the present state of natural pigments research, chemical structure and classification, bioactivities or therapeutic activities, and their application in the development of functional and nutraceutical food formulations. In addition to offering a roadmap for future study in the field of natural pigments and their application in the sustainable development of functional foods, this chapter provides a methodical summary of the current literature on the subject.

15.2 Natural Pigments

Before the dawn of the modern age, natural organic pigments had an essential and significant role in various therapeutic activities. In most cases, the fruits, vegetables, and roots of plants are the sources that are utilized in the process of extracting natural pigments. In the area of food science, interest in natural pigments has recently increased because they are derived from biological substances, occasionally called bio-colours. Natural pigments have a variety of purposes besides just imparting aesthetic beauty to food formulations.[9] They have sparked renewed interest due to their potential health advantages and concerns about the safety of synthetic colourants. Numerous pigments such as chlorophylls, carotenoids, anthocyanins, and betalains have shown various therapeutic activities.[11] The classification and different types of natural pigments are given in Figure 15.1. The plant-based natural pigments can be extracted from leaves, stems, bark, and flowers of plants, or vegetable and fruit waste and by-products. These pigments possess potential biological roles in the protection of human health, like anti-inflammatory, antioxidant, anticancer, antimicrobial, cardioprotective and antithrombotic properties, improving ocular and neural health, and the prevention of non-communicable-diseases.[6,12,13]

15.2.1 Carotenoids

Carotenoids are tetraterpenoid compounds involved in many biological roles with various health benefits.[14,15] Carotenoid serves as a precursor of vitamin A, with carotenoids containing β-rings having the greatest provitamin A activity.[16] Based on their functional groups, carotenoids are classified into xanthophylls (oxygen-containing groups) such as β-cryptoxanthin, lutein, astaxanthin and zeaxanthin, and those having just carbon chains called carotenes (α-carotene, β-carotene, and lycopene).[17] Many chronic diseases, such as heart disease, neurodegenerative disorders, type 2 diabetes, and

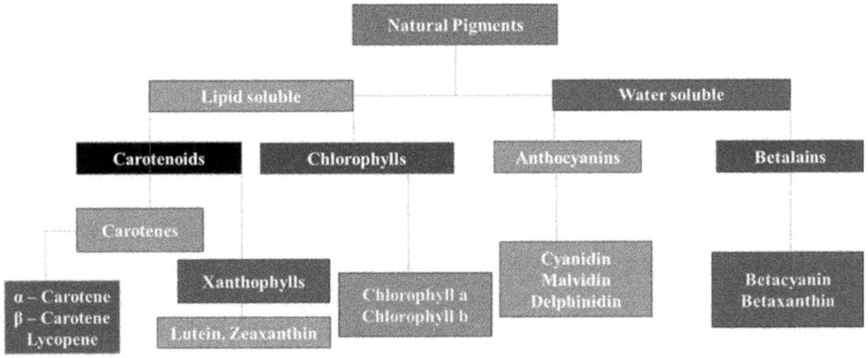

Figure 15.1 Classification of agri-food-based natural pigments found in different food sources (fruits and vegetables).

various types of cancer, such as prostate and gastric cancer, can be improved by involving various mechanisms of provitamin A (carotenoids), such as antioxidant activity, regulation of gene transcription, enhancement of gap junction, phase II enzyme-inducing activity, and the capacity to boost the immune system.[15,18] The lipo-solubility of nutrients determines the phases of their release, transport, and bio-assimilation from food. These are associated with prostaglandin E_2 (PGE2), interleukin (IL)-8, -6, and -1, oxidised phospholipids (such as OxLDL), tumour necrosis factor-α (TNF-α), nuclear factor kappa-light-chain-enhancer of activated B cells (NF-κB), and C-reactive protein (CRP). Oxidative stress modulation mediated through nuclear factor erythroid 2-related factor 2 (Nrf2) and peroxisome proliferator-activated receptor (PPAR) helps in control of the levels of mediators by raising the expression of antioxidant and cytoprotective phase II enzymes. The primary purpose of carotenoid-mediated Nrf2 signalling is to reduce inflammatory reactions and oxidative damage. As a result, each of these impacts the lowering of inflammation in the human body.[19]

15.2.2 Anthocyanins

Anthocyanins (ANCs) are water-soluble pigments that belong to the flavonoids group and are the most spectacular plant-based pigments widely spread in nature.[12,13] These are responsible for most of the red to blue colours in flowers and fruits and a variety of physiological functions in plant tissues.[20] Plants include twenty anthocyanidins with a C_{15} structure but no skeleton extension; however, only six of them are found in practically all fruits and vegetables: pelargonidin, cyanidin, delphinidin, peonidin, petunidin, and malvidin. Anthocyanins are found in practically all plants as anthocyanidin glycosides. The sugar moieties (often mono-, di-, or tri-glycosides) can be acylated with aliphatic and aromatic acids.[21] The anthocyanidins (delphinine, malvin, marvidin, pelargonin, pelargonidin and cyanidin) are unstable aglycones, and the presence of the glycoside part (glucose, galactose, rhamnose, and arabinose) increases their stability.[22–24] ANCs are identified into two forms: anthocyanidin glycosides and acylated anthocyanins. ANCs have received a lot of attention for their colour and ability to fight free radicals, which is closely linked with their chemical structure.[25,26] The multiple health benefits of these pigments support their roles as natural food colourants, functional ingredients, and dietary components in various food applications.[27,28] They can be absorbed throughout the digestive system after intake, but the lower intestine acts as the main site of absorption and metabolism. Due to the site of the absorbance of the anthocyanins, hybrid metabolites are created, which boost the bioavailability (12–15%) of anthocyanins after being ingested.[28,29] Although there has been significant progress in understanding the metabolic pathways, absorption, and bioavailability of anthocyanins concerning cardiovascular disease, no clear studies have been established. These are associated with treating obesity due to their ability to increase lipoprotein metabolism and thus reduce the body's fat production. Clinical studies have shown certain anthocyanin extracts to reduce abdominal obesity, cholesterol, triacylglycerol, and low-density lipoprotein.[30] These involve a potential protective effect against oxidative stress-induced apoptosis.

According to a previous study, ANCs are particularly good at neutralising free radicals and are crucial in reducing mutagenesis and carcinogenesis in cells.[28] They have demonstrated efficacy in inhibiting various cancers' initiation, development, and progression, including metastatic melanoma, prostate, colorectal, liver, breast, cervical, and lung cancers.[31] Several aggregative mechanisms account for the anti-carcinogenic activity in the human body, including inhibition of oxidative DNA damage,[28] anti-angiogenesis,[32] and cell cycle delay.[33] Several studies have shown that anthocyanins can help to reduce inflammatory diseases like gum disease, diarrhoea, inflammation after eating, and laryngeal reflux. Anti-inflammatory medications and anthocyanins may be combined to create novel therapeutic approaches for diseases brought on by inflammation. More and more research suggest anthocyanins may slow or stop age-related health issues and mental decline. Multiple studies have found evidence that anthocyanins safeguard the brain and enhance memory, cognitive, and motor function.[28]

15.2.3 Betalains

Betalains are an important group of natural, water-soluble, nitrogen-containing indole-based pigments and show different colours over a wide range of pH, from 3 to 7, which makes them acceptable for use in food colouring in the pH range of 3 to 7.[24,34] Their nitrogenous core structure is called betalamic acid. The reaction of betalamic acid with imino compounds or amino acids/derivatives results in the formation of different types of betacyanins, which are primarily red-violet (betanin), and betaxanthins (yellow-orange colour: vulgaxanthin and indicaxanthin, predominantly).[34,35] Depending on the pH of the environment and the food matrix, the betalains take different tones of colours, such as tawny ($pH > 8.5$), blue-violet (pH 6.5–8.0), red-purple (pH 3.0–6.5) or red-violet ($pH < 3.0$).[23] Jurić *et al.*[12] mentioned that not all types of betalains have the same biological antioxidant activities and stability. The health benefits of betalains includes their preventive and anticancer properties, anti-lipidemic effects, high antimicrobial activity, and preventive oxidative-stress-related disorders due to their high antioxidant activity.

15.2.4 Chlorophylls

Chlorophylls are lipid-soluble natural pigments with a porphyrin structure that are widely distributed in green fruits and vegetables and are characterized by a brilliant green colour range from olive-green to dark green and significant biological activity. They are identified into six different types (*a*, *b*, *c*, *d*, *e*, and *f*), although only two types (chlorophyll-*a* and -*b*) represent the major chlorophylls used in the food industry as natural colourants.[12,36] Chlorophyll-*a* has a methyl group and gives a green-blue colour, while chlorophyll-*b* has an aldehyde group and gives a yellow-green colour.[12,23] Due to the chlorophylls' bioactive properties, they have a significant influence on human health, as they can contribute to rebalancing the gut microbiota, and, due to their unique chemical structure, the chlorophylls exhibit antioxidant and antimicrobial properties. This group of pigments have anti-inflammatory, anticancer, and anti-obesity properties.[37] They are also involved in the rebalance the gut bacteria and lower the risk of colorectal cancer. Since chlorophylls can be converted to haemoglobin or work as a substitute for it when there isn't enough, they have been recommended as an add-on treatment for thalassemia and haemolytic anaemia.[38,39] Janiszewska-Turak *et al.*[23] reported that chlorophyll has a similar structure to haemoglobin, therefore it is able to regenerate or act as a substitute for haemoglobin in haemoglobin deficiency conditions. Other health benefits of chlorophyll include its ability to rejuvenate and energize the body, detoxification of the liver, cleaning of the intestine, normalize blood pressure, and combat bad odour, bad breath, and body odour due to the magnesium it contains.[24] Recovering chlorophylls from different sources is an exciting way to make food production more environmentally friendly.[37]

15.3 Therapeutic Activities of the Natural Pigments

15.3.1 Anti-inflammatory Activity of Pigments

Inflammation is an immune response generated after an infection or injury to the body. Several biochemical processes and cells are involved in this process, which restores damaged tissues and physiological function by eliminating harmful substances. A recent study has found a link between inflammation and how diseases like cancer and arthritis start. Signals that initiate, sustain, and terminate inflammation exert close control over the process. Damage to cells and tissues and autoimmune diseases result from an imbalance between these signals. Acute inflammation is a condition that can last for minutes, hours, or several days. It is characterized by fluid leakage and the movement of leukocytes, especially neutrophils, out of the affected area. In contrast, chronic inflammation is associated histologically with lymphocytes, macrophages, fibrosis, and tissue necrosis and has a longer duration.[40]

Pigments improve human health through various bioactivity pathways, as shown in Figure 15.2. The natural pigments found in fruits and vegetables have better anti-inflammatory activities than other dietary compounds. When there is acute or long-term inflammation, high amounts of inflammatory chemicals are released, which causes an oxidative imbalance. Scientists have discovered that products from plants can help reduce inflammation. For example, campferol-3-*O*-methyl, which is found in ginger extract as the active ingredient. It inhibits nitric oxide and prostaglandin

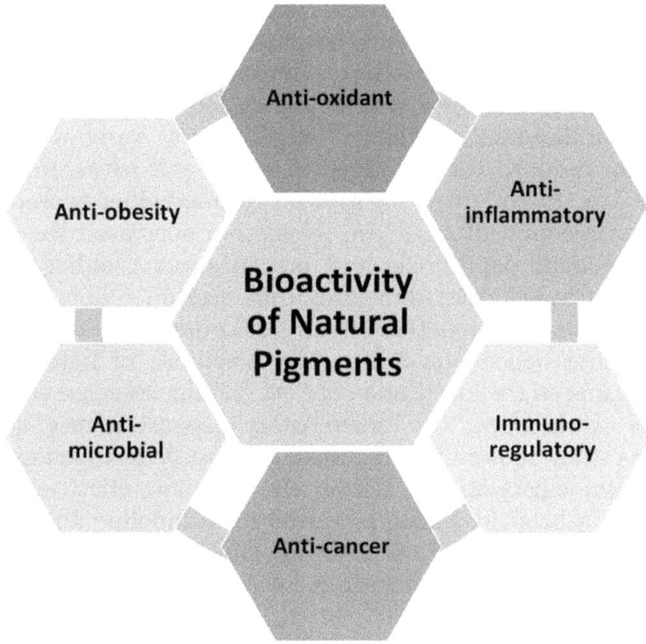

Figure 15.2 Different types of bioactivities shown by natural pigments.

E_2 synthesis. Inhibiting the activity and synthesis of a variety of proinflammatory chemicals and enzymes, including tumour necrosis factor (TNF-α), cyclooxygenase (COX), and lipoxygenase (LOX), may also give anti-inflammatory properties.[41] By inhibiting the activation of nuclear factor kappa light-chain-enhancer of activated B cells (NF-κB), which regulates DNA transcription, cytokine production, and cell survival, and by suppressing prostaglandin E_2 (PGE2), which impairs T cell receptor signalling, various pigments may aid in the resolution of inflammation.[42,43]

Many carotenoids, anthocyanins, and flavonoids are involved in reducing inflammation. The anti-inflammatory activity of these phytochemicals is well-known, however their precise mechanisms of action are not completely understood. The impact of carotenoids on intracellular signalling cascades, affecting gene expression and protein translation, has been connected to their various characteristics. Carotenoids limit the synthesis of inflammatory cytokines like interleukin-8 and prostaglandin E_2 by preventing NF-κB's translocation to the nucleus. Carotenoids can prevent oxidative stress by increasing the nuclear factor erythroid 2-related factor 2 pathway and activating phase II enzymes and antioxidants such glutathione-S-transferases. *In vitro* studies have demonstrated that β-carotene, lycopene, and lutein can inhibit the generation of reactive oxygen species (ROS), which can induce inflammation.[44] Lutein (20 μM) and β-carotene (20 μM) significantly reduced ROS (by 20% and 10%, respectively) in AGS cells (human gastric epithelial cells) stimulated with H_2O_2.[45] Similarly, β-carotene (10 and 20 μM) dramatically decreased ROS in AGS cells after *Helicobacter pylori* activation.[46] After studying the anti-inflammatory qualities of pectinolytic preparations of raspberry anthocyanins, researchers found that the anthocyanins could suppress the activities of lipoxygenase and cyclooxygenase 2, which in turn produced anti-inflammatory effects.[47] According to Yang *et al.*,[48] betanin administration reduced the amounts of NF-κB and iNOS proteins in the heart of AMI rats. In rats treated with *N*-nitrosomethylbenzylamine (NMBA), subcutaneous injection of E162 dye (betalains) decreased the rates of inflammation of both papillomas and precancerous esophageal lesions.[49] Researchers concluded from this study that betanin reduce the paraquat-induced acute damage in rats through anti-inflammatory pathways in several tissues, with dose dependent effects. The potentials of natural and semi-synthetic betalains on the inactivation of COX and lipoxygenase were examined in an *in vitro* investigation. The inflammatory response is mediated by these enzymes. This study showed that the most effective cyclooxygenase inactivator is phenethyl-amine-betaxanthin. Conversely, the most effective lipoxygenase inactivators are a betalain analog produced from indoline and the naturally occurring pigment betanidin.[50] Diets rich in pigments are also linked to various age-related health benefits including lowered age-related inflammatory diseases.[51]

15.3.2 Antioxidant Activities of Pigments

Antioxidants protect human health from infections and degenerative diseases by inhibiting and eliminating free radicals. They can remove free radicals such as reactive oxygen species (ROS) and reactive nitrogen species (RNS).[52] In addition, they serve as chain disruption agents or chain initiation inhibitors (second line of defence mechanisms). These antioxidants can keep cells from getting damaged by oxidation, which is a cause of many diseases.[52,53] Many synthetic antioxidants reduce or prevent oxidation, but they might have harmful impacts on human health. That is why scientists are trying to find natural pigments which act as antioxidants as well as colouring food formulations.[54] Antioxidant activity that involves inhibition or reduction of oxidative stress is the primary mechanism of action of natural pigments, and due to this reason they are classified as anti-carcinogenic and immuno-active therapeutic agents.

The primary recognized role of carotenoids in supporting health is as an essential dietary supply of vitamin A, mostly as β-carotene, β-cryptoxanthin, and α-carotene.[55] In conjunction with this function, their antioxidant capacity is described as the most representative and is extended to all carotenoids, regardless of whether they have provitamin A activity.[28] The physiological role of carotenoids includes their role as antioxidants in cell membranes, their role in the oxidative cycle of the cell with other main antioxidants, and controlling cell growth and differentiation. Vitamin A aids cell growth, immune system function, prenatal development, and eye health.

Numerous biological activities of ANCs, such as antioxidant, anti-inflammation and antimutagenic activity, induction of differentiation, inhibition of proliferation through modulation of signal transduction pathways, induction of cell cycle arrest, and stimulation of apoptosis or autophagy of cancer cells, as well as anti-invasion and anti-metastasis, are reported to underpin their potential anti-tumour effects.[56] Anthocyanins have the capability to eliminate free radicals,[57] prevent cancer cells,[58] and protect eyesight due to their antioxidant activity. In addition, *in vitro* and *in vivo* investigations confirmed their low toxicity and safety. Consequently, anthocyanins are extensively used in the food, pharmaceutical, and cosmetic industries.[59]

Different acylation mechanisms increased the antioxidant potential of betacyanins. Red beets can stop lipid degradation and haem breakdown very well. The longevity of the pigments can affect both the breakdown of the antioxidants (phytochemicals) in the natural plant pigments and the saturation of the colour.[60,61] According to Yang *et al.*,[48] betanin treatment decreased oxidative damage in a rat model of acute myocardial infarction (AMI) by inhibiting GSH (glutathione) activity, catalase (CAT), and superoxide dismutase (SOD), and increasing the malondialdehyde (MDA) and myeloperoxidase (MPO) levels.

15.3.3 Anti-carcinogenic Activities of Pigments

Natural pigments with antioxidant and anti-inflammatory properties are hypothesised to possess anticancer properties. Some carotenoids are precursors to vitamin A, which is necessary for cell differentiation, growth, eyesight, and other physiological activities. Carotenoids have antioxidant qualities that protect against a weaker immune system, premature ageing, heart disease, and arthritis. Some carotenoids have also been shown to be anti-inflammatory (like astaxanthin), protect against oxidation (like lycopene), or help people lose weight (like fucoxanthin).[18] Pigments are also getting more attention in cancer studies[18,62] due to their more anti-carcinogenic properties like stopping the cell cycle, which is closely linked to apoptosis,[63] being anti-metastatic, anti-invasive, anti-angiogenic, and anti-proliferative, stopping tumour growth,[18] and lowering the risk of cancer. Also, a lot of preclinical cancer studies have shown that carotenoids can help cells die by targeting signalling pathways involved in extrinsic and intrinsic pathways,[64,65] metabolic apoptotic processes,[66] and epigenetic modulation.[67] Some clinical research on anticancer also focuses on fenretinide or bexarotene, two synthetic retinoids belonging to the apocarotenoid family. Although fenretinide exerted anticancer effects in preclinical cancer models, the results of more recent clinical studies are inconclusive.[18] On the other hand, it was found that mixing bexarotene with chemotherapeutic drugs was safe and well-accepted and that all-*trans* retinoic acids (tretinoin) had a safe and effective effect in cancer treatment.[68] Above all, preclinical evidence that carotenoids are linked to apoptotic molecules and signalling pathways is the basis for more research and the possible future use of carotenoids that target apoptotic processes to improve cancer management in terms of specific targeting and personalized cancer therapy.[18] On the other hand, there is not enough data to suggest that carotenoids have any pro-apoptotic effects in clinical settings. Carotenoids such as fucoxanthin extracted from marine sources showed strong anticancer effects on a wide range of cancer cells through apoptosis with *in vitro* and *in vivo* studies. But their effects were very different on different types of cancer cells, and the metabolites (halocynthiaxanthin and fucoxanthinol) were more effective at fighting cancer than fucoxanthin. Foods with lycopene had effects that both sped up the growth and differentiation of osteoblast cells and slowed down the growth and breaking down of osteoclasts. Consumption of dietary lycopene is associated with a lower incidence of prostate cancer and a potential protective effect against developing liver tumours.[69]

Anthocyanins have also been studied and reviewed extensively for their ability to fight cancer.[56] Various studies demonstrated the anticancer properties and mechanisms of anthocyanins. Cyanidin-3-glucoside (C3G) prevented the activation of the ErbB2/cSrc/FAK pathway caused by ethanol. The ability of C3G to inhibit cell migration/invasion was touted as advantageous for preventing breast cancer metastasis induced by ethanol. The explanation for anthocyanins' anticancer properties is that they can inhibit angiogenesis and promote apoptosis.[70] *In vitro* cell culture studies and animal models showed potential anticancer and anti-angiogenesis effects. The process of angiogenesis is essential to the progression of cancer since it is one of the critical steps in transforming tumours from benign to malignant. Anti-angiogenesis is a way to avoid cancer by stopping the growth of new blood vessels that bring oxygen to tumour cells. With anticancer properties, anthocyanins have the ability to stop or reduce the tumorigenesis and disintegrate the cells. Fimognari *et al.*[71] discovered that cyanidin-3-*O*-β-glucopyranoside (Cy-g) could stimulate the differentiation of human acute promyelocytic leukemia cell line HL-60 in a dose-dependent manner by activating PI3K and PKC through the detection of markers and kinase inhibitors in the cell differentiation process. Cy-g (200 $mg\,mL^{-1}$) treatment resulted in the reduction of oncogene *c-Myc* expression and the presentation of differentiation traits in HL-60 cells, including increased adhesion and enhanced esterase activity. Nevertheless, after PI3K and PKC inhibitor therapy, Cy-g's ability to stimulate HL-60 differentiation was greatly diminished. According to research by Longo *et al.*,[72] anthocyanin-treated human hepatocellular carcinoma (HCC) cell line PLC/PRF/5 resulted in down-regulated expressions of downstream Bcl-2 family and rapamycin target proteins, but up-regulated expressions of eukaryotic initiation factor 2α and autophagy-related gene *LC3-II*. These findings suggest that anthocyanins can trigger autophagy in human HCC cells. MDA-MB-231 and MCF7 breast cancer cell lines showed that anthocyanins and anthocyanin–pyruvic acid adduct extracts (250 $g\,mL^{-1}$) could prevent cancer cells from spreading. The extracts reduced cancer cell proliferation by acting as chemoinhibitors. In MDA-MB-231, the anthocyanin–pyruvic acid adduct extract has a better impact, indicating an action independent of oestrogen receptors. In addition to blueberry anthocyanins, anthocyanin-rich extracts (50 $g\,mL^{-1}$ monomeric anthocyanin) from chokeberry inhibit the growth of human HT-29 colon cancer cells by 60% after 24 hours, increase expression of tumour suppressor genes (*p21WAF1* and *p27KIP1*), and reduce cyclooxygenase 2 gene expression by 35%. According to data from a prior study, 5% whole freeze-dried black raspberries and the anthocyanin-rich fraction supplemented to *N*-nitrosomethylbenzylamine-induced F344 rats have the potential to prevent the development of cancer by reducing angiogenesis, inflammation, and cell proliferation in both preneoplastic and papillomatous oesophageal tissues, and by inducing apoptosis. As anticipated, the extracts do not inhibit the growth of normal colonic cells.[73]

In a different study, betanin/isobetanin treatment significantly decreased the proliferation and viability of cancer cells, changed the potential of the mitochondrial membrane (especially through intrinsic and extrinsic apoptotic pathways), and caused autophagosome vesicles to form in MCF-7-treated cells. Additionally, it caused a significant increase in the expression of the apoptotic Bad, TRAILR4, FAS, and p53 proteins, which led to autophagic cell death. The authors draw the conclusion that betanin/isobetanin therapy may be helpful in the treatment of cancer, particularly in tumours with functioning p53. Normal cell lines were unaffected by the betanin-enriched extract.[74]

15.3.4 Antimicrobial Activity of Pigments

Natural pigments may be interesting to use as co-preservatives to lower the amounts of synthetic preservatives in food matrices, because they provide disease-preventative measures to human health by killing microbes. More than forty plant extracts are involved in antibacterial activity. Pigments, due to their antimicrobial properties, may be intriguing additives for use as co-preservatives to reduce synthetic preservatives in food matrices, and they may also provide a disease-preventing action that is beneficial to human health.[75] A key potential barrier to the widespread use of natural plant pigments is the lack of stability, an important quality in the food sector.[52,61] Various natural plant pigments are used in the modern food industry, but relatively few studies have examined their antimicrobial properties.

Various pigments are responsible for inhibiting tryptophan catabolism, as they may act by detoxifying reactive oxygen species, thereby reducing the effects mediated by interferon and reducing reactive oxygen species formation.[40] The antimicrobial qualities of commercial water-soluble annatto extracts come from *cis*- and *trans*-norbixin.[75] Dietary flavonoids may have important uses in the food industry as natural antimicrobial agents according to their mechanism.

Carotenoids extracted from solid coffee waste have been demonstrated to have antimicrobial properties against toxic fungi such *Aspergillus flavus*, *A. parasiticus*, *A. carbonarius*, and *A. ochraceus*, as well as pathogenic bacteria like *Salmonella colorless*, *Escherichia coli*, *Staphylococcus aureus*, and *Listeria monocytogenes*.[76] In an another study, it was discovered that carotenoid extract worked incredibly well against yeast and Gram-positive and Gram-negative types of bacteria.[77] The researchers found that the carotenoid extract was more effective against *Pseudomonas* sp. and *Escherichia coli* than it was against *Staphylococcus aureus*, *Micrococcus*, *Candida* sp., and other bacteria based on a comparison of the diameters of the inhibitory zones. In a study, the antimicrobial activity of fucoxanthin was evaluated using the agar disc-diffusion and micro-dilution techniques. According to the agar disc-diffusion method, fucoxanthin had the maximum efficacy against *Streptococcus agalactiae*, followed by *Staphylococcus epidermidis* and *Staphylococcus aureus*.[78]

Pomegranate anthocyanins have exhibited antibacterial activity against *Salmonella*, with a MIC of 10.75–12.5 mg mL^{-1}. This MIC greatly suppressed *Salmonella* development in chicken refrigerated at 4 °C.[79] Similarly, anthocyanins isolated from lowbush blueberries had a minimum bactericidal concentration (MBC) of 69.50 mg mL^{-1} and a MIC of 34.75 mg mL^{-1} against *S. typhimurium*.[80] Llivisaca *et al.*[81] found that the anthocyanin extract of vaccinium fruit has a comparable zone of inhibition to ampicillin against *S. typhimurium*. Anthocyanins isolated from *Curtisia dentata* have been demonstrated to have good antitoxic effect, inhibiting the expression of the Vtx1 and Vtx2 genes in *E. coli*.[82]

Betalains have antimicrobial activity against a wide range of microbes, including multidrug-resistant bacteria and single-species and dual-species biofilm-producing bacteria. This is significant given the current antimicrobial resistance issue reported by the World Health Organization. The variety of betalains' antimicrobial capabilities is highlighted by research showing antiviral effectiveness against dengue virus, *in silico* studies involving SARS-CoV-2, and anti-fungal effects. Betacyanin has been shown to have antimalarial and anti-infective properties in living infection models, despite the paucity of *in vivo* research on the subject. Although the exact cellular mechanisms underlying betalains' antibacterial activity are yet unknown, previous studies have established the groundwork for a potentially unique antimicrobial agent.[83] According to Vijaya and Thangaraj, betalains from red beetroot ethanol extract reduced *E. faecalis* growth in a dose-dependent manner, and they measured inhibitory zones of about 8, 10, 12, and 14 mm after treatment with 1, 2, 4, and 8 mg of extract, respectively.[84]

In a study, extract of chlorophylls from plant materials using an ethanol–water mixture was developed, followed by their conversion to a micellar form using Pluronic F127, a polyethylene glycol and polypropylene glycol block copolymer, to check their antimicrobial properties.[85] The authors calculated the extract's photoinduced cytotoxicity against *Staphylococcus aureus* bacterium cells. The experiments demonstrated that the bacterial cells were completely inactivated, whereas a comparable experiment with healthy cells did not reveal any appreciable cytotoxic effects.

15.4 The Trend of Natural Pigments in Food Applications

Food pigments play a crucial role in food production, masking unpleasant attributes or enhancing the food products' natural properties.[17] Over the past 35 years, there has been a growing trend towards replacing synthetic colourants with natural pigments due to consumer demand for more natural products and the safety and health benefits of natural pigments.[86] Thus, plant-based natural pigments are considered functional food ingredients with a wide range of applications in the development of novel functional and nutraceutical food products (Table 15.1). Due to the significant low stability of natural pigments under food processing and storage conditions, the encapsulation technology has a promising effect on protecting the natural colourings against deterioration and enhancing their stability within the processing environment of a model food system.[6,87] Food products are naturally coloured by four main groups of natural pigments: red-blue-purple anthocyanins, red betalains, green chlorophylls, and yellow-orange-red carotenoids.[12]

15.4.1 Carotenoids in Functional Foods

Carotenoids are a group of naturally derived food colourants with growing demand, mainly due to their excellent colouring abilities, bioactive properties, and other health-promoting characteristics. The regular intake of carotenoids in diets is related to lower risks of chronic diseases such as cardiovascular diseases (CVDs), cancer, neurological disorder, and eye-related diseases.[88] They are known to be responsible for many of the brilliant red, orange, and yellow colours in a wide range of vegetables, fruits, and food products.[13,36] Carotenoids are also popular for their antioxidant potential and have been used as colouring compounds or natural dyes in many food applications such as frozen desserts, salad dressings, butter, margarines, spreading cheese, roasted foods (popcorn), non-alcoholic drinks, fruit juices, confectionery, baked goods, ice-cream, yoghurt, jams, pastries, jellies, *etc.*[23,88] Moreover, these pigments are also widely used for cosmetic, pharmaceutical and nutraceutical purposes due to their promising bioactive ingredients.[13]

The best-known carotenoids in food industry applications are β-carotene, annatto (bixin, norbixin), capsanthin, lutein, and canthaxanthin with β-carotene as the most common plant colourant used in foods. The utilization of carotenoids in the food industry as colourant additives and functional ingredients is challenging due to their water insolubility, instability, and quick degradability in the presence of reactive oxygen species or exposure to light, acidic pH and high temperature, in addition to their low bioavailability.[89,90] Previous studies (Table 15.1) have developed suitable alternatives for carotenoid delivery, such as in water-dispersible products, colloidal suspensions, emulsions, and colloidal dispersions, which boost the application of carotenoids in food sectors producing functional food products.[89] Other approaches have been proposed to solve these disadvantages,

Table 15.1 Technological application of encapsulated natural pigments in food industries.

Pigment	Source	Mode of application	Type of functional food	Finding	Ref.
Carotenoid	Melons	Added in the form of nanoencapsulations with porcine gelatin	Yoghurt	The nanoencapsulation improved β-carotene stability during storage for 60 days, promoted carotenoid solubility, and contributed a positively perceived yellow colour to the yoghurt.	92
Lycopene	Tomato waste	Added in the form of microencapsulation with sucrose and gelatin	Wheat noodles	The study results showed that encapsulation of lycopene reduced its degradation by 37.14% in noodles and improved sensory attributes, particularly sensory score for the colour of noodles, compared to the control noodles with free lycopene.	93
Carotenoids (lutein, zeaxanthin, β-cryptoxanthin, α-carotene, β-carotene)	Yellow bell pepper	Added in the form of microencapsulation with β-cyclodextrin	Isotonic beverages	Isotonic beverages with added extract exhibited dose-dependent luminosity and redness increase but decreasing yellowness. No differences during the storage time (21 days) were observed among the beverages for lightness, but yellowness significantly decreased ($p < 0.05$).	94
β-Carotene	(No data)	Added in the form of microencapsulation with soy protein isolate	Ice cream	The colorimetric evaluations of the ice creams confirmed the importance of β-carotene encapsulation in solid lipid microparticles to obtain a product with longer stability.	95
Carotenoids	Carrot waste	Added in the form of microencapsulation with alginate beads	Yoghurt	The yoghurt was stored for 28 days at 4 °C and showed improved antioxidant activity whilst preserving the microbial and physiochemical properties of the yoghurt.	96

Anthocyanins	Blueberry extract	Added in the form of nanoencapsulation with chitosan	Beverage	The encapsulation improved the stability of anthocyanins during storage compared to the free anthocyanins, where the stability achieved 4.5% *vs.* 71.2% for 35 days at refrigeration temperature (4 °C) and 68.4% *vs.* 49.7% after 12 days at room temperature (25 °C).	98
Anthocyanins	Italian black rice (*Oryza sativa* L., var. Artemide)	Added in the form of microencapsulation with maltodextrin–gum Arabic mixture	Baked cookies/biscuits	The fortified biscuits showed a higher antioxidant capacity than the control biscuits (482–595 *vs.* 283 mg TE per g of biscuit). Cyanidin-3-*O*-glucoside was the most abundant anthocyanin in all fortified biscuits (3220 mg per g of biscuit).	99
Procyanidins	(No data)	Added in the form of microencapsulation with extruded cornstarch	Chicken sausages	The chicken sausage containing MPS had a more stable secondary structure of protein compared to the control sample. The authors noticed that MPS addition significantly reduced the hardness and chewiness of the samples, and inhibited the texture change caused by protein oxidation ($p < 0.05$).	100
Anthocyanins	*Solanum melongena* L. bark	Added in the form of microencapsulation with gum Arabic	Yoghurt	The yoghurt content of phenolic compounds ranged from 106.01–239.90 mg GAE per 100 g of yoghurt prepared with non-encapsulated extract, and from 225.17–291.98 mg GAE per 100 g of yoghurt prepared with encapsulated extract.	101
Anthocyanins	Black carrot extract	Added in the form of microencapsulation with gum Arabic–maltodextrin	Ice cream	The highest total phenolic content (545.38 ± 4.34 mg GAE per 100 g) was observed in ice cream containing 9% microencapsulated anthocyanin powder. The quality attributes of prepared ice cream samples containing microencapsulated ANCs were also found acceptable till the end of the study (60 days).	102

Table 15.1 (*Continued*)

Pigment	Source	Mode of application	Type of functional food	Finding	Ref.
Betacyanin	Beetroot pomace	Added in the form of microencapsulation using soy protein isolate	Pseudocereal-enriched einkorn wheat biscuits	The authors observed that the microencapsulate-enriched water biscuits had the highest content of betanin, isobetanin, total phenolics and highest antioxidant capacity, thus demonstrating the protective effect of microencapsulation during the production process.	105
Betalain	Cactus fruit pulp (*Opuntia ficus-indica*)	Added in the form of microencapsulation using calcium alginate	Gummy candies	The betanin stability in the food system was assessed during 30 days of storage at 4 °C and there was no significant variation in total colour parameters for all samples (ΔE^* values less than 4).	106
Betalain	Beetroot extract	Added in the form of microencapsulation using maltodextrin	Whole milk stirred yoghurt	The addition of encapsulated beetroot extract did not modify ($p<0.05$) the rheological characteristics but significantly reduced syneresis and increased protein content and antioxidant activity ($p<0.05$); moreover, it has the most acceptable sensorial criteria.	108
Chlorophyll	Alfalfa extract	Added in the form of microencapsulation using agar and gelatin mixture	Jelly and gummy candy	The microencapsulated chlorophyll green pigment showed a lower degree of change in the colour factors (L^*, a^*, and b^*) and higher heat stability than the non-encapsulated green pigment.	110
Chlorophyll	*Nannochloropsis oculata* microalga extract	Added in the form of microencapsulation using maltodextrin	White chocolate	The chlorophyll-*a* content of the obtained white chocolate samples was between 9.60–22.9 $\mu g\, g^{-1}$. It was found that the sensory properties of the samples, such as appearance, texture, mouth melt and odour, were not affected by the level of microalgae use ($p>0.05$).	111

such as micro- and nanoencapsulation by host molecules, of which nano-encapsulation by formation of inclusion complexes has been highlighted as a straightforward and effective alternative, with possible benefits in terms of increased stability, solubility, and bioavailability of the carotenoids.[91]

Medeiros *et al.*[92] studied yoghurt fortification with a porcine gelatin-encapsulated carotenoid extract from melons. The authors noticed that the nanoencapsulation improved β-carotene stability during storage for 60 days, promoted carotenoid solubility and contributed a positively perceived yellow colour to the yoghurt. Ranveer and Sahoo developed wheat noodles enriched with microencapsulated lycopene extracted from tomato waste and encapsulated with sucrose and gelatin.[93] The study results showed that the encapsulation of lycopene decreased its degradation by 37.14% in noodles and improved sensory attributes, particularly sensory score for the colour of the noodles, compared to the control noodles with free lycopene. Lobo *et al.*[94] encapsulated yellow bell pepper pigments, which consist of the main carotenoids lutein, zeaxanthin, β-cryptoxanthin, α-carotene, and β-carotene, with β-cyclodextrin and evaluated their stability in isotonic beverages (pH 2.9; 0.02%, 0.05%, and 0.06% extract addition). Isotonic beverages with added extract exhibited dose-dependent increase in luminosity and redness but decreased yellowness. No differences were observed among the beverages for lightness during the storage time (21 days), but yellowness significantly decreased ($p<0.05$).

The feasibility of incorporating β-carotene-loaded solid lipid microparticles (BCSLM) into vanilla ice creams was investigated by Lima *et al.*,[95] through the physico-chemical characterization and evaluation of sensory acceptability of the products. The colorimetric evaluations of the ice cream samples showed that the ice cream produced with partial substitution of artificial additives by BCSLM containing α-tocopherol presented a more intense colour, while in the product with non-encapsulated β-carotene fast degradation of carotenoid was confirmed, highlighting the importance of the encapsulation techniques. The results of the sensorial assessment of the products were highly satisfactory and showed that the panellists preferred the ice creams produced with BCSLM containing α-tocopherol and with partial substitution of artificial additives by BCSLM containing α-tocopherol, confirming the feasibility of incorporating BCSLM into ice creams to reduce the application of artificial dyes to the product. Šeregelj *et al.*[96] have recently reported the encapsulation of carotenoids from carrot waste in alginate beads as part of a yoghurt product at concentrations of 2.5 and 5 g per 100 g. The yoghurt was stored for 28 days at 4 °C and showed improved antioxidant activity whilst preserving the microbial and physiochemical properties of the yoghurt.

15.4.2 Anthocyanins in Functional Foods

Anthocyanins (ANCs) are structurally diverse and their colour range is affected by the pH of the surrounding food matrix, thus providing a wide array of colouring properties (from red-orange to blue-purple) for different food applications. Different pH of the solution makes this pigment appear pink-purple (pH 1–2), violet-indigo (pH 3–5), blue (pH 6–7), mint green (pH 8–9), and green (pH 10–13).[13,17]

Echegaray *et al.*[97] reported the potential use of anthocyanins as functional compounds, effective anti-oxidative stress agents and natural food colouring alternatives to artificial dyes. Among the major health properties associated with ANCs are their anticancer activities linked to chemopreventive and chemoprotective effects *in vivo* and *in vitro* in several cancer cell lines, their antioxidant and anti-inflammatory effects, decreased arterial hypertension, improved metabolism, their inhibitory effect on mutagenesis and carcinogenesis, and prevention of cardiovascular diseases.[12,17,23] The above health properties prove how important it is for people to consume anthocyanins daily. Application of ANCs by the food industry is a common practice and they can be met, for example, in fruit, alcoholic and non-alcoholic drinks, sauces, cheeses, milk desserts, jellies, jams, and candies.[23] However, anthocyanins are easily degraded due to their highly reactive nature and relatively low stability in food processing conditions.[13]

ANCs are weakly stable compounds and their decomposition may be caused by the presence of oxygen, temperature increase, UV light, acidic pH, and a longer period of heating or storage. These factors bring about formation of yellow or colourless chalcones, and finally cause browning.[23] Colour loss of ANCs may be prevented by storing them at low temperatures, in dark containers or under oxygen-free packaging.[36] Awolu and Oladeji[24] have reported that the stability of ANCs increased with reduced water activity, therefore they are suitable for colouring dried and intermediate-moisture foods. Due to the instability of anthocyanins, the extracts obtained after their recovery and purification are often subjected to stabilization processes using many techniques (utilization of controlled atmospheres, use of metal ions, addition of –SH group-containing compounds and polyphenolic compounds with stabilizing effect, and encapsulation) in order to be used in the food industry.[97] Among the preceding stabilizing techniques, the encapsulation method has been hailed and taken a special interest of researchers for further investigations.

Many research studies have demonstrated the effectiveness of encapsulation in stabilizing anthocyanins within the matrix of different functional food. He *et al.*[98] studied the stability of a blueberry-derived mixture rich in anthocyanins nanoencapsulated with chitosan in a model beverage system. In this work, the authors noticed that encapsulation improved the stability of anthocyanins during storage compared to the freely-added anthocyanins. They achieved stability of 84.5% *vs.* 71.2% for 35 days at refrigeration temperature (4 °C) and 68.4% *vs.* 49.7% after 12 days at room temperature (25 °C). A study was conducted by Papillo *et al.*[99] that demonstrated the potential of incorporating encapsulated anthocyanin-rich extract of Italian black rice (*Oryza sativa* L., var. Artemide) into the preparation of functional baked cookies/biscuits. The authors stabilized the hydroalcoholic ANC-rich extract by both spray-drying (with and without a coating agent, maltodextrin–gum Arabic mixture) and freeze-drying to obtain ANC ingredients that were more stable during the baking of the biscuit at 180 °C for 25 min. Under these conditions, the authors obtained a biscuit with a higher anthocyanin content (between 194 and 230 mg per g biscuit, according to the stabilization method of the extract) compared to the control, where anthocyanins were not detected. At the same time, the fortified biscuits showed a higher antioxidant capacity than the control biscuits (482–595 *vs.* 283 mg Trolox equivalents (TE) per g of biscuit). Cyanidin-3-*O*-glucoside was the most abundant anthocyanin in all fortified biscuits (3220 mg per g of biscuit). According to the authors, the encapsulation of the black rice ANC-rich extracts had a mild protective effect on their anthocyanins.

The functions of microencapsulated procyanidins by extruding corn starch (MPS) in maintaining the quality of chicken sausages stored at 4 °C for 28 days were investigated by Song and co-workers.[100] The MPS added to chicken sausages had a strong inhibitory effect on the formation of disulfide bonds (46.76 μmol per g protein), and a slight impact on the amino acid content. The chicken sausages containing MPS had a more stable secondary structure of protein compared to the control sample. The authors noticed that the MPS addition significantly reduced the hardness and chewiness of the samples, and inhibited the texture change caused by protein oxidation ($p < 0.05$). Moreover, the microstructure of chicken sausages suggested that the MPS addition did not damage the structure of chicken sausages. Therefore, MPS is a potential natural antioxidant as it delayed the oxidative deterioration and quality degradation of chicken sausages. In the dairy sector, a study was carried out by Barretto *et al.*[101] to evaluate the stability of non-encapsulated and spray-dried gum Arabic-encapsulated anthocyanin from *Solanum melongena* L. bark as a natural dye in yoghurts. With non-encapsulated extract the anthocyanin content decreased during storage for 30 days in a refrigerator in the absence of light, while it remained constant when encapsulated extract was added to yoghurt. The content of phenolic compounds in yoghurt ranged from 106.01–239.90 mg gallic acid equivalents (GAE) per 100 g of yoghurt prepared with non-encapsulated extract, and from 225.17–291.98 mg GAE per 100 g of yoghurt prepared with encapsulated extract. The authors proved the effectiveness of the gum Arabic encapsulation in protecting the ANC-rich yoghurt. In the same field, Shamshad *et al.*[102] examined the effect of storage on quality attributes of ice cream enriched with anthocyanin-rich black carrot extract encapsulated in gum Arabic–maltodextrin. The authors explained that during storage the ANC content declined due to oxygen exposure and interaction with other food ingredients. The highest total phenolic content (545.38 mg GAE per 100 g) was observed in the ice cream containing 9% microencapsulated anthocyanin powder. The quality attributes of prepared ice cream samples containing microencapsulated ANCs were also found acceptable till the end of the study (60 days). This research study identified the potential effect of addition of microencapsulated anthocyanin powder during ice cream preparation to stabilize the black carrot ANCs during the processing and storage of ice cream and to contribute positively to the sensory characteristics of ice cream.

15.4.3 Betalains in Functional Foods

Betalains show a potential health benefit which validates the requirement of their regular consumption in different food products. Betacyanins are a common type of betalains, used in the food industry due to their better stability at high temperatures and acidic pH and being less prone to oxidation than betaxanthins, in addition to their potential effect in reducing the risk of coronary diseases.[12,24] Beet powder, with EEC No. E162, is recognized as a permitted colourant under Title 21 of the Code of Federal Regulations of the Food and Drug Administration (FDA) in the United States. It is currently used to colour a range of fabrics, foods, pharmaceuticals, and other products with compatible physicochemical properties, in addition to creating dye-sensitized solar cells.[103]

Due to bioactive potential, betalains have been introduced in different food products such as confections, ice cream, yoghurt, flavoured milk drinks, jellies, jam, candies, ready-made frostings, cake mixes, syrups and beverages.[17,24] Nevertheless, the application of betalains in the food industry is limited due to their low stability to light and high-temperature, and their distinctive unappealing earthy flavour.[17,34] Several factors may influence the stability of betalains, which must be taken into account to ensure the pigment properties. The main factors are exposure to high temperatures above 70 °C, pH lower than 3 or higher than 7, light, oxygen and the presence of high water activity.[12,23] Based on the chemical structures of betalains, they have the highest stability in the range of pH 4–6.

On the basis of these properties, betalains are recommended to be used in foods with a short shelf-life, produced with minimal heat treatment, and packaged and marketed in a dry state under reduced levels of light, oxygen, and humidity.[36] Zaki and Jai[13] have mentioned that among the promising techniques for stabilising the betacyanins (betalains) are complex formation, co-pigmentation, and encapsulation, which would boost the application of betalains in the food industry by diversifying the functional food products in the market (Table 15.1). Betalains were successfully encapsulated and protected by ionic gelation, with higher stability observed especially at low relative humidity. Good encapsulation efficiency and better stability of betalain were obtained by using a double emulsion (water-in-oil-in-water or w/o/w emulsions) with the betalains dispersed in the inner aqueous phase.[104] Remarkably, the w/o/w emulsion structure acts as a barrier to protect the hydrophilic bioactive embedded in the inner aqueous phase of the capsule, while allowing the gradual release of the payload when oil digestion occurs. Different encapsulation matrices (polysaccharides such as maltodextrin, different combinations of other polysaccharides, or protein–polysaccharide complexes) show great potential for increasing the stability of betalain.[12]

Beetroot (*Beta vulgaris* L.) pomace was used as a source of betacyanin and betacyanin derivatives extract further encapsulated using soy protein isolate and used in pseudocereal (amaranth, buckwheat, and quinoa)-enriched einkorn wheat (*Triticum monococcum*) water biscuits.[105] The authors observed that the microencapsulated-enriched water biscuits had the highest contents of betanin, isobetanin, total phenolics and antioxidant capacity, thus demonstrating the protective effect of microencapsulation during the production process. Gummy candies were selected as a model food system by Otálora *et al.*[106] to incorporate betalain-rich capsules obtained by ionic gelation with calcium alginate from a betalain-rich extract of *Opuntia ficus-indica* fruit. The betanin stability in the food system was assessed during 30 days of storage at 4 °C and there was no significant variation in total colour parameters for all samples (ΔE^* values less than four). Thus, the stability of the encapsulated betalain colour in the gummy candy was confirmed, giving this food product a vivid red-purple colour, and representing a promising application for these natural pigments in the food industry. To enhance the stability of betalains, Kaur *et al.*[107] encapsulated betalains extracted from the pomace of red beetroot (*Beta vulgaris* L.) to produce a functional kulfi. When compared to the control, they found that the betalain-fortified functional kulfi had a lower microbial profile, a lower melting rate, and more antioxidant activity. Flores-Mancha *et al.*[108] developed yoghurt to which encapsulated lyophilized beetroot extract was added, which had higher sensory quality and antioxidant activity ($p < 0.05$). The addition of encapsulated beetroot extract did not modify the rheological characteristics ($p < 0.05$) but significantly reduced syneresis and increased protein content and antioxidant activity ($p < 0.05$); moreover, it has the most acceptable sensorial criteria (colour, taste, and general appearance). These results indicated that encapsulated beetroot extracts represent an option to manufacture functional yoghurt with added antioxidant activity.[108]

15.4.4 Chlorophylls in Functional Foods

Chlorophylls are a group of naturally occurring pigments responsible for green colour. Scientific studies revealed that the intake of chlorophyll with foods may beneficially act on human health through its antioxidant, antimutagenic, antigenotoxic and anticancer effects, and it supports healthy intestinal flora and strengthens immunity.[17,36] Due to their medicinal properties and their improvement to the human metabolism, chlorophylls have been used in various food and beverage industries such as jellies and jams, sweets, sugar confectionaries, chewing gums, soup concentrates, sauces, condiments, soft drinks, cheese spreads, ice cream, yoghurts, custards, puddings, preserves and vegetable pickles.[17,23] However, the commercial-scale application of chlorophylls in functional food production is limited because, under unfavourable conditions, the degradation and discoloration of chlorophylls takes place rapidly.

Chlorophyll colourants show the lowest stability of all the natural pigments and have low resistance to high temperature, light, oxygenation, and pH change (low pH conditions). They also undergo degradation in the presence of oxygen, pro-oxidants, catalytic metals, oxidative enzymes, and unsaturated lipids in food and even during storage under exposure to light and low temperatures.[12] A historical approach to chlorophyll stabilization has been established to form complexes with zinc or copper ions to form the more stable green metallo-chlorophylls. However, the need for concentrations of Zn^{2+} or Cu^{2+} salts above the limits established by the US FDA encouraged the use of physical encapsulation for chlorophyll stabilization.[109] Jurić *et al.*[12] explained the remarkable potential effect of using maltodextrin alone as a wall material to encapsulate chlorophyll which results in its highest storage stability (95–98%), because of the minimized moisture uptake and maximized encapsulation efficiency. The authors highlighted the promising effect of whey protein isolates in the successful encapsulation of chlorophylls because they can contribute to improving the water solubility and biological stability of chlorophylls. The previous studies showed that different types of microparticles have potential to stabilize the chlorophyll and improve its antioxidant activity and physicochemical properties as a natural food additive in functional food formulations, as shown in Table 15.1.

Recently, Raei and co-authors microencapsulated chlorophyll by freeze-drying extract from alfalfa and analysed its application in heated food (jelly and gummy candy).[110] In that study, the authors used a mixture of agar and gelatin as wall materials and reported the optimized formula to produce microcapsules with the highest encapsulation efficiency. Furthermore, they reported that during heating the microencapsulated chlorophyll green pigment showed a lower degree of change in the colour factors (L^*, a^*, and b^*) and higher heat stability than the non-encapsulated green pigment. Chlorophyll degradation was observed from a pH decrease in the a^* factor, which represents the red colour grade and microencapsulation significantly prevented this phenomenon. These differences demonstrate the higher resistance of the microencapsulated dye samples, which encountered less damage during heating processes, and introduce the application of microencapsulation for the green pigments of alfalfa as a good marker to prevent their destruction in the food industry. Genc Polat *et al.*[111] applied *Nannochloropsis oculata* microalga extract rich in chlorophyll spray-dried-encapsulated with maltodextrin in white chocolate as a colouring agent.

The chlorophyll-*a* content of the obtained white chocolate samples was between 9.60–22.9 $\mu g\, g^{-1}$. The obtained chocolate enriched with chlorophyll extract presented varied luminosity (61.6–78.0) and increased a^* and hue compared to control chocolate samples. It was found that the sensory properties of the samples, such as appearance, texture, mouth melt and odour, were not affected by the level of microalgae use ($p > 0.05$). However, it was notable that the taste and overall acceptability decreased ($p < 0.05$) depending on the level of the dried and encapsulated microalgae used.

The authors explained that these disadvantages can be overcome by the use of different encapsulation agents and encapsulation conditions.[111]

15.5 Conclusion and Future Prospectives

In recent years, the significance of investigating and expanding the use of natural colourants in unique and appealing food matrices has become prominent. Anthocyanins, carotenoids, and chlorophylls are pigments that not only contribute colour to food but also provide consumers with therapeutic benefits such as antioxidant, antibacterial, anticancer, and anti-inflammatory properties. However, in order for natural food colourants to become popular in the food industry, their high cost and low stability, as well as the FDA's and the European Union's strict regulations, standards, and lengthy toxicological evaluations, must be addressed through additional research efforts. Although natural plant-based colours are known to be highly sensitive to food processing conditions, an eco-friendly standard that puts safety above reevaluating processing parameters and focuses on how those changes affect colour stability is advised and could be a suitable alternative to synthetic dyes. More research is needed to include cutting-edge technology for making environmentally friendly substitutes for synthetic colourants in a variety of areas of our life, in order to contribute to a greener future. Finally, future studies should attempt to improve our understanding of the biochemical properties of natural pigments, not only to develop techniques to address their cost and stability challenges, but also to uncover their potential as functional food ingredients and nutraceuticals that can directly contribute to a sustainable functional food production chain.

References

1. ReportLinker, 2023, retrieved from https://www.globenewswire.com/news-release/2023/08/10/2723022/0/en/Functional-Food-Market-Size-Share-Analysis-Growth-Trends-Forecasts-2023-2028.html (accessed on May 8, 2024).
2. N. Shen, J. Ren, Y. Liu, W. Sun, Y. Li, H. Xin and Y. Cui, *Food Chem.*, 2023, **403**, 134422.
3. M. Sharma, K. Sridhar, V. K. Gupta and P. K. Dikkala, *Curr. Res. Green Sustainable Chem.*, 2022, **5**, 100340.
4. Natural Food Colors Market, 2022, retrieved from https://www.thebrainyinsights.com/report/natural-food-colors-market-13584#:~:text=The%20global%20natural%20food%20colors,quickly%20throughout%20the%20projected%20period (accessed on May 8, 2024).
5. A. Ahmadi, S.-A. Shahidi, R. Safari, A. Motamedzadegan and A. G.-H. Saraei, *Food Chem. Toxicol.*, 2022, **163**, 112980.
6. M. Sharma, Z. Usmani, V. K. Gupta and R. Bhat, *Crit. Rev. Biotechnol.*, 2021, **41**(4), 535–563.
7. M. Ghaffarzadeh and M. Edrisi, *J. Stud. Color World*, 2016, **6**(1), 63–82.

8. M. Eksiri, S. Shahidi and L. Nateghi, *Iran. J. Chem. Chem. Eng.*, 2022, **41**(9), 2936–2945.
9. R. S. Parmar and C. Singh, *Biochem. Biophys. Rep.*, 2018, **13**, 22–26.
10. T. Singh, V. K. Pandey, K. K. Dash, S. Zanwar and R. Singh, *J. Agric. Food Res.*, 2023, **12**, 100628.
11. S. Ghosh, T. Sarkar, A. Das and R. Chakraborty, *LWT*, 2022, **153**, 112527.
12. S. Jurić, M. Jurić, Z. Król-Kilińska, K. Vlahoviček-kahlina, M. Vinceković, V. Dragović-Uzelac and F. Donsì, *Food Rev. Int.*, 2020, **38**(8), 1735–1790.
13. N. A. M. Zaki and J. Jai, *Malays. J. Chem. Eng. Technol.*, 2020, **3**(1), 44–49.
14. P. Swapnil, M. Meena, S. K. Singh, U. P. Dhuldhaj, Harish and A. Marwal, *Curr. Plant Biol.*, 2021, **26**, 100203.
15. R. K. Saini, P. Prasad, V. Lokesh, X. Shang, J. Shin, Y.-S. Keum and J.-H. Lee, *Antioxidants*, 2022, **11**, 795.
16. D. B. Rodriguez-Amaya, Natural Food Pigments and Colorants, in *Bioactive Molecules in Food*, ed. J.-M. Mérillon and K. G. Ramawat, Springer, Cham, 2018, pp. 867–901.
17. I. Luzardo-Ocampo, A. K. Ramírez-Jiménez, J. Yañez, L. Mojica and D. A. Luna-Vital, *Foods*, 2021, **10**, 634.
18. L. Koklesova, A. Liskova, M. Samec, K. Zhai, M. Abotaleb, M. Ashrafizadeh, A. Brockmueller, M. Shakibaei, K. Biringer, O. Bugos, M. Najafi, O. Golubnitschaja, D. Büsselberg and P. Kubatka, *Biomolecules*, 2020, **10**, 1653.
19. Z. Kohandel, T. Farkhondeh, M. Aschner and S. Samarghandian, *Biomed. Pharmacother.*, 2021, **137**, 111374.
20. J. Martín, E. M. Kuskoski, M. J. Navas and A. G. Asuero, in *Flavonoids: From Biosynthesis to Human Health*, ed. G. C. Justino, Antioxidant capacity of anthocyanin pigments, IntechOpen, London, 2017, p. 55.
21. F. Blando, N. Calabriso, H. Berland, G. Maiorano, C. Gerardi, M. A. Carluccio and Ø. M. Andersen, *Int. J. Mol. Sci.*, 2018, **19**, 169.
22. J. C. Espin, C. Solar-Rivas, H. J. Wichers and C. Garcia-Viguera, *J. Agric. Food Chem.*, 2000, **48**, 1588–1592.
23. E. Janiszewska-Turak, A. Pisarska and J. B. Królczyk, *Nauka Przyr. Technol.*, 2016, **10**(4), 1–13.
24. O. O. Awolu and O. A. Oladeji, *Eurasian J. Food Sci. Technol.*, 2021, **5**(1), 25–40.
25. V. O. Silva, A. A. Freitas, A. L. Maçanita and F. H. Quina, *J. Phys. Org. Chem.*, 2016, **29**, 594–599.
26. J. Martín, M. J. Navas, A. M. Jiménez-Moreno and A. G. Asuero, in *Phenolic compounds – Natural Sources, Importance and Applications*, ed. M. Soto-Hernandez, M. Palma-Tenango, and M. R. Garcia-Mateos, Anthocyanin pigments: Importance, sample preparation and extraction, IntechOpen, London, 2017, p. 35.
27. J. M. Silván, M. Reguero and S. de Pascual-Teresa, *Food Funct.*, 2016, **7**(2), 1067–1076.
28. J. A. Fernández-López, V. Fernández-Lledó and J. M. Angosto, *RSC Adv.*, 2020, **10**, 24669.

29. I. A. Ludwig, P. Mena, L. Calani, G. Borges, G. Pereira-Caro, L. Bresciani, D. Del Rio, M. E. Lean and A. Crozier, *Free Radical Biol. Med.*, 2015, **89**, 758–769.
30. M. Lee, S. R. Sorn, Y. Park and H.-K. Park, *J. Med. Food*, 2016, **19**, 995–1003.
31. S. Li, X. Wei, J. He, X. Tian, S. Yuan and L. Sun, *Biomed. Pharmacother.*, 2018, **105**, 83–94.
32. M.-K. Kang, S. S. Lim, J.-Y. Lee, K. M. Yeo and Y.-H. Kang, *PLoS One*, 2013, **8**, e79823.
33. M.-C. López de las Hazas, J. I. Mosele, A. Macià, I. A. Ludwig and M.-J. Motilva, *J. Agric. Food Chem.*, 2017, **65**, 6477–6487.
34. J. Miller, *J. Food Drug Res.*, 2022, **6**(1), 3–4.
35. P. Rahimi, S. Abedimanesh, S. A. Mesbah-Namin and A. Ostadrahimi, *Crit. Rev. Food Sci. Nutr.*, 2019, **59**(18), 2949–2978.
36. M. J. Shetty, P. R. Geethalekshmi and C. Mini, *Trends Biosci.*, 2017, **10**(21), 4057–4064.
37. P. Ebrahimi, Z. Shokramraji, S. Tavakkoli, D. Mihaylova and A. Lante, *Plants*, 2023, **12**, 1533.
38. Y. Sun, L. Gu, J. Wen, C. van der Tol, A. Porcar-Castell, J. Joiner, C. Y. Chang, T. Magney, L. Wang, L. Hu, U. Rascher, P. Zarco-Tejada, C. B. Barrett, J. Lai, J. Han and Z. Luo, *Global Change Biol.*, 2023, **29**, 2893–2925.
39. L. Q. Zepka, E. Jacob-Lopes and M. Roca, *Curr. Opin. Food Sci.*, 2019, **26**, 94–100.
40. C. M. Moreno-Ley, G. Osorio-Revilla, D. M. Hernández-Martínez, O. A. Ramos-Monroy and T. Gallardo-Velázquez, *Hum. Nutr. Metab.*, 2021, **25**, 200126.
41. Y. Marunaka, R. Marunaka, H. Sun, T. Yamamoto, N. Kanamura, T. Inui and A. Taruno, *Molecules*, 2017, **22**(2), 209.
42. T. Jurikova, J. Mlcek, S. Skrovankova, D. Sumczynski, J. Sochor, I. Hlavacova, L. Snopek and J. Orsavova, *Molecules*, 2017, **22**, 944.
43. W. Kozłowska, A. Matkowski and S. Zielińska, *Front. Plant Sci.*, 2022, **13**, 888509.
44. J. Bouayed and T. Bohn, *Oxid. Med. Cell. Longevity*, 2010, **3**, 228–237.
45. Y. Kim, J. H. Seo and H. Kim, *J. Nutr. Sci. Vitaminol.*, 2011, **57**, 216–223.
46. S. H. Jang, J. W. Lim and H. Kim, *J. Physiol. Pharmacol.*, 2009, **60**, 131–137.
47. U. Szymanowska and B. Baraniak, *Antioxidants*, 2019, **8**(8), 299.
48. B. Yang, F. Cao, H. Zhao, J. Zhang, B. Jiang and Q. Wu, *Int. J. Clin. Exp. Pathol.*, 2016, **9**(3), 2777–2786.
49. J. F. Lechner, L.-S. Wang, C. M. Rocha, B. Larue, C. Henry, C. M. McIntyre, K. M. Riedl, S. J. Schwartz and G. D. Stoner, *J. Med. Food*, 2010, **13**(3), 733–739.
50. P. J. Vidal, J. M. López-Nicolás, F. Gandía-Herrero and F. García-Carmona, *Food Chem.*, 2014, **154**, 246–254.

51. A. Patel, U. Rova, P. Christakopoulos and L. Matsakas, *Front. Mar. Sci.*, 2022, **9**, 1015419.
52. H. Y. Leong, P. L. Show, M. H. Lim, C. W. Ooi and T. C. Ling, *Food Rev. Int.*, 2017, **34**(5), 463–482.
53. A. Guaadaoui, S. Benaicha, N. Elmajdoub, M. Bellaoui and A. Hamal, *Int. J. Nutr. Food Sci.*, 2014, **3**(3), 174–179.
54. E. J. Lee, D. An, C. T. T. Nguyen, B. S. Patil, J. Kim and K. S. Yoo, *J. Agric. Food Chem.*, 2014, **62**, 1324–1331.
55. M. Eggersdorfer and A. Wyss, *Arch. Biochem. Biophys.*, 2018, **652**, 18–26.
56. B. W. Lin, C. C. Gong, H. F. Song and Y. Y. Cui, *Br. J. Pharmacol.*, 2017, **174**(11), 1226–1243.
57. X. Zhao and Z. Yuan, *Chem. Biodiversity*, 2021, **18**(10), e2100399.
58. E. E. Ozkan, M. F. Seyhan, O. K. Sirin, T. Yilmaz-Ozden, E. Ersoy, S. D. H. Cakmar, A. C. Goren, H. Y. Aydogen and O. Ozturk, *J. Food Biochem.*, 2021, **45**(9), e13904.
59. L. Zhang, G. Fan, M. A. Khan, Z. Yan and T. Beta, *Food Chem.*, 2020, **323**, 126714.
60. H. Y. Leong, P. L. Show, M. H. Lim, C. W. Ooi and T. C. Ling, *Food Rev. Int.*, 2018, **34**, 463–482.
61. R. E. Wrolstad and C. A. Culver, *Annu. Rev. Food Sci. Technol.*, 2012, **3**, 59–77.
62. M. Chatterjee, K. Roy, M. Janarthan, S. Das and M. Chatterjee, *Curr. Pharm. Biotechnol.*, 2012, **13**, 180–190.
63. M. Gao, F. Dang and C. Deng, *Eur. J. Pharmacol.*, 2019, **859**, 172528.
64. Y. Jeong, J. W. Lim and H. Kim, *Nutrients*, 2019, **11**, 762.
65. X. Gong, J. Smith, H. Swanson and L. Rubin, *Molecules*, 2018, **23**, 905.
66. N. Ishijima, K. Kanki, H. Shimizu and G. Shiota, *Cancer Sci.*, 2015, **106**, 567–575.
67. H. Mollaei, R. Safaralizadeh, E. Babaei, M. R. Abedini and R. Hoshyar, *Biomed. Pharmacother.*, 2017, **94**, 307–316.
68. M. Lanhez, S. A. Pinto, H. A. Miot and E. Bagatin, *Int. J. Dermatol.*, 2019, **58**, 365–373.
69. J. L. Rowles, K. M. Ranard, J. W. Smith, R. An and J. W. Erdman, *Prostate Cancer Prostatic Dis.*, 2017, **20**, 361–377.
70. B. Alappat and J. Alappat, *Molecules*, 2020, **25**(23), 5500.
71. C. Fimognari, F. Berti, M. Nusse, G. Cantelli-Forti and P. Hrelia, *Biochem. Pharmacol.*, 2004, **67**, 2047–2056.
72. L. Longo, F. Platini, A. Scardino, O. Alabiso, G. Vasapollo and L. Tessitore, *Mol. Cancer Ther.*, 2008, **7**, 2476–2485.
73. H. E. Khoo, A. Azlan, S. T. Tang and S. M. Lim, *Food Nutr. Res.*, 2017, **61**(1), 1361779.
74. L. Nowacki, P. Vigneron, L. Rotellini, H. Cazzola, F. Merlier, E. Prost, R. Ralanairina, J. P. Gadonna, C. Rossi and M. Vayssade, *Phytother. Res.*, 2015, **29**(12), 1964–1973.
75. A. Shakeri, V. Soheili, M. Karimi, S. A. Hosseininia and B. S. F. Bazzaz, *J. Food Meas. Charact.*, 2018, **12**, 356–361.

76. M. D. Moreira, M. M. Melo, J. M. Coimbra, K. C. dos Reis, R. F. Schwan and C. F. Silva, *Waste Manage.*, 2018, **82**, 93–99.
77. S. Sinha, S. Das, B. Saha, D. Paul and B. Basu, *Front. Microbiol.*, 2023, **13**, 1088477.
78. T. M. Karpiński and A. Adamczak, *Antioxidants*, 2019, **8**(8), 239.
79. B. A. Wafa, M. Makni, S. Ammar, L. Khannous, A. B. Hassana, M. Bouaziz, N. E. Es-Safi and R. Gdoura, *Int. J. Food Microbiol.*, 2017, **241**, 123–131.
80. A. Lacombe, V. C. H. Wu, J. White, S. Tadepalli and E. E. Andre, *Food Microbiol.*, 2012, **30**(1), 124–131.
81. S. Llivisaca, P. Manzano, J. Ruales, J. Flores, J. Mendoza, E. Peralta and J. M. Cevallos-Cevallos, *Food Sci. Nutr*, 2018, **6**(4), 934–942.
82. J. H. Doughari, P. A. Ndakidemi, I. S. Human and S. Benade, *J. Ethnopharmacol.*, 2012, **141**(3), 1041–1050.
83. V. N. Wijesinghe and W. S. Choo, *J. Appl. Microbiol.*, 2022, **133**(6), 3347–3367.
84. D. Vijaya and N. Thangaraj, *J. Pharm. Sci. Res.*, 2019, **11**, 2422–2425.
85. K. D. S. S. Campanholi, J. M. Jaski, R. C. da Silva Junior, A. B. Zanqui, D. Lazarin-Bidóia, C. M. da Silva, E. A. da Silva, N. Hioka, C. V. Nakamura and L. Cardozo-Filho, *J. Photochem. Photobiol., B*, 2020, **203**, 111763.
86. S. N. H. M. Azmin, N. S. Sulaiman, M. S. Mat Nor, P. S. Abdullah, Z. Abdul Kari and S. Pati, *Appl. Biochem. Biotechnol.*, 2022, **194**(10), 4655–4672.
87. J. L. D. Antigo, A. P. Stafussa, R. de Cassia Bergamasco and G. S. Madrona, *J. Food Eng.*, 2020, **285**, 110101.
88. M. Sharma and R. Bhat, *Foods*, 2021, **10**(4), 787.
89. D. B. Rodriguez-Amaya, *Forum Nutr.*, 2003, **56**, 35–37.
90. S. Jurić, M. Jurić, Ż. Król-Kilińska, K. Vlahoviček-Kahlina, M. Vinceković, V. Dragović-Uzelac and F. Donsì, *Food Rev. Int.*, 2022, **38**(8), 1735–1790.
91. C. A. Fuenmayor, O. G. Baron-Cangrejo and P. A. Salgado-Rivera, *Polysaccharides*, 2021, **2**(2), 454–476.
92. A. K. O. C. Medeiros, C. C. Gomes, M. L. Q. A. Amaral, L. D. G. Medeiros, I. Medeiros, D. L. Porto, C. F. S. Aragão, B. L. L. Maciel, A. H. A. Morais and T. S. Passos, *Food Chem.*, 2019, **270**, 562–572.
93. R. C. Ranveer and A. K. Sahoo, *South Asian J. Manage. Res.*, 2014, **6**(1), 461–470.
94. F. A. T. Lobo, V. Silva, J. Domingues, S. Rodrigues, V. Costa, D. Falcão and K. G. de Lima Araújo, *J. Sci. Food Agric.*, 2018, **98**(7), 2665–2671.
95. J. G.-de Lima, T. C. Brito-Oliveira and S. C.-de Pinho, *Food Sci. Technol.*, 2016, **36**(4), 664–671.
96. V. Šeregelj, L. Pezo, O. Šovljanski, S. Lević, V. Nedović, S. Markov, A. Tomić, J. Čanadanović-Brunet, J. Vulić and V. T. Šaponjac, *LWT–Food Sci. Technol.*, 2021, **138**, 110732.

97. N. Echegaray, P. E. S. Munekata, P. Gullón, C. K. O. Dzuvor, B. Gullón, F. Kubi and J. M. Lorenzo, *Crit. Rev. Food Sci. Nutr.*, 2022, **62**(6), 1553–1567.
98. B. He, J. Ge, P. Yue, X. Yue, R. Fu, J. Liang and X. Gao, *Food Chem.*, 2017, **221**, 1671–1677.
99. V. A. Papillo, M. Locatelli, F. Travaglia, M. Bordiga, C. Garino, M. Arlorio and J. D. Coïsson, *Food Chem.*, 2018, **269**, 603–609.
100. J. Song, L. Jiang, M. Qi, W. Suo, Y. Deng, C. Ma, H. Li and D. Zhang, *J. Food Sci.*, 2022, **87**(3), 1184–1196.
101. F. J. de F. P. Barretto, H. A. Clemente, A. L. B. D. Santana and M. A. de S. Vasconcelo, *Rev. Bras. Frutic.*, 2020, **42**, 1–13.
102. A. Shamshad, Iahtisham-Ul-Haq, M. S. Butt, G. A. Nayik, S. Al Obaid, M. J. Ansari, I. K. Karabagias, N. Sarwar and S. Ramniwas, *Food Sci. Nutr.*, 2023, **11**, 3976–3988.
103. G. Richhariya, A. Kumar, P. Tekasakul and B. Gupta, *Renewable Sustainable Energy Rev.*, 2017, **69**, 705–718.
104. M. C. Otalora, J. G. Carriazo, L. Iturriaga, C. Osorio and M. A. Nazareno, *Food Chem.*, 2016, **202**, 373–382.
105. A. Hidalgo, A. Brandolini, J. Čanadanović-Brunet, G. Ćetković and V. T. Šaponjac, *Food Chem.*, 2018, **268**, 40–48.
106. M. C. Otálora, H. de Jesús Barbosa, J. E. Perilla, C. Osorio and M. A. Nazareno, *LWT–Food Sci. Technol.*, 2019, **103**, 222–227.
107. N. Kaur, A. Kaur, K. Sridhar, M. Sharma, T. P. Singh and S. Kumar, *Int. J. Food Sci. Technol.*, 2021, **56**(10), 5362–6370.
108. M. A. Flores-Mancha, M. G. Ruiz-Gutiérrez, A. L. Rentería-Monterrubio, R. Sánchez-Vega, J. Juárez-Moya, E. Santellano-Estrada and A. Chávez-Martínez, *J. Food Process. Preserv.*, 2021, **45**(1), e15628–e15660.
109. I. Viera, A. Perez-Galvez and M. Roca, *Molecules*, 2019, **24**(1), 154–171.
110. A. Raei, S. A. Yasini Ardakani and M. Daneshi, *J. Food Process Eng.*, 2017, **40**(5), e12529–e12538.
111. D. Genc Polat, Y. Durmaz, N. Konar, O. S. Toker, I. Palabiyik and M. Tasan, *J. Appl. Phycol.*, 2020, **32**(5), 3077–3088.

CHAPTER 16

Functional Food Production Utilizing Microbial Vitamins and Carotenoids

PRAVEEN KUMAR DIKKALA,[a] JAHNAVI KAKARLAPUDI,[b] PREETHI ROKALLA,[c] KANDI SRIDHAR,[d] PRINCE CHAWLA,[e] MINAXI SHARMA*[f] AND VIJAI KUMAR GUPTA*[g]

[a] Department of Food Technology, Koneru Lakshmaiah Education Foundation, Vaddeswaram, Andhra Pradesh 522303, India; [b] Department of Microbiology, College of Basic Sciences and Humanities, Punjab Agricultural University, Ludhiana, Punjab-141004, India; [c] Department of Food Science and Technology, Punjab Agricultural University, Ludhiana, Punjab-141004, India; [d] Department of Food Technology, Karpagam Academy of Higher Education Deemed to be University, Coimbatore 641021, India; [e] Department of Food Technology and Nutrition, Lovely Professional University, Phagwara 144411, Punjab, India; [f] Research Centre for Life Science and Healthcare, Nottingham Ningbo China Beacons of Excellence Research and Innovation Institute (CBI), University of Nottingham Ningbo China, Ningbo-315000, China; [g] School of Biotechnology, Dublin City University, Glasnevin, Dublin D09 K20V, Ireland
*Emails: minaxi86sharma@gmail.com; vijaikumar.gupta@dcu.ie

16.1 Introduction

As the population is increasing, the requirement for bioactive components is increasing day by day. The majority of industrial techniques used these days are not sustainable since they rely on non-renewable resources and

Food Chemistry, Function and Analysis No. 44
Functional Foods of the Future
Edited by Vijai Kumar Gupta, Minaxi Sharma, Smriti Gaur and Ramesh Chander Kuhad

Published by the Royal Society of Chemistry, www.rsc.org

frequently result in the generation of hazardous wastes. To replace chemical processing and its methods, the scientific community has developed numerous disciplines and methods to extract different bioactive components, including metabolic engineering, bioinformatics, metabolomics, systems biology, synthetic biology, *etc.*, to accelerate developments in 'white' biotechnology.[1] The production of various bioactive compounds, especially vitamins and carotenoids because of their health beneficial properties, from different sources has been increased in the last few decades. Microbial pigments show associated biological functions, which may benefit humans more than just as colouring compounds.[1] In nature, different microbes spontaneously synthesize vitamins and natural pigments, but their related metabolic pathways should be strictly regulated in lab-scale experiments and need to be increased to industry level.[1] Application of microorganisms in production of microbial vitamins and pigments is gaining tremendous attention from researchers and scientists, due to their rapid growth, weather-independence, scalability, and lack of competition.[2]

From the white biotechnological aspect, it is feasible to create strains that are capable of producing a significant quantity of vitamins and pigments (carotenoids) by making use of different innovative techniques that are available for metabolic engineering. The most crucial step in the metabolic process that is involved in the extraction of vitamins is fermentation. The purification of the fermentation fluid must be accomplished with a high yield and at an industrially acceptable cost. The development of bioprocesses has been driven chiefly by commercial considerations rather than environmental concerns. Furthermore, it is crucial to examine the significance of environmental and regulatory factors in this context.[3–5]

Approximately 90% of the total market share is presently held by synthetic carotenoids. The production of carotenoids through organic synthesis or extraction from plant-based sources is constrained by low yields, resulting in increased product prices. As a result, there is considerable interest in the microbial synthesis of various carotenoids, such as astaxanthin, β-carotene, lutein, lycopene, and zeaxanthin, as a more favourable alternative that has shown superior productivity. Furthermore, by using agro-industrial wastes as low-cost substrates, microbial manufacture of commonly used carotenoids may be a preferable option for producing unique C_{50} carotenoids at a low price.[6] There is a significant increase in demand for naturally generated carotenoids with various microbial pigments, driven mainly by expanding health concerns associated with synthetic carotenoids. Fungal strains such as *Monascus*, *Cordyceps*, *Serratia*, *Penicillium*, *Aspergillus*, *Fusarium*, and *Talaromyces* are used to extract biopigments such as carotenoids, riboflavin, lycopene, melanins, quinones, and betalains.[1]

Fungal carotenoids possess notable attributes beyond their use as dyes, including antibacterial, antioxidant, anticancer, gene regulation and cytotoxic effects. These traits have expanded their application across other domains.[7–9] In order for novel bioprocesses to become commercially available, it is necessary to solve a number of additional technological and regulatory problems.[3] Throughout the course of this chapter, we will discuss and investigate the process of extracting vitamins and carotenoids from a variety of microbiological sources using contemporary biotechnological methods, as well as their applicability in a variety of food processing businesses.

16.2 Vitamins and Carotenoids from Microbial Cell Factories

16.2.1 Vitamins

Vitamins are either water-soluble or fat-soluble. As the name suggests, a water-soluble vitamin dissolves in water easily and is insoluble in organic solvents. Fat-soluble vitamins are dissolved in fats but not in water and are stored in the liver or fatty tissues for future use. While vitamins are essential nutrients for all living things, many plants and microorganisms can synthesize them naturally by themselves.[10] This chapter primarily explores water-soluble vitamins, such as vitamin C and the vitamin B complex (thiamine, riboflavin, niacin, pantothenic acid, pyridoxine, biotin, folate, and cobalamin), along with fat-soluble vitamins E and K. The discussion includes insights into producing microorganisms using advanced techniques in metabolic engineering.

16.2.1.1 Water-soluble Vitamins

In the realm of water-soluble vitamins, microorganisms and their enzymes play a pivotal role in their production through intricate metabolic pathways. Within the bacterium *Escherichia coli* (*E. coli*), the biosynthesis of thiamine (vitamin B_1) is orchestrated by the enzymes ThiC/ThiD, ThiE, and ThiL.[11,12] Along with *E. coli*, several other microorganisms like *Salmonella typhimurium* and *Aspergillus oryzae* have been observed to produce thiamine.[13,14] Similarly, in *Bacillus subtilis* the synthesis of riboflavin (Vitamin B_2) involves the concerted action of RibA, RibB, RibD, RibH, RibF, and RibC enzymes.[15] Fungi such as *Candida famata* and *Ashbya gossypii* were also commercially used in riboflavin production.[16,17] Vitamin B_9 (folate) biosynthesis in *B. subtilis* or *A. gossypii* relies on AroH, FolE, NudB, FolB, FolK, and DfrB enzymes.[18,19] In *E. coli* and yeast, vitamin B_3 (niacin) is produced through the activity of NadB, NadA, NadC, NadD, and NadE/NadF enzymes.[20,21] Vitamin B_6, produced *de novo* in various organisms such as *Sinorhizobium meliloti, E. coli, B. subtilis* engages enzymes like PdxB, SerC, PdxA, PdxJ, and Dxs.[22] The synthesis of vitamin B_5 (pantothenic acid) in microorganisms like *B. subtilis* and *Corynebacterium glutamicum* involves ilvBHCD and panDBEC enzymes.[23,24] *E. coli*'s biosynthesis of Vitamin B_7 (biotin) is a complex process orchestrated by BioC, BioH, BioF, BioA, BioD, BioB, FabD, FabG, FabA, FabF, and FabI enzymes. The aerobic pathway for cobalamin (vitamin B_{12}) synthesis encompasses an extensive array of enzymes, including HemA, ALA, HemB, HemC, HemD, CobA, CobI, CobG, CobJ, CobF, CobK, CobL, CobH, CobB, cobNST,

CobR, CobO/CobP, CobQ, CobS/CobV, CobC, CobD, and CobU/CobT. Bacterial cultures like *Gluconobacter oxydans*, *Bacillus megaterium* and *Ketogulonigenium vulgare* were reported to produce vitamin C (L-ascorbic acid).[25]

16.2.1.2 Fat-soluble Vitamins

Among the fat-soluble vitamins, vitamin D, a group of fat-soluble secosteroids, is divided into two primary compounds including vitamin D_2 (ergocalciferol) and vitamin D_3 (cholecalciferol). Vitamin D_2 can be produced from ergosterol, primarily through yeast fermentation [26] and, on the other hand, researchers are exploring microbial biosynthesis for vitamin D_3, employing microorganisms like *Rhodococcus*, *Streptomyces*, *Pseudonocardia* sp., and *Mycobacterium*.[27] Vitamin E, a group of lipid-soluble antioxidants, comprises tocopherols and tocotrienols that can be produced by photosynthetic organisms, such as algae, green plants, and cyanobacteria like *Synechocystis*. Researchers have also explored microorganisms like *Euglena gracilis*, using precursors like L-tyrosine as substrates in the growth media to increase α-tocopherol accumulation.[28] Furthermore, engineering yeast strains using heterologous genes from photosynthetic organisms have enabled the production of tocotrienols, another form of Vitamin E.[29] Vitamin K exists in two types: vitamin K_1 (phylloquinone/phytomenadione) and vitamin K_2 (menaquinone, MK). Plants synthesize vitamin K_1, while microorganisms, including *B. subtilis, E. coli*, lactic acid bacteria, and *Flavobacterium* sp., are responsible for vitamin K_2 production (Figure 16.1).[30]

16.2.2 Microbial Carotenoids

Carotenoids, a subgroup of isoprenoids (also known as terpenoids), represent a highly diverse category of secondary compounds found in plants, fungi, algae, and bacteria. Over the past decade, significant advancements have been made in the microbial production of carotenoids due to their increasing demand for their antioxidant properties and vitamin A activity. Microalgae species from the genera *Chlorella*, *Dunaliella*, *Scenedesmus*, and *Haematococcus* are recognized as primary producers of commercially significant carotenoids. These microorganisms have the capacity to synthesize carotenoids like astaxanthin, lutein, and β-carotene. Notably, *Dunaliella salina*, under specific extreme environmental conditions, can accumulate as much as 10% β-carotene based on dry cell weight. Similarly, *Haematococcus pluvialis* stands out as a valuable natural source of astaxanthin, comprising 80–99% of the total carotenoids (Figure 16.2).[33,34]

A soil bacterium such as *Corynebacterium glutamicum*, offers a unique platform for producing novel C_{50} carotenoids, including 4-decaprenoxanthin and its glycosylated forms, which are particularly valuable for medical and cosmetic applications due to their UV light-protective properties.[36] Genetically engineered *C. glutamicum* strains have demonstrated remarkable pigmentation with more than 80-fold increase in lycopene content.[37]

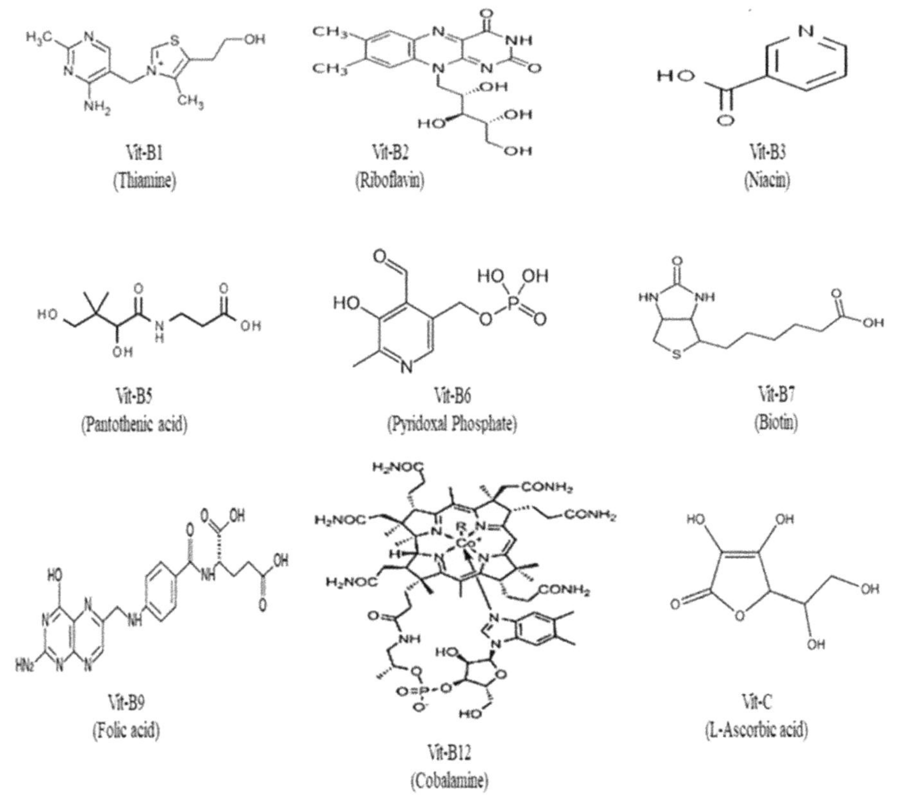

Figure 16.1 Chemical structures of water-soluble vitamins.[31,32] Reproduced from ref. 31, https://doi.org/10.3390/nu14193940, under the terms of the CC BY 4.0 license, https://creativecommons.org/licenses/by/4.0/. Reproduced from ref. 32, https://doi.org/10.3390/nu12092550, under the terms of the CC BY 4.0 license, https://creativecommons.org/licenses/by/4.0/.

A filamentous fungus, *Blakeslea trispora*, is extensively employed for commercial β-carotene and lycopene production. Fungi offer advantages for carotenoid production, including rapid mycelial growth and responsiveness to metabolic stimulators and inhibitors in large-scale industrial settings. Additionally, the utilization of agro-industrial waste materials can reduce production costs and contribute to environmental sustainability (Figure 16.3).[38]

16.3 Advanced Biotechnological Techniques for Production of Vitamins and Carotenoids

16.3.1 Mutagenesis and Genetic Engineering

This approach involves introducing genetic mutations or engineering specific genes in microorganisms to enhance the production of vitamins. By modifying the genetic makeup of the microorganism, researchers can increase the yield.

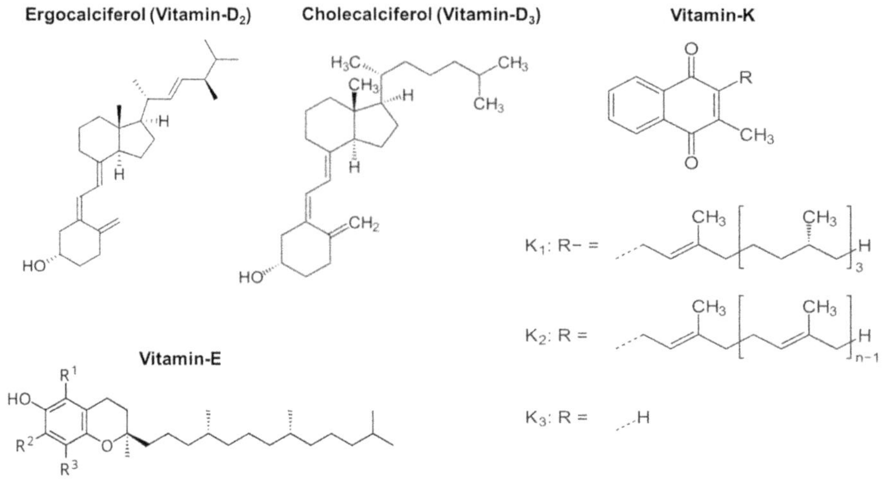

Figure 16.2 Chemical structures of fat-soluble vitamins.[35] Image in the Public Domain.

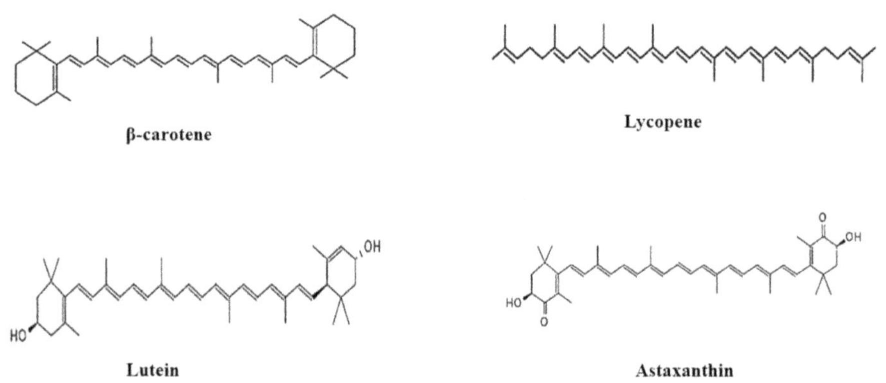

Figure 16.3 Chemical structures of carotenoids.[39,40] Reproduced from ref. 39 with permission from Oxford University Press, Copyright 1998. Reproduced from ref. 40, https://doi.org/10.1371/journal.pone.0135779, under the terms of the CC BY 4.0 license, https://creativecommons.org/licenses/by/4.0/.

16.3.2 Insertion and Deletion

In this method, genetic manipulation is used to insert or delete specific genes in microorganisms. This approach can lead to increased production by optimizing the metabolic pathways within the microorganism.

16.3.3 Gene Cloning

Gene cloning involves the insertion of genes responsible for vitamin production from various sources into a microorganism. This approach allows microorganisms to synthesize products by harnessing genes from other organisms with efficient production pathways.

16.3.4 Inverse Metabolic Pathway Engineering

This technique focuses on modifying the metabolic pathways within microorganisms to enhance the production of carotenoids. By providing specific substrates or modifying metabolic modules, researchers can increase the yield.

16.3.5 Genetic Manipulation

Genetic manipulation involves transferring genes responsible for carotenoid production from one organism (the source) to another (the host).

16.3.6 RNA Interference and Antisense Constructs

This approach uses RNA interference or antisense constructs to down-regulate specific genes within microorganisms.

These biotechnological approaches leverage genetic and metabolic engineering to optimize microorganisms for the efficient production of vitamins and carotenoids. Each method is tailored to specific compounds and microorganism–host systems, aiming to achieve higher yields and cost-effective production. Table 16.1 discusses the various research done using different biotechnological approaches to increase the yield of various vitamins and carotenoid production.

16.4 Microbial Vitamin and Carotenoid Production: Role in Sustainable Economy

16.4.1 Impact on the Environment

The synthesis of microbial vitamins and carotenoids as well as their potential ecological advantages can have a substantial impact on the environment. The commercial manufacturing of vitamins and carotenoids frequently uses microorganisms like bacteria, yeast, and microalgae. Fermentation is frequently used in these processes, which may have an impact on the environment. Depending on the microbe, substrate, and production scale used, the environmental impact of fermentation-based manufacturing methods might vary. It might need energy for waste management, nutrition supply, and temperature control, all of which might increase greenhouse gas emissions. In photosynthetic organisms, carotenoids function as photoprotectants to guard against harm from excessive light exposure. The stability of terrestrial and aquatic ecosystems can be preserved in this way.[49]

Reactive oxygen species (ROS) that are damaging to organisms can be scavenged by carotenoids, which lowers oxidative stress in a variety of

Table 16.1 Various biotechnological approaches for vitamin and carotenoid synthesis.

Biotechnological method	Type of vitamin/carotenoid produced	Type of microorganism used	Yield	Medium, precursor used and genes responsible	Ref.
Microbial vitamins					
Mutagenesis and genetic engineering	Vitamin B1	*B. subtilis* TH95	1.27 $mg L^{-1}$	Minimal medium Mutation in thiamine pyrophosphate kinase activity (*thiN*) and thiamine-related transport protein (*ykoD* and *yuaJ*) genes	41
Mutagenesis and genetic engineering	Vitamin B_2	*Candida famata*	1026 $mg L^{-1}$	Yeast extract peptone dextrose medium, fluorophenylalanine precursor Overexpression of *sef1* and *imh3* genes	16
Mutagenesis and genetic engineering	Vitamin B_{12}	*P. denitrificans*	214 $mg L^{-1}$	Beet molasses medium with choline chloride precursor Overexpression of *cobF–cobM* gene cluster and *cogA* and *cobE* genes	42
Insertion and deletion	Vitamin B_{12}	*S. meliloti* (MC5-2)	156 $mg L^{-1}$	Minimal medium with cobalt chloride as precursor Deletion of *cobl* gene and overexpression of *hemE*	43
Insertion and deletion	Vitamin B_3	Yeast	8 $mg L^{-1}$	Yeast extract peptone dextrose medium with nicotinic acid Knockout NR importer Nrt1 in the NR-non-salvaging genotype *nrkl, urhl, pmpl* (strain PAB038)	20
Insertion and deletion	Vitamin B_5	*C. glutamicum*	1000 $mg L^{-1}$	Minimal medium Deletion of *ilvA* gene and overexpression of *ilvBNCD* and *panBC* genes	23
Insertion and deletion	Vitamin D	*Rhizobium erythropolis*	573 $mg L^{-1}$	MM medium Insertion of the gene-expression cassette encoding *Bacillus megaterium* glucose dehydrogenase-IV into the chromosome of *R. erythropolis*	27
Gene cloning	Vitamin E	*S. cerevisiae*	320 $mg L^{-1}$	Synthetic dextrose medium Gene cloning from various photosynthetic organisms; codon optimization and protein truncation	29

Table 16.1 (*Continued*)

Biotechnological method	Type of vitamin/ carotenoid produced	Type of microorganism used	Yield	Medium, precursor used and genes responsible	Ref.
Microbial carotenoids					
Inverse metabolic pathway engineering	β-Carotene	*S. cerevisiae*	2.24 $mg L^{-1}$	Supplementation of oleic acid and palmitoleic acid as substrate	44
Inverse metabolic pathway engineering	β-Carotene	*E. coli*	2100 $mg L^{-1}$	Engineering methylerythritol phosphate (MEP) module to increase yield	45
Genetic manipulation	Lutein	Host: *Chlamydomonas reinhardtii* Source: *Dunaliella salina*	1.75 $mg g^{-1}$ dry weight of cells	Transfer of *PSY gene* from source to host to increase yield by 2.6-fold	46
Genetic manipulation	Astaxanthin	Host: *Synechococcus* sp. Source: *Haematococcus pluvialis*	11.4 $mg g^{-1}$ dry weight of cells	Transfer of phytoene desaturase (*Pds*) genes into host to upregulate astaxanthin production	47
RNA interference and antisense constructs	Lycopene	*Dunaliella salina* V-101	108.34 $mg g^{-1}$ dry weight of cells	Tris-acetate-phosphate medium Downregulation of phytoene desaturase (*Pds*) gene expression	48

organisms, including plants and animals. This may improve the condition of the ecology. Some microorganisms produce vitamin B_{12}, which can have an impact on the development and metabolism of aquatic phytoplankton. Aquatic food webs may be impacted in a cascade manner by changes in vitamin B_{12} availability. Vitamin K_2, which is necessary for animal blood clotting, is produced by certain bacteria. These bacteria play a part in the nutrient cycling in ecosystems and have an effect on animal health.[50]

16.4.2 Impact on Food Sustainability

By boosting the nutritional value, lowering waste, and reducing the overall environmental effect of food production, microbial vitamins and carotenoids can have a substantial positive impact on the sustainability of food. The production of vitamins like vitamin K and the B vitamins (including B_2, B_{12}, and folic acid) can be done by microbial fermentation techniques. These vitamins are crucial for human health yet are frequently deficient in some diets. Microbes can also manufacture carotenoids like β-carotene.[51,52]

A precursor of vitamin A, β-carotene is crucial for the immune system and vision. To address nutrient shortages in populations, it is now possible to fortify staple foods with important nutrients by integrating microbially generated vitamins and carotenoids into food items. Food waste and byproducts can be turned into useful goods using microbial fermentation. For instance, microorganisms can ferment waste from the processing of fruits and vegetables to create carotenoids and other beneficial substances.[53] Microbial processes help to reduce food waste, which is important for ensuring the sustainability of the food supply. This is done by transforming food waste into useful ingredients. Compared to conventional agricultural practices, microbial production of vitamins and carotenoids frequently uses fewer resources (such as water and arable land). Additionally, it may lead to lower energy use and greenhouse gas emissions, resulting in a more sustainable food production system.[54] Various food items, such as baked goods, dairy goods, drinks, and plant-based foods, can use microbial vitamins and carotenoids. These nutrients can be included in a variety of diets and culinary cultures thanks to their adaptability (Table 16.2).[53]

Table 16.2 Various food applications of microbial vitamins and carotenoids.

Vitamin/carotenoid	Microbe used	Food formulation	Findings	Ref.
Vitamin B_{12}	*Propionibacterium*	Fortified dairy products	Increased vitamin B_{12} content in the product	55
β-Carotene	*Blakeslea trispora*	Fortified fruit juices	Enhanced color and vitamin A content	56
Riboflavin (vitamin B_2)	*Ashbya gossypii*	Baked goods	Improved riboflavin levels and nutritional value	57
Folic acid (vitamin B_9)	*Saccharomyces cerevisiae*	Fortified cereals	Increased folic acid content	58
Vitamin K_2	*Bacillus subtilis*	Fermented soybean products	Higher vitamin K_2 content and enhanced taste	59
Lycopene	*Blakeslea trispora*	Tomato-based sauces	Enhanced color and antioxidant properties	60
Astaxanthin	*Haematococcus pluvialis*	Nutraceutical supplements	Improved stability and bioavailability	61
Vitamin D_2	Yeast *(Saccharomyces)*	Fortified plant-based milk	Increased vitamin D_2 content	62
Ascorbic acid	*Acetobacter species*	Fermented foods	Increased vitamin C levels	63
Biotin (vitamin B_7)	*Escherichia coli*	Baked goods	Enhanced biotin content and nutritional value	64

16.4.3 Impact on Bio-based Economy

Microbial vitamins and carotenoids contribute to a variety of industries, including food, medicine, cosmetics, and biofuels, and hence play a vital part in the bio-based economy. When compared to conventional chemical synthesis techniques, these chemicals have a number of advantages, including sustainability, cost effectiveness, and reduced environmental impact. They are created by microbes through fermentation processes. In comparison to conventional agriculture for plant-based sources, the production of vitamins and carotenoids by microorganisms often uses less land, water, and energy. This environmentally friendly production technique lessens the impact.[65]

In comparison to synthetic manufacture or extraction from plant sources, microbial fermentation procedures can be tuned for high-yield production of vitamins and carotenoids.[66] In the food and beverage business, microbial vitamins are crucial additives for fortification and colour improvement. Examples of these vitamins include B vitamins (such as B_2, B_{12}) and carotenoids (such as β-carotene).[52] Because of their health advantages, such as their antioxidant capabilities and involvement in disease prevention, microbial vitamins and carotenoids find applications in pharmaceuticals and nutraceuticals. Because of the skin-protective and colouring benefits that carotenoids have, they are employed in cosmetics and personal care products as natural substitutes for synthetic pigments.[52,67] Microbial carotenoids can be used as starting materials for the synthesis of biofuels and bioplastics, aiding in the advancement of clean, sustainable energy sources.[68]

16.5 Conclusion

Different bioactive components have been effectively produced in industrial bioprocesses using both rational and traditional metabolic engineering techniques. In the market, these are currently dominated by chemical synthesis, and thus can be replaced with natural ones synthesized through a microbiology route. Through this way, bioactive component production is transitioning from chemical synthesis to microbial production on a significant scale. A number of factors suggest pigments derived from various microorganisms will eventually be extracted with different techniques. Microbiological creation of bioactive compounds that are good for business would be a good choice. Different microbes of all types, *i.e.* bacteria, yeast, and fungi, have the potential to produce several vitamins and natural pigments (carotenoids). Certain bacteria exhibit the capacity to produce vitamins and carotenoids of substantial economic value due to their notable size. Nevertheless, there needs to be more information regarding the methodologies employed to attain optimal production of these vitamins and carotenoids at a large-scale industrial level; there is a need to increase the quantity produced. There are numerous process parameters for enhanced accumulation of vitamins and carotenoids in microorganisms, which is an essential step towards the economically feasible production of carotenoids and vitamins. Along with different advancements in biotechnological sciences, they are indirectly improving microbial production of vitamins and carotenoids. The comprehensive establishment of novel pathways to extract different bioactive components like carotenoids and vitamins still needs effort to be developed at the industrial level.

References

1. M. A. Toma, M. H. Rahman, M. S. Rahman, M. Arif, K. H. M. N. H. Nazir and L. Dufossé, *J. Fungi.*, 2023, **9**(4), 454.
2. H. Meruvu and J. C. dos Santos, *Crit. Rev. Biotechnol.*, 2021, **41**, 1153–1177.
3. C. G. Acevedo-Rocha, L. S. Gronenberg, M. Mack, F. M. Commichau and H. J. Genee, *Curr. Opin. Biotechnol.*, 2019, **56**, 18–29.
4. R. K. Saini and Y.-S. Keum, *Indian. J. Microbiol.*, 2017, **57**(1), 129–130.
5. R. Anand, L. Mohan and N. Bharadvaja, *Rev. bras. farmacogn.*, 2022, **32**, 491–501.
6. C. Li, C. A. Swofford and A. J. Sinskey, *Metab. Eng. Commun.*, 2020, **10**, e00118.
7. S. H. Choi, S.-W. Kim, Z. Ku, M. A. Visbal-Onufrak, S.-R. Kim, K.-H. Choi, H. Ko, W. Choi, A. M. Urbas and T.-W. Goo, *Nat. Commun.*, 2018, **9**, 452.
8. C. Ramesh, N. V. Vinithkumar, R. Kirubagaran, C. K. Venil and L. Dufossé, *Microorgan.*, 2019, **7**, 186.
9. A. C. Lagashetti, L. Dufossé, S. K. Singh and P. N. Singh, *Microorgan.*, 2019, **7**, 604.
10. C. J. Blake and E. J. M. Konings, *J. AOAC Int.*, 2019, **88**, 325–330.
11. S. Hwang, B. Cordova, M. Abdo, F. Pfeiffer and J. A. Maupin-Furlow, *J. Bacteriol.*, 2017, **199**(7), e00810-16.
12. S. Cardinale, F. G. Tueros and M. O. A. Sommer, *Cell Rep.*, 2017, **20**(5), 1029–1037.
13. D. M. Downs and L. Petersen, *J. Bacteriol.*, 1994, **176**, 4858–4864.
14. M. Tokui, T. Kubodera, K. Gomi, N. Yamashita and A. Nishimura, *J. Biosci. Bioeng.*, 2011, **111**, 388–390.
15. Z. Shi, S. Zhen, G. A. Wittert, B. Yuan, H. Zuo and A. W. Taylor, *PLoS One.*, 2014, **9**, e88862.
16. K. V. Dmytruk, V. Y. Yatsyshyn, N. O. Sybirna, D. V. Fedorovych and A. A. Sibirny, *Metab. Eng.*, 2011, **13**, 82–88.
17. T. Sugimoto, S. Kanamasa, T. Kato and E. Y. Park, *Appl. Microbiol. Biotechnol.*, 2009, **83**, 529–539.
18. C. Serrano-Amatriain, R. Ledesma-Amaro, R. Lopez-Nicolas, G. Ros, A. Jimenez and J. L. Revuelta, *Metab. Eng.*, 2016, **38**, 473–482.
19. M. Jagerstad and J. Jastrebova, *J. Agric. Food. Chem.*, 2013, **61**, 9758–9768.
20. P. Belenky, R. Stebbins, K. L. Bogan, C. R. Evans and C. Brenner, *PLoS One.*, 2011, **6**, e19710.

21. Z. Wang, Z. Liu, W. Cui and Z. Zhou, *Appl. Biochem. Biotechnol.*, 2017, **182**, 1458–1466.
22. F. M. Commichau, A. Alzinger, R. Sande, W. Bretzel, D. R. Reuss and M. Dormeyer, *et al.*, *Metab Eng.*, 2015, **29**, 196–207.
23. R. Leonardi and S. Jackowski, *EcoSal Plus*, 2007, **2**, DOI: 10.1128/ecosalplus.3.6.3.4.
24. H. P. Hohmann, J. M. V. Dijl, L. Krishnappa and Z. Prágai, *Host Organisms: Bacillus Subtilis*, John Wiley & Sons, Ltd, Hoboken, NJ, 2016.
25. Y. Wang, L. Liu, Z. Jin and D. Zhang, *Front. Bioeng. Biotechnol.*, 2021, **17**(9), 661562.
26. T. Tan, M. Zhang and H. Gao, *Enzyme Microbial. Technol.*, 2003, **33**, 366–370.
27. Y. Yasutake, T. Nishioka, N. Imoto and T. Tamura, *ChemBioChem*, 2013, **14**, 2284–2291.
28. Y. Tani and H. Tsumura, *Agric. Biol. Chem.*, 1989, **53**, 305–312.
29. B. Shen, P. Zhou, X. Jiao, Z. Yao, L. Ye and H. Yu, *Nat. Commun.*, 2020, **11**, 5155.
30. G. K. Schwalfenberg, *J. Nutr. Metab.*, 2017, **2017**, 6254836.
31. M. A. Ali, H. A. Hafez, M. A. Kamel, H. I. Ghamry, M. Shukry and M. A. Farag, *Nutrients*, 2022, **14**(19), 3940.
32. J. Thomas, A. Stephen, I. Nader, J. Zita, S. Tarassoli, D. Thomas, H. Patrick, T. Catherine and W. Iain, *Nutrients*, 2020, **12**, 2550.
33. L. C. Mata-Gómez, J. C. Montañez, A. Méndez-Zavala and C. N. Aguilar, *Microb. Cell. Fact.*, 2014, **13**(1), 1–11.
34. R. R. Ambati, S. M. Phang, S. Ravi and R. G. Aswathanarayana, *Mar. Drugs*, 2014, **12**(1), 128–152.
35. J. A. P. Kiran, *Transl. Anim. Sci.*, 2019, **3**, 903–910.
36. S. A. Heider, P. Peters-Wendisch, R. Netzer, M. Stafnes, T. Brautaset and V. F. Wendisch, *Appl. Microbiol. Biotechnol.*, 2014, **98**, 1223–1235.
37. S. A. Heider, P. Peters-Wendisch and V. F. Wendisch, *BMC Microbiol.*, 2012, **12**, 1–11.
38. Y. Q. Shi, X. L. Xin and Q. P. Yuan, *Biotechnol. Lett.*, 2012, **34**, 849–852.
39. S. Gradelet, L. B. Anne, B. Raymond, S. Marc and A. Pierre, *Carcinogenesis*, 1998, **19**, 403–411.
40. A. Ana-Andreea, F. Mathias, G. Denise, M. Maria, S. José, P. Serge, R. Bruno and P. Andrew, *PLoS ONE.*, 2015, **10**(8), e0135779.
41. G. Schyns, S. Potot, Y. Geng, T. M. Barbosa, A. Henriques and J. B. Perkins, *J. Bacteriol.*, 2005, **187**, 8127–8136.
42. K. T. Li, D. H. Liu, J. Chu, Y. H. Wang, Y. P. Zhuang and S. L. Zhang, *Bioprocess Biosyst. Eng.*, 2008, **31**, 605–610.
43. Y. Cai, M. Xia, H. Dong, Y. Qian, T. Zhang, B. Zhu, J. Wu and D. Zhang, *BMC Biotechnol.*, 2018, **18**, 27.
44. X. Sun, L. Liu, Y. Zhao, T. Ma, F. Zhao, W. Huang and J. Zhan, *Food Chem.*, 2016, **192**, 43–52.
45. J. Zhao, Q. Li, T. Sun, X. Zhu, H. Xu, J. Tang, X. Zhang and Y. Ma, *Metab. Eng.*, 2013, **17**, 42–50.

46. I. Couso, M. Vila, H. Rodriguez, M. A. Vargas and R. León, *Biotechnol. Prog.*, 2011, **27**(1), 54–60.
47. J. Steinbrenner and G. Sandmann, *App. Environ. Microbiol.*, 2006, **72**(12), 7477–7484.
48. R. Srinivasan, S. Babu and K. M. Gothandam, *Bioresour. Technol.*, 2017, **242**, 311–318.
49. K. Joshi, P. Kumar and R. Kataria, *Process. Biochem.*, 2023, **128**, 190–205.
50. K. Jomova, R. Raptova, S. Y. Alomar, S. H. Alwasel, E. Nepovimova and K. Kuca, *Arch. Toxicol.*, 2023, **97**, 2499–2574.
51. H. Papapostolou, V. Kachrimanidou, M. Alexandri, S. Plessas, A. Papadaki and N. Kopsahelis, *Antioxid.*, 2023, **12**, 1030.
52. S. Nazir, M. Habib, S. Ahmad, Z. R. A. A. Azad and F. M. Allai, *Microbial Vitamins and Carotenoids in Food Production and Processing System*, ed. S. A. Ashraf and M. Kuddus, Apple Academic Press, 2024, vol. 4, pp. 105–123.
53. F. Mozzi, M. E. Ortiz, J. Bleckwedel, L. De Vuyst and M. Pescuma, *Food Res. Int.*, 2013, **54**(1), 1152–1161.
54. M. Riaz, F. M. Anjum, N. Huma and T. Zahoor, *Crit. Rev. Food Sci. Nutr.*, 2013, **53**(10), 962–971.
55. F. Watanabe, Y. Yabuta and T. Bito, *J. Agric. Food Chem.*, 2013, **61**(28), 6769–6775.
56. Y. Ma, A. Kilaru, J. Li, M. Schreiber, M. Wright and S. Kushinsky, *MicrobiolOpen*, 2018, **7**(4), e00575.
57. K. Y. Choi, J. H. Kim and S. H. Lee, *J. Microbiol. Biotechnol.*, 2017, **27**(4), 646–652.
58. A. Tamura, T. Tamura, Y. Tanaka, Y. Watanabe and K. Saito, *J. Biosci. Bioeng.*, 2014, **117**(6), 722–727.
59. K. Y. Yoon, E. E. Woodams and Y. D. Hang, *Food Microbiol.*, 2016, **25**(3), 323–328.
60. M. I. Khan, J. H. Shin, J. D. Kim and M. H. Cho, *J. Agric. Food Chem.*, 2019, **67**(10), 2722–2730.
61. R. R. Ambati, S. M. Phang, S. Ravi and R. G. Aswathanarayana, *Mar. Drugs*, 2019, **12**(3), 128–152.
62. M. S. Calvo, S. J. Whiting and C. N. Barton, *Amer. J. Clin. Nutri.*, 2019, **88**(2), 491S–499S.
63. H. K. Park, S. H. Lee and D. H. Lee, *J. Agric. Food Chem.*, 2018, **66**(21), 5280–5288.
64. T. Friedrich, M. Aramaki, K. Tzschoppe and A. C. Illing, *Biochim. Biophys. Acta, Proteins Proteomics*, 2017, **1865**(11), 1487–1493.
65. B. I. Morinaka and H. B. Bode, *Nat. Rev. Chem.*, 2020, **4**(8), 476–480.
66. M. Gouda, S. H. Omar and S. A. Mohamed, *Egypt. J. Aquat. Res.*, 2018, **44**(1), 31–38.
67. M. J. Kiki, *Cosmetics*, 2023, **10**, 47.
68. K. Samadhiya, R. Sangtani, R. Nogueira and K. Bala, *Front. Microbiol.*, 2022, **12**, 674864.

Subject Index